PRAISE FOR

TRAPPED UNDER THE SEA

AN AMAZON.COM BEST NONFICTION BOOK OF 2014

"While corporate boardrooms are the usual point of entry for dramas involving big money and technological hubris, Swidey, a journalist and author, works instead from the bottom up in his impressively reported account. . . . His is a skillful examination into the basic fragility of such huge infrastructural projects and a lesson in how worker fatalities result not so much from single catastrophic mistakes as from 'a series of small, bad decisions made by many individuals.'"

—*New York Times Book Review*

"Perhaps Swidey's greatest accomplishment is how through it all—the bravery, the bungling, and the loss—he manages to attain a level of suspense akin to that accomplished by Sebastian Junger in *The Perfect Storm* . . . A masterfully crafted saga."

—*Boston Globe*

"Intense . . . A *Perfect Storm* of public works: the great, awful narrative about the building of a ten-mile tunnel that ends in a very dark place beneath the Atlantic. Maybe not for claustrophobes; definitely for everyone else."

—*New York*

"This book will take you on a journey into a fascinating but little-known world—it's the anatomy of a tragedy, a dramatic tale with a cast of vividly drawn characters, superbly written and researched."

—**Jonathan Harr, author of *A Civil Action* and *The Lost Painting***

"Dramatic . . . Through his meticulous reporting, Swidey sheds light on how the largest monuments to our collective genius are also the most likely to be seriously flawed."

—**Chris Jones, *Esquire***

"A harrowing account of how commercial divers risk their lives to improve ours. After reading Neil Swidey's engrossing *Trapped Under the Sea*, you will never look at a bridge or tunnel in the same way."

—*Men's Journal*

"*Trapped Under the Sea* is a heartbreaking tale of real-life bravery, real-life bungling, and real-life tragedy. Neil Swidey is a terrific storyteller."

—**Elizabeth Kolbert, author of *Field Notes from a Catastrophe* and *The Sixth Extinction***

"*Trapped Under the Sea* is extraordinary. It bears comparison with *The Perfect Storm* in its brilliant evocation of everyday, working-class men thrust into a harrowing, at times heroic confrontation with death and disaster."

—**Dennis Lehane, author of *Live By Night* and *Shutter Island***

"[*Trapped Under the Sea*] transcends narrow geography in many ways: as exemplary investigative reporting, as superb narrative writing, as a cautionary tale of capitalistic greed, as a case study of how government agencies can protect or harm, and as a rare glimpse into the scary world of underwater dive crews. . . . [Swidey] masterfully portrays the lives of the five divers, their loved ones, their work colleagues, and their supervisors. It is a rare book that portrays blue-collar skilled laborers so thoroughly and compellingly."

—**Steve Weinberg, *Dallas Morning News***

"Thrilling and beautifully told, *Trapped Under the Sea* delivers us into a dangerous and mysterious world, a place that speaks to our darkest fears and where heroes work, as Swidey so masterfully shows us, just beneath the surface of our everyday lives."

—**Robert Kurson, author of *Shadow Divers***

"A harrowing account of one of the largest engineering projects in U.S. history and of the hubris and ignorance that led to tragedy . . . [and] a cautionary tale, which Mr. Swidey writes with splendid heart."

—*Wall Street Journal*

"A fascinating, sympathetic, and suspenseful look at a doomed, high-risk engineering job, the working-class men who dared to undertake it, and its ripple effect on the survivors. Claustrophobic and compelling."

—**Chuck Hogan, author of** *Devils in Exile* **and** *The Town*

"A marvel of masterful reporting and suspenseful writing. Neil Swidey has delivered a gripping, action-filled account of the human costs deep inside a feat of modern engineering. He has a remarkable knack for bringing to life indelible characters and making readers hold our breath as these brave men enter the claustrophobic world of their undersea lives."

—**Mitchell Zuckoff, author of** *Frozen in Time* **and** *Lost in Shangri-La*

"Captivating . . . Swidey brands the disaster with a human face by introducing the men to the reader and extracting lessons learned through a careful examination that he passes along in a narrative nonfiction piece that would no doubt make his glorious predecessors in the investigative magazine genre of the early 20th century proud."

—**Fort Worth** *Star-Telegram*

"Neil Swidey's detail-rich account of this unlikely disaster is a stirring tribute to the men, how they lived, and how they died."

—*Mother Jones*

"Neil Swidey's *Trapped Under the Sea* combines rich characters with a thrilling and tragic story that offers something for readers of all stripes . . . At once tragic and ironic, insightful and enraging."

—*The Blaze*

"*Trapped Under the Sea* offers vital insights into how organizations work—or fail to work—and how very smart people can make very bad decisions. Neil Swidey's riveting account of the Deer Island disaster should be essential reading for anyone in a position of leadership. I couldn't put it down."

—**Amy Edmondson, Harvard Business School Novartis Professor of Leadership and Management and author of** *Teaming*

"Swidey's book is, at its core, a story about people: the people who risked their lives. The people who loved them. And the people who should have seen the disaster to come."

—*Maclean's*

"A gripping (and true) tale . . . told in a you-are-there narrative style that recalls Jon Krakauer's *Into Thin Air*."

—*Civil Engineering*

"Gripping . . . This virtuoso performance combines insights into massive engineering projects, corporate litigation, environmental science, and cutthroat free-market behavior with vivid personal stories."

—*Publishers Weekly*, **starred review**

"The pacing and feel of a special-ops adventure and the insight of a public-policy investigation."

—*Booklist*, **starred review**

"Unforgettable . . . Seems destined to become a nonfiction classic."

—*Engineering News-Record*

"A riveting, tragic true story . . . Fascinating."

—*Parade*

TRAPPED UNDER THE SEA

ONE ENGINEERING MARVEL,
FIVE MEN, AND A DISASTER TEN MILES
INTO THE DARKNESS

NEIL SWIDEY

B \ D \ W \ Y
BROADWAY BOOKS
| NEW YORK |

Published in the United States by Broadway Books, an imprint of the Crown Publishing
Group, a division of Random House LLC, a Penguin Random House Company, New York.
www.crownpublishing.com

BROADWAY BOOKS and its logo, B \ D \ W \ Y, are registered trademarks of Random
House LLC.

Originally published in hardcover in the United States by Crown Publishers, an imprint
of the Crown Publishing Group, a division of Random House LLC, New York, in 2014.

Neil Swidey based this book on articles that originally appeared in *The Boston Globe*.
Inquiries concerning permission to reprint any article or portion thereof should be directed
to Brian McGrory, Editor at *The Boston Globe*, 135 Morrissey Boulevard, Boston, MA 02125.

Library of Congress Cataloging-in-Publication Data
Swidey, Neil.
Trapped under the sea: one engineering marvel, five men, and a disaster ten miles into the
darkness / Neil Swidey.—First edition.
Includes bibliographical references and index.
1. Sewage disposal plants—Accidents—Massachusetts—Boston Metropolitan Area.
2. Survival—Massachusetts—Boston Metropolitan Area. 3. Deer Island (Mass.:
Cape)—Buildings, structures, etc. I. Title.
TD524.M4S95 2014
363.11'9628390974461—dc23   2013013657

ISBN 978-0-307-88673-6
eBook ISBN 978-0-307-88674-3

Printed in the United States of America

Book design by Phil Mazzone
Illustrations copyright © by Javier Zarracina
Cover design by TK
Cover photograph by TK

10 9 8 7 6 5 4 3 2 1

First Paperback Edition

To my true north—Denise, Sophia, Nora, and Susanna
—and to my brothers and sisters,
who taught me how to find my way

"Wouldst thou,"—so the helmsman answered,
"Learn the secret of the sea?
Only those who brave its dangers
Comprehend its mystery!"

    —Henry Wadsworth Longfellow, "The Secret of the Sea"

The work an unknown good man has done
is like a vein of water flowing hidden underground,
secretly making the ground green.

    —Thomas Carlyle

CONTENTS

CONTENTS

PROLOGUE

JULY 21, 1999

He was about to begin a miles-long trek into the dank, dark intestines of the earth, but DJ Gillis wouldn't go until he'd found a piece of twine.

Two of the other divers had already packed themselves into the "man cage," a tubular basket of yellow metal that would be lowered by crane down a 420-foot shaft. But they couldn't go without DJ, and he wasn't about to be hurried.

"C'mon, DJ," one of the guys shouted. "Let's go!"

DJ ignored the calls as he searched an equipment trailer, sifting through piles of rain gear and crates of tools. The summer sun was beginning its climb above Deer Island, a peninsula that hangs down from the north like a comma into Boston Harbor, curling in front of Logan International Airport.

DJ's boss, a hard-charging guy by the name of Tap Taylor, was standing near the man cage and losing his patience. "Let's go!" Tap yelled.

The two of them had a close if combustible relationship. Tap's singular focus was building his small New Hampshire commercial diving business into something bigger, and he thought nothing of logging fourteen-hour shifts every day of the week. Still, he had a soft spot for DJ, treating the breezy twenty-nine-year-old more like a kid brother than an employee.

Over the years, they had developed a rhythm as predictable as the banter between an anchorwoman and weatherman on the eve of a big storm. DJ, a six-foot-two, solidly built charmer, would show up late to the job site, occasionally dropped off by some blonde or brunette he'd been partying with the night before. As DJ peeled off yesterday's clothes and put on his dive gear, Tap would curse, threatening to kick him off the job. But those outbursts always ended the same way. Before long, Tap would calm down and begin pumping DJ for details from his latest adventure hopping bars and hopping beds.

"C'mon!" Tap shouted again.

"If you're in that much of a hurry," DJ barked, "then go without me!"

It was the morning of July 21, 1999, a Wednesday, and the tension was thick, mainly because so many problems had surfaced on the project that Monday and Tuesday. Getting down the thirty-foot-wide shaft would be the easy part. The challenge would come when the divers had to make their way to the end of a pitch-black tunnel that began at the base of the shaft and kept going and going, under the sea, for nearly ten miles.

Tap would be monitoring the divers' progress from topside on Deer Island. He was in no mood for DJ's usual antics. In reality, neither was DJ. The only woman he had on his mind right now was the Virgin Mary. He needed the piece of twine to tie a small religious medal to the underside of his hard hat. The oval medal had once belonged to his grandfather, a carpenter who helped build the Prudential Tower that defined Boston's skyline.

DJ had asked his mother for it the night before, remembering the story of how his grandfather had kept the Miraculous Medal in his pocket the whole time he spent erecting that skyscraper, taking comfort in the Blessed Virgin's protection. Seeking comfort himself, DJ had gingerly asked his mom, "Is that still around?"

"Yes," she had replied cautiously. "Why?"

"I'm a little concerned about the job."

As much as he had tried to downplay his growing uneasiness with the tunnel project, DJ hadn't been surprised to see fear flash across his mother's face. He had just broken one of the cardinal rules he'd learned early on in his career as a commercial diver, when he'd seen oil rigs capsize and cranes collapse: Never tell your family the truth about the dangers of the job. It isn't fair to dump that kind of worry on them.

But this wasn't like any job DJ had worked on before. Hell, it wasn't like any job *anyone* had worked on before. And that challenge—to make history in his field, to do the seemingly undoable—was what had sold DJ on the tunnel assignment in the first place. With everything that had gone down in the last few days, though, he was feeling some buyer's remorse.

Finally, he found the twine, fastened the Virgin Mary, and put on his hard hat as he strode over to the man cage.

Tap was still heated. "What the *hell* were you doing?" he snapped. "We've got a job to do here."

DJ took off his hat and turned it over, so his boss could see the medal dangling from it. "I'm taking care of myself," he said.

Tap's steam instantly lifted. "It's getting that bad, huh?"

The man cage spun slowly as it moved down the shaft, like a gentle whirlpool of water circling around a drain. DJ was a talker, but now he felt no urge to speak. He stared across the cage at the other two divers, guys he'd known for only a couple of weeks. He had an especially tough time getting a bead on the shorter of the two, Dave

Riggs. While divers tended to be a rowdy bunch, pounding beers and swapping stories after their shift, Riggs didn't drink and kept to himself. During one of their first days working together, DJ had been doing what he always did—telling lots of tales, peppered with colorful words—when Riggs turned to him and said, "I'd appreciate it if you didn't use foul language within my earshot." DJ narrowed his eyes as he stared back at Riggs in disbelief. Then he said the first thing that came into his head: "Are you fucking kidding me?" He'd been a diver for seven years, and he'd never once come across anyone who confused a job site with a church pew.

He had a better connection with the other guy in the cage, Donald Hosford, known as Hoss to everyone except his mother. Hoss had a ropy, six-foot-five build and the rugged looks of someone who might appear in a magazine ad for the Copenhagen chew he always kept wadded under his lip. He was only twenty-four years old, but he worked with the confidence and sure movements of a seasoned veteran. To DJ, Hoss embodied the cowboy spirit of the experienced divers he'd always looked up to: take-charge guys who could be rough and even crude on the job, but who always addressed a lady as "ma'am" and reflexively pulled out a chair for her. Still, DJ thought to himself, the guy seemed so steely that, if need be, he could put a bullet in another man's forehead and then go right back to eating his supper.

As the man cage neared the bottom, DJ fixed his eyes on a fat ventilation duct that transported air from Deer Island, down the shaft, and into the tunnel. For years, a "bag line" had run all nine and a half miles of the tunnel, providing plenty of ambient air to the subterranean workers known as sandhogs, who were responsible for burrowing down under. Now the duct ended right where the shaft did, at the very start of the tunnel.

DJ had heard that the sandhogs had spent nearly a decade mining the tunnel—twice as long as planned—and that the contractor was tens of millions of dollars in the red. The fact that he and the other divers were being called in during the final hour was itself evidence that something had gone seriously wrong. After all, the divers were being asked to finish the job even though the ventilation, electricity, and transportation systems—the infrastructure that had kept the sandhogs alive—had already been removed from the tunnel.

DJ couldn't tell if the project's bosses viewed his crew as their cavalry or as the equivalent of a Hail Mary pass. The few sandhogs still hanging around on Deer Island certainly seemed to resent the divers' arrival, as though it signified a failing on their part. When the divers suffered a delay after one of their equipment trailers was damaged, a veteran sandhog had reacted with exasperation, asking, "How long is it going to take to fix it?" Hoss hadn't missed a beat in putting the guy in his place. "Well, I'll tell you what," he had said, smiling with a wad of chew bulging under his lip. "It ain't gonna take nine years."

At the base of the shaft, it was cold—about fifty degrees—and misty. There was a decent amount of light and air there, but when DJ let his eyes wander east into the tunnel, they quickly got lost in the dark. To him, it looked like the very center of a black hole.

High above that hole, Deer Island was in the final stages of its conversion into one of the largest and most sophisticated sewage treatment plants in the country, the destination for every toilet flush in the eastern half of Massachusetts. On the southern end of the island stood the plant's most unmistakable component, a cluster of giant digester tanks that looked like mutant eggs from some 1950s sci-fi movie set. Each egg was fourteen stories high, and fittingly, there were a dozen of them. But the most complicated part of the project, the construction of the Deer Island Outfall Tunnel, had taken place far from view.

Imagine you are venturing into a tunnel that's been bored into the bedrock underneath the ocean and that continues straight out, hundreds of feet below the seafloor, for almost ten miles. There is no light, besides the faint glow coming from the bulb on your helmet. There is no sound, besides the water dripping overhead or sloshing around your boots. Most important, there is no breathable air, besides what you brought in with you, a lifeline pumping through a hose and into your facemask. At the end of the tunnel, you don't even have enough room to stand up straight, since it chokes down to just five feet in diameter before ending abruptly. It's the world's longest dead-end tunnel, so there's no way out other than turning around and making the hazardous trek back to where you started.

For perspective, consider that the deepest ocean point on earth, the Challenger Deep valley in the Pacific Ocean's Mariana Trench, lies 6.8 miles below the surface of the water—a distance nearly three miles *less* than the Deer Island tunnel's total length. Granted, the character of Challenger Deep's remoteness is vertical, and the tunnel's is mostly horizontal. But because of the tunnel's forbidding conditions, its endpoint effectively makes it farther from the surface than the base of that trench.

All the sewage treated at Deer Island's new plant would ultimately make its way to the end of that horizontal tunnel, then travel through a series of vertical pipes climbing up to the ocean floor, before being released into the sea. But that couldn't happen unless DJ and the other divers managed to get all the way out there first. They had been dispatched on a hazardous, high-stakes mission to fix a problem that had confounded some of the world's top engineering and construction companies for years. If the divers were successful, the empty tunnel could be flooded, allowing billions of gallons of treated sewer water to begin flowing out to sea. Left unsolved, the problem threatened to turn the new tunnel into a $300 million white elephant, if not render the entire court-ordered cleanup of Boston Harbor a multibillion-dollar waste of taxpayer money.

Boston Harbor was once a national embarrassment, a waterway indivisible from American history but blackened by the smothering amounts of sewage and sludge dumped into it every day. Thanks largely to the Deer Island treatment plant and tunnel, it is now considered one of the country's cleanest urban harbors, an unambiguous environmental success story, with no shortage of academic papers extolling it as a national model. The part of the story that is seldom discussed, however—and the part that offers critical lessons for understanding how very bright people can make very bad decisions—is how a vast engineering marvel of a project ended with a handful of divers being sent into the darkness with an improvised, untested plan. And how their mission turned into a harrowing race to get out alive.

DJ and the other divers were used to danger. They were Navy SEAL–type guys who ran toward it when everybody else was running away. Unlike their usual assignments, this one wouldn't require them to work while submerged in water. But the conditions they would face in the tunnel were far more hazardous. They had been hired because they knew how to do construction work in dicey settings where they had to supply their own breathing air. And by this point, the long, empty tunnel had become an oxygen starved, toxic tube.

It wasn't hard for the divers to detect the grumbling and finger-pointing coming from the project's major players. Yet DJ and the others were completely in the dark about the intensity of the battles that had preceded their arrival. They didn't know that brinksmanship between the contractor building the tunnel and the government agency that owned it had triggered the fateful decision to strip the tunnel of all its lighting and breathable air. And they certainly had no idea the contractor had warned, in an eerily prescient memo one year earlier, that if a dive team were to be sent in after the tunnel's ventilation had been removed, "the risk of catastrophe would be exponentially higher."

As it turned out, the divers were asked to do something so experimental, requiring them to be so utterly cut off from civilization, that they might as well have been working on the surface of the moon.

In other ways, though, they were simply continuing a tradition where blue-collar workers are the ones expected to transform the dazzling dreams of engineers and the promises of politicians into concrete reality. True, they weren't working in the days of the Hoover Dam's construction, when regulation was scant and the loss of about one hundred worker lives was viewed as an actuarially unremarkable toll for such an ambitious undertaking. The threshold for acceptable human loss had been lowered dramatically over the course of the twentieth century. Nonetheless, the divers were working at a time when designers and engineers—emboldened by new technology and pressured by governments and corporations to address a growing population's rapacious need for more fuel, transportation, and waste management—were pushing the limits of the possible. And no matter how impressive an engineer's solution might have looked on a computer screen, it still required a bunch of workers in hard hats to carry it out.

These are the largely invisible laborers who build our bridges and repair our pipelines, dig our tunnels and erect our towers. While the designers and developers take the bows at the banquets and ribbon cuttings, the workers are the ones who assume the real risks. Yet they tend to flash into public consciousness only when groups of them are killed or imperiled. In 2010 alone, eleven oil rig workers lost their lives on the Deepwater Horizon in the Gulf of Mexico, twenty-nine coal miners were incinerated in the Upper Big Branch explosion in West Virginia, and thirty-three miners were trapped for sixty-nine days in the San Jose copper and gold mine in Chile. A year earlier, the *Las Vegas Sun* won a Pulitzer Prize for exposing the deaths of twelve construction workers within eighteen months during a frenzied building boom along the Vegas Strip—a worker death rate of one every six weeks.

Still, few people notice the casualties when they come, as they typically do, in increments of one or two. That's almost always the

pattern in the world of commercial divers, a fraternity of just a couple thousand full-time construction workers who happen to do their
building and welding underwater. At the time of the Deer Island mission, a fresh government analysis zeroing in on diver fatalities gar-
nered little public attention even as its findings rippled through the
fraternity. The report from the Centers for Disease Control found
that commercial divers were killed on the job at a rate forty times
higher than that of the average worker.

The Deer Island tunnel was twenty-four feet in diameter but became
progressively narrower along its final stretch. DJ took comfort in the
fact that at least the divers wouldn't have to walk the entire nine and
a half miles. Instead, they would travel in two souped-up Hummer
Humvees. They needed two because deep inside the tunnel there
wouldn't be enough room to turn a vehicle around. So they used a
two-way trailer to tow a second Humvee facing in the opposite direction, which they would use for the return trip. DJ liked the military-
special-ops feel to the mission, even if the unorthodox breathing
system at the center of it made him nervous.

Over by the Humvees, he found the other two divers who would
round out their team. Tim Nordeen was a husky guy who had recently shaved down his bushy beard into a goatee so it wouldn't interfere with the seal on his facemask. At thirty-nine, Tim was the
oldest guy on the crew and had the air of someone who'd experienced
enough in life to know not to sweat its little stresses.

Standing on the bumper of one of the Humvees was foreman Billy
Juse, who happened to be DJ's close friend and the only member of
the team he knew well. Billy had a slim, five-foot-nine frame and a
dark mustache with edges as squared off as tape cut with scissors.
DJ loved to tease Billy about that thick, dense landing strip under his
nose. "Hey, Billy," he would say. "I forgot to bring my lunch today.

You mind if I clean off your mustache when you're done eating? I'm sure I can find enough left over in there to make a sandwich."

The team would drive the Humvees as far into the tunnel as they could go. At that point, two divers would camp out while the other three trudged on foot to the very end of the tunnel. The assignment called for DJ and Tim to remain in the Humvee. But before the journey began, Billy pulled his buddy aside. He said his back was hurting and asked DJ to switch places with him.

"You're a pussy," DJ said.

"Maybe so," Billy shot back, "but I'm your boss."

That was only partially true, and they shared a knowing laugh. They both usually answered to Tap, but it had already become painfully clear that on this job, even Tap wasn't in charge.

After a slow, two-hour drive, the lead Humvee reached the nine-and-a-quarter-mile mark, beyond which it could go no farther. Billy and Tim remained planted inside the vehicle, monitoring the breathing system for all five divers, while DJ, Riggs, and Hoss began their trek on foot.

The three guys remained connected, through about fifteen hundred feet of hose, to the air system back in the Humvee. The hose was called an umbilical because a diver could not survive without it any longer than a baby in the womb could survive without hers.

It took DJ, Riggs, and Hoss twenty minutes to trudge, with their gear, about eleven hundred feet. Hoss positioned himself outside the final stretch of the tunnel, just before its width shrank to five feet. His job was to watch over their connections to the main breathing system and use a communications wire to call in periodic updates to Tim back in the Humvee.

DJ and Riggs slogged onward, duck-walking several hundred feet more to the tunnel's end. Once there, Riggs took on the toughest assignment, shimmying deep into a slimy, skinny connector pipe to remove a heavy safety plug. With DJ's help, he would need to repeat that task in a series of pipes. To most people, the claustrophobia inside that cramped, black space so far from civilization would be paralyzing. The pipe Riggs had to crawl through measured just thirty inches across—not much wider than his shoulders. And he had to be careful not to jostle the cumbersome breathing gear that was keeping him alive.

Not long after Riggs emerged from the pipe, Hoss noticed that the divers' umbilicals had become dangerously tangled. He instructed DJ and Riggs to get their hoses straightened out before doing anything else.

As Hoss began his own untangling effort, he looked up to see something startling. DJ was collapsing onto the tunnel floor, in a strange kind of involuntary slow motion.

"Are you okay, DJ?" Hoss yelled, muffled, through his facemask.

Before he could finish his sentence, Riggs went down next, falling on one knee in front of DJ.

Hoss suddenly felt light-headed himself, in a warm and fuzzy way, as though he had just tossed back a few cocktails. He had enough presence of mind to know that warm and fuzzy was not how you wanted to be feeling when you were buried under the sea, miles from land. "I need to call Tim," he said.

When he reached Tim on the comm wire, Hoss pressed him for what the oxygen content was in their main breathing supply. Anything below 19 percent meant trouble. Tim said he and Billy would check. A few seconds later Tim called back. "Shit, it's 8.9!" he shouted, frantically.

Then the line went dead.

1

DJ

DJ pulled into the driveway, got out of his Ford Bronco, and stepped into what felt like a 1980s music video. Straight ahead was a sun-tanned brunette washing her car while wearing ripped jean shorts and a wet half-shirt. As he trained his eyes on her, DJ could practically hear the thumping hair-metal-band soundtrack playing in his head. Actually, it wasn't all in his head. There was music coming from around the back of the house, where someone had placed a speaker facing out of a first-floor window.

At a picnic table, three attractive women in their early twenties sat in Daisy Duke cutoffs and tight tops, drinking wine coolers and taking in the sun on a late summer afternoon. It was a Friday in August 1993, and DJ, a month shy of his twenty-fourth birthday, had just returned to Massachusetts after more than two years working as an offshore diver in the Gulf of Mexico. During his time away, his mother and younger brother had moved into the upstairs apartment of this two-family house in Waltham, a former mill city west of Boston. They were away on vacation now, so DJ was on his own as he saw the new place for the first time.

A guy around his age approached him, explaining that he lived in the downstairs apartment with his girlfriend, the dark-haired car washer. He invited DJ to grab a beer and join the party that was just getting started. DJ didn't need much convincing and plopped himself down at the picnic table. Given his muscular build, broad smile, and easy conversational skills, he never had much trouble getting noticed by girls, even if he was shier than he let on.

A blonde named Lisa, in between drags on her cigarette, began chatting him up. She had a big laugh to match a big personality. When she told a story, she used her chin, not just her hands, for emphasis. DJ immediately liked her. But his eyes were more drawn to the black-haired woman sitting next to her, who introduced herself as Donna. At least that's what it sounded like to DJ's ears. But when he called her that a few minutes later, she quickly corrected him. It may have sounded like Donna, but her name was actually spelled Dana. To nail the correct pronunciation—*DAH-nah*—you needed to contort your mouth into a horizontal line as exaggerated as a mailbox slot. *Seems like a lot of trouble for a name,* DJ thought to himself. But he was so smitten that he didn't mind. Dana had bronze skin, big alluring eyes, and milky teeth that lit up her face when she smiled.

DJ could sense that all the girls were fascinated by his tales of adventure as a diver in the Gulf. He explained how he would get helicoptered way out to sea, onto a giant oil platform the size of a village, so he could do complicated work hundreds of feet underwater. When he'd left Waltham a few years earlier, he'd been just another construction worker hanging out at the bar. Now, having turned his childhood love of the water into a thrilling career, he could claim a deep well of true stories. He knew to prune from his anecdotes all the unglamorous realities of life as an offshore diver—the smelly sleeping quarters and grunt work bordering on hazing—and stick to the exciting stuff.

The party grew as the night wore on. When DJ noticed at one

point that Dana had disappeared, he turned his attention to blond-haired Lisa. As night turned to morning, they made their way upstairs to his mother's apartment, where they hooked up. She took off early the next morning, explaining that she had to head out of town.

It didn't take long for day two of the party to get going. Once again Dana was there, and this time DJ didn't let her out of his sight. She told him she was a hairstylist at a high-end Boston salon. She clearly liked to have a good time, but DJ detected something classy and almost exotic about her, with her dark hair falling around her face, hiding one of her eyes.

Late that night, after most of the partyers had cleared out, DJ realized he had lost the key to his mother's apartment. "Don't worry," his downstairs neighbor told him. "You can crash on our couch."

Things heated up between DJ and Dana once they found themselves alone. Eventually they moved from the couch to the dark kitchen and then, for the crescendo, onto the kitchen counter. Suddenly someone flicked on the overhead light. Dana yelled. DJ turned to see a groggy guy who was clearly startled to find he had produced such drama—or more accurately, interrupted it. That was all the motivation DJ needed to head outside, climb up to the second floor, and crawl through a window. He then let Dana in, and they spent the night together with no further interruptions.

Even as the fun began to wind down on Sunday, DJ had something else to look forward to. While he had been working as a nonunion diver in the Gulf, his hope in returning to Boston was to join Local 56, the union for pile drivers and commercial divers. The pay and benefits were a lot better for union divers, which is why DJ worried it would be hard to break into the local. But just a few days earlier, he'd received instructions from the union hall to show up at the beginning of the week for a job rebuilding a ship terminal in South Boston. In one weekend, everything in his personal and professional life seemed to come together.

Late that afternoon Dana made a call. After she hung up, she mentioned her sister would be coming over.

"Oh," DJ said. "You got a sister?"

"Yes," Dana replied, tilting her head in surprise. "You met her."

DJ racked his brain but couldn't recall meeting her sister. Dana insisted. "You know, Lisa—the blond girl."

The words were a punch to his gut. When Lisa arrived, DJ noticed her flinch as she saw Dana run her hand through his light brown hair. DJ tried the only move he could think of, a last-minute call for clemency in the form of a pained look shot directly at Lisa. With it, he was wordlessly saying: *I had no idea.*

Lisa, to his eternal relief, returned a forgiving look.

Later, when her sister was out of the room, Lisa said to him, "I see you and Dana are getting along well."

"Yeah," DJ replied nervously. "Listen, Lisa, I didn't know—"

She cut him off. "You don't have to explain anything. I can tell you didn't know we were sisters."

DJ couldn't have been more relieved. He was attracted to Dana in a way he hadn't felt before, though there were few signs of the role she would come to play in his life.

There was, however, more fallout from his fantasy weekend. All that partying caught up with him to the point where, on his first day of his first union job, he showed up to the worksite late. The supervisor told DJ that if he thought he could just waltz in whenever he felt like it, he should save everybody some time and go find another job. "Don't even think about being late again," the guy barked, "unless you've got a really good story."

DJ flashed an impish grin. "Have I got a story for you!" he said, launching into the tale of his weekend with two sisters. By the end, the supervisor was the one grinning. He let DJ's tardiness slide.

. . .

DJ didn't get into it then, but his full life story was just as interesting.

As a kid, he'd lived in eleven states in a dozen years. Tennessee, Maryland, Florida, Ohio, Indiana, Illinois, Michigan, New York, Georgia, North Carolina, Oklahoma. He was forever the new kid, the outsider who would never be around long enough to make real friends. In every new town, he carried an unfamiliar accent with inflections from the last stop, setting him up for taunting from the other kids.

His mother, Lorraine, had grown up on a farm near the rocky coastline of Cape Breton Island in Nova Scotia. Her father regularly traveled to the States to find work as a carpenter, and at age twenty-one, Lorraine had followed him. In Boston she fell for an Irish construction worker, got pregnant, and eventually gave birth to a boy she named Donald James Gillis, after her two brothers. Soon everyone took to calling him DJ.

After a few years raising DJ alone, Lorraine married a pipe fitter who helped build power plants, specializing in nuclear facilities. She had a son with him, and as he chased work across the country, she followed along with her two young boys. She was fond of her husband but felt that his weakness for alcohol made life a struggle. When he drank his wages, Lorraine had to provide for her sons with the small paychecks she earned working in a series of service jobs: waitress, hospital worker, clerk at a fireworks stand.

In this sea of uncertainty, Lorraine came to rely heavily on DJ. Even as a young boy, he had shown himself to be cool in a crisis. When he was eight years old, he appeared at their Oklahoma door carrying his bloodied five-year-old brother, David. DJ was holding a towel down on his brother's forehead and squeezing his hand tightly. Lorraine shrieked. "Oh my God! What happened?"

But DJ's first words to his mother were reassuring. "Don't be scared, Mom. He's gonna be okay."

A neighborhood punk had thrown a piece of pipe at DJ's brother, producing an enormous gash in his head. DJ had carried David several blocks home, like a firefighter calmly removing a child from a house in flames. "From that day on," Lorraine would later say, "DJ was like a father, constantly protecting his brother—and me."

When Lorraine finally tried to steel herself to leave her husband, DJ was the one she confided in. "Don't be scared, Mom," he told her. "You're tough. We can make it together." She drew strength from those words, even though they came from the mouth of a twelve-year-old boy.

After bouncing around for a few years, Lorraine and her sons landed back in the Boston area in 1984, on DJ's fifteenth birthday. They arrived in Newton, a small city bordering Boston, and settled into a gloomy, almost petrified house with plastic-encased sofas. It was owned by an invalid woman whom Lorraine would care for in exchange for housing. Newton was an affluent, education-obsessed city, so the transition was not easy for DJ, a high school freshman with a spotty transcript. Kids made fun of his cowboy boots and curious southern accent. They called him a redneck, a dagger of an insult if there ever was one in such a brainy town. When he visited classmates' homes, he found a degree of wealth that would have been inconceivable in the sticks of Tennessee and Oklahoma. It made him only more embarrassed about his borrowed space in an old lady's dusty house.

By his sophomore year, his family had moved to the neighboring city of Waltham, whose complexion turned quickly from gritty industry and tired triple-deckers to sleek high-tech offices and green suburbia. When he attended his first teen house party, he found himself being stared down by one of the toughest kids in school, a beefy pot dealer. DJ was tall but at that point very thin, so everyone expected him to get pummeled. Yet all those years moving around and constantly having to defend himself had turned him into an agile fighter.

He ended up throwing the pot dealer through a window. After that, no one made fun of DJ's accent anymore.

Before long he was getting into so many fistfights that his guidance counselor called him into his office and asked, "Is everything all right at home?" He figured DJ's string of black eyes and busted lips were signs of abuse, not the emblems of his new identity as someone who was both cool and fearless.

The next year he transferred to the city's vocational high school, where he discovered a talent for welding. Before long, he picked up a part-time job with a local welder. After graduating from high school, his mother helped him buy a brand-new black Firebird Formula 350. He treasured that car, keeping it immaculate. He cruised the downtown strip with pride, as though his Firebird were announcing to the world the arrival of the kid who had always been ashamed of how little he had.

A life of the party who seemed just dangerous enough to be attractive, DJ had become a guy that girls wanted a piece of and other guys wanted to be around. He cemented his appeal by adding muscle to his build through weight lifting. After a lonely and turbulent childhood, he savored the newfound attention. He moved quickly and partied vigorously, making up for lost time.

Still, for all the fun he was having, DJ realized that on some level he was lost. He had struggled just to get through high school, so college didn't seem viable. He worked several blue-collar jobs but couldn't glimpse much of a future. When a relative offered him the chance to work with him doing commercial construction in Syracuse, New York, DJ said yes.

Cut off from their normal social circles, they filled some of their downtime taking scuba-diving lessons at a local pool. As a kid, DJ had always loved swimming, refusing to get out of the water until he had reached that blue-lipped, body-shivering state that only children

and shipwreck victims seem able to tolerate. Before one class, as he
flipped through a scuba magazine, he came across an article describing commercial diving as a career. It explained how these divers did
underwater welding and construction and traveled the world, finding
adventure and making good money. It dawned on DJ: *This is something I could be good at.*

In February 1991, twenty-one-year-old DJ moved to Houston to
enroll at the Ocean Corporation dive school. After a month in school,
around the time when many wannabe divers realize how punishing
the job is and quit, DJ called home to Lorraine.

"Mom," he said, "this is *exactly* what I want to be doing."

DJ grabbed his yellow fiberglass diving helmet out of his truck and
headed over to the job site. The SuperLite 17 was his prized possession. He'd bought it used, when he was just getting started in the
Gulf, and even then it had set him back nearly three grand. Despite its
name, the helmet still weighed about thirty pounds out of the water.
This job in the fall of 1993 had brought DJ to a hydroelectric plant in
southeastern Vermont. Given his experience as a deep-sea diver in
the Gulf, where he'd done underwater welding at depths of 250 feet,
he figured working in the waters of a New England river one-tenth as
deep would be like a dip in the pool.

When he was working offshore, he'd had no choice but to own his
own helmet—which divers call their "hat." Without it, you weren't
considered a diver. But as DJ looked around at the rest of the Vermont
crew, he noticed that very few of them had their own hats. He'd heard
that the standard practice for these "inland" divers, who plied their
trade in rivers and harbors, was to take turns using a helmet owned
by the employer.

When one of the inland guys told him he was about to learn how real diving was done, DJ let out a big laugh. "I know you guys work your asses off," he cracked, "but I don't know if you're really *divers*." He invoked another term the offshore guys had for their shallow-diving counterparts—*mud bugs*—and then said, "If anything goes wrong in the water, all you gotta do is take your hat off and stand up."

The inland guy made some quip about DJ being yet another off-shore prima donna, then said, "You'll learn."

DJ knew full well that he had chosen to come back east and do this kind of inland work. But he'd done that more for reasons of family. He doubted the assignments in New England would ever be able to give him the adrenaline rush he'd regularly gotten from his deep dives off the coast of Louisiana and Texas.

Yet it wouldn't take long for him to see that the inland guys had a point. Back in the Gulf, after he'd done a dive, he'd be required to sit in a chamber for a long period to avoid decompression sickness. Known as the bends, this potentially deadly affliction involves the formation of gas bubbles inside the body when divers make their ascent to the surface too rapidly from the high-pressure environment of the deep sea. DJ would then be forbidden from going back into the water for up to eighteen hours. Once he had been promoted from his job as a tender—the rookie who feeds hoses and checks equipment for full-fledged divers—he essentially never did that lowly work again. But on this Vermont job, he quickly learned that after he finished his dive, he was expected to keep working from the surface, serving as a tender for another diver. Because they were diving in depths of less than forty feet for no more than a few hours at a time, there was no need to decompress. That allowed for much longer workdays.

After suiting up, DJ splashed into the murky gray-green waters of the upper Connecticut River. He quickly realized the conditions were worse than he expected. Not only was there much less visibility than in the Gulf, but it was so much colder. In the relatively temperate waters down south, DJ had kept warm simply by feeding a hot-water hose into his dive suit and letting the liquid circulate. But that solution wasn't adequate for the frigid waters of Vermont in autumn.

One diver who had experience both offshore and inland showed DJ the work-around. Called a spider, it involved taking a thick, nearly foot-long hose—which the guys naturally called a donkey dick—and sliding the hot-water hose into one end of it. At the other end, there were half a dozen quarter-inch-diameter silicone tubes, which resembled the water hoses used by dental hygienists during cleanings. After DJ snaked one of these tubes down to each of his feet and hands, and others to his groin and back, the hot water would instantly circulate throughout his body. That was enough to make the bitter cold bearable.

There were two main strains of rivalry among commercial divers: inland versus offshore, and East Coast versus West Coast. Basically, the divisions boiled down to bragging rights about who did better work under tougher conditions. Regardless, there was one current that was always powerful enough to get them to close ranks. Whenever some scuba-diving hobbyist tried to invite himself into the diving fraternity, he'd be swiftly reminded that people who made their living as divers, no matter what type or on which coast, never got to spend their workdays frolicking around with starfish in the azure bathwater of the Caribbean. That kind of diversion might have made for a fun vacation, but it was *not* real diving.

Recreational divers, with their self-contained breathing apparatus, usually go no deeper than 130 feet. Commercial divers, attached by umbilicals to a surface supply of compressed air or mixed gas, can go to depths of a couple hundred feet. For shallower dives, they typically use a premixed oxygen-nitrogen combo called nitrox, and for deeper ones, a helium and oxygen blend called heliox. Going still deeper, from 300 to 1,000 feet, requires what's called saturation diving. Highly trained divers are lowered by crane in a special chamber called a bell, whose inside pressure precisely matches that of the deep water outside it. Those "sat" divers typically don't surface for about a month. They work down below from this bell and sleep in bunk beds in a slender attached vessel filled with a helium-rich air supply that makes the divers all sound like Donald Duck.

At this point, climbing the ranks to become a sat diver remained a distant goal for DJ. Still, he'd already spent countless hours making heliox dives in the 200-to-300-foot depth range. So even if he was still adjusting to the cold and the murkiness of this river water in Vermont, he knew he'd have no trouble shining when the real work began.

He inserted a waterproof electrode, or rod, into a rubber stinger, fired up the DC power, and watched the familiar arc form. This was underwater welding. The very idea of it seems to violate the public service announcements that every child has heard about how electricity and water don't mix. But when done with the right equipment and under the right circumstances, underwater welding is a safe and essential part of marine construction. In the Gulf, DJ might go through fifty rods in one dive, either welding parts together to repair something or "burning" to break down a structure, such as an old oil platform. The scale was quite a bit smaller on this job in Bellows Falls, a village of restored brick buildings nestled between the river and the green mountains that gave Vermont its name.

Native Americans used to gather at "Great Falls," a gorge at a narrow point of the otherwise wide Connecticut River, to catch migrating salmon. But the migration had largely ended two centuries earlier, when the waters were dammed and a canal was built. The waterfalls were used first to power multiple mills and later to generate electricity. Now, after forging an agreement with environmental groups, the big power company in town was working to make things right again for the salmon. The job involved building an elaborate concrete fish "ladder," which would allow the migrating young salmon to get past the hydroelectric plant without getting mangled in its turbines.

To do construction in water, you must either work submerged or you must move the water. The plan here mostly called for the latter, through the construction of a cofferdam. This temporary dam structure, the kind often used on bridge construction or repair work, resembled an enormous bread loaf pan, twenty feet wide by a hundred feet long. Once it had been assembled, out of steel sheet piles and gaskets, all the water inside it would be pumped out, allowing workers to work "underwater" in an entirely dry space, as they built the actual fish ladder. Still, constructing the cofferdam—everything from drilling anchor holes to cutting steel—required real underwater work, a challenging task that naturally fell to divers.

As the months-long job wore on, DJ found himself bonding with the inland guys. They were clearly impressed with his burning skills, and he had gained a new appreciation for the tough conditions they had to endure. Not that he would ever admit to that publicly. Commercial divers exist in a hardworking, hard-drinking, itinerant world where they must constantly prove themselves to one another. At the bar after one shift, DJ continued to press his chauvinism. "Gulf divers," he crowed, "are more experienced, smarter, better looking, get more women, drink more beer, and drive better trucks than inland divers."

Outside the construction trailer, a short guy with his sandy hair pulled back in a ponytail charged up to a much taller, much younger diver and clocked him. The kid had been running his mouth, trying to take credit for some burning work the older guy had actually done, and he'd also been tossing his wet gear onto the other divers' dry stuff. So the ponytailed guy had administered some riverside justice.

DJ looked on in admiration. Dive work was often like a relay race, where one guy would get as much of a task done as he could and then hand things off to the next guy. Sometimes divers would burn their initials or a mark to note their progress, before coming up to the surface. But it was essentially an honor system. This young diver had violated the code, so he needed to be taught a lesson.

In no time, that ponytailed guy, who was two decades older than DJ, became his closest friend on the job. Ron Kozlowski seemed to have packed a couple of lifetimes of adventure into his forty-four years. As a Navy SEAL, he'd served in Vietnam and been injured in a firefight near Chu Lai. As a commercial diver, he'd worked both offshore and inland jobs all over the globe. As hard as he worked, the cocksure guy was just as serious about his surfing, photography, and mountain biking, and he spent the bulk of the winter at his place in Costa Rica. While most of the other Vermont divers stayed in one motel, DJ followed Kozlowski to another one ten miles away. They both worked the overnight shift, where Kozlowski taught him, among other things, how to do that spider work-around to keep warm. They began hanging out together during the day, whether snowboarding or just relaxing over a few beers back at the motel. In Kozlowski, DJ spied the kind of future he wanted.

Kozlowski cautioned DJ that the diving life had its drawbacks. No matter how seasoned you were, danger was always lurking. A year earlier he'd been repairing electrical cables that were buried just under the seafloor in the 130-foot waters of Long Island Sound. After crawling along a lengthy stretch, with very little visibility, he finally located the bad splice that had to be repaired. But when he went to stand up, he was shocked to find he couldn't lift his head up more than a few feet. When enough mud had cleared, the headlamp on his helmet revealed that he was trapped under a very large fishing net. He was horrified, but experience told him the fear coursing through him would put him at risk for hyperventilating. To guard against that, he turned up the air in his hat. He used the comm wire in his umbilical to tell a coworker on the boat above him that he was in trouble. With his heart racing, he lay flat on his stomach and began crawling in reverse. After he'd shimmied himself some fifty feet, he reached up his hand to feel for the net. It wasn't there. He was in the clear. The next day the boat pulled up the discarded fishing net, which turned out to be two hundred feet long.

Kozlowski knew DJ was a young, ambitious diver who had a mother he was close to and a girlfriend he was crazy about. Dana, the hairstylist, had come to visit DJ in Vermont more than once. Kozlowski was middle-aged and divorced. Still, he'd been through enough to warn DJ never to come clean with his loved ones about how hazardous diving could be. He told him about a T-shirt that had been popular in the trade when he was starting out. It featured the image of a helmeted diver playing a piano that had a scantily clad woman lying on top of it. Underneath that picture was the caption DON'T TELL MY MOTHER I'M A DEEP SEA DIVER. TELL HER I'M A PIANO PLAYER IN A WHOREHOUSE.

DJ first met Tap Taylor when Tap came to Bellows Falls to pick up some extra hours. Tap was working in Vernon, a Vermont town just down the river, on another project being run by the same employer, the respected construction giant Kiewit. DJ later learned that Tap's full name was George Tapley Taylor III, a handle that made him sound more like a Harvard legacy than a sturdy, down-to-earth diver. So Tap it was.

Like DJ, Tap had just recently joined the union. The hardworking guy had meaty hands like vise grips, and they always seemed to be in motion. A thirty-year-old with receding brown hair, Tap had a wife and one-year-old daughter at home back in New Hampshire. He had dropped out of college after one semester and logged some fun years as a ski instructor in Colorado, then some less fun years helping his father build houses. In commercial diving, he told DJ, he'd finally found his calling.

What Tap lacked in training—he had never gone to commercial dive school—he made up for in drive. Not long after he'd earned his scuba certification in high school, he had become fascinated by a 1710 shipwreck off Maine's Boon Island. Not only did he dive to search for wreckage near the coast, but he arranged with the Boston Public Library to view the captain's log, donning white gloves over those rugged hands of his and tenderly turning each page. It was clear to DJ that once Tap got interested in something, he went all in.

As Tap began working more shifts in Bellows Falls, he and DJ grew closer. Although Tap was a family man hustling to earn enough money to make a down payment on a house, he found DJ's footloose style a welcome reminder of his own freewheeling bachelor days. And DJ found in Tap a straightforward guy who liked to have fun but who also had a clear sense of direction in life.

In short order, DJ had formed two close friendships on the Vermont job. In the beginning, all three guys got along well. Before long, though, tension developed between Kozlowski and Tap. Some of it had to do with personality differences, but more of it was about how to deal with management.

Tap soon tired of all Kozlowski's stories, in which the veteran always seemed to cast himself in the lead. More than that, he saw Kozlowski as a chronic complainer, someone who made lots of demands of the bosses and kept chiseling away until he got what he wanted. When Kozlowski tried to rile up the diver rank-and-file and insist that their Kiewit managers give them more holiday time off, Tap wanted no part of it. He could only shake his head and wonder, *What will it take to keep this guy happy?* Tap had no problem admitting how grateful he was to be working for Kiewit, which, despite being based in off-the-radar Omaha, Nebraska, was one of the biggest contractors in the country.

Kozlowski, meanwhile, saw himself as wise to the ways of the world. He liked how receptive DJ was to his advice, but he sensed growing resistance from Tap. And he came to see Tap as a naïve guy too concerned with trying to please management. Kozlowski had lasted in the field a lot longer than most divers, but he knew that hadn't been by accident. Over the years, he'd seen friends injured and even killed, often because of money. Sometimes an employer had been trying to save a buck and insisted on shortcuts that put lives at risk. Other times it was the diver who was reckless. "Most guys don't make it ten years," he told DJ. "They get hurt seeing dollar signs, taking chances, trying to be a hero to impress the bosses."

Kozlowski told DJ about a big job in Boston that had consumed most of his time over the last two years. It was related to another Kiewit project, the construction of the miles-long Deer Island sewer tunnel. But Kozlowski was working for a different contractor on a jack-up barge parked in Massachusetts Bay, above the tunnel's eventual end point. In the fall of 1991, a storm raced up the East Coast so fast that the contractor missed its window for bringing the three-thousand-ton barge safely back to shore. The swells were lapping at the barge, even though it had been jacked up ninety feet above the water. Still, Kozlowski said, the rig's supervisor actually instructed him to make a dive to inspect a drill shaft. "Are you freakin' crazy?" Kozlowski shot back. He refused to go.

The supervisor backed down, and soon that guy's bosses ordered the entire crew to evacuate the barge. Even though Kozlowski had survived chopper rides under gunfire in Vietnam, he said his helicopter ride from the rig back to land that day was the scariest of his life. During the storm, the only way the contractor knew whether the evacuated jack-up barge was still standing was by repeatedly calling the fax machine aboard the rig and hearing it beep. After that weather event on October 30, 1991, passed, it was given a new name: The Perfect Storm.

Kozlowski used the story to support his contention that divers had to look out for themselves and each other, even if that made them unpopular with the higher-ups. If they didn't put their own safety first, who would? Kozlowski told DJ that the contractor on that job had actually not been a reckless one. And some hazards are simply the nature of the business. Still, he said, on virtually every job, no matter how well run it was or how much the employer talked about safety, a diver was going to be asked by someone—at some level of authority—to take an unnecessary risk. Having the good sense and strong stomach to stand up and say no, Kozlowski said, was a muscle that every diver needed to develop.

DJ tried to avoid choosing sides between Kozlowski and Tap, though he put more stock in what a seasoned ex–Navy SEAL like Kozlowski had to say. Still, he had no way of knowing then that Kozlowski would one day give him the most important advice of his life—advice that DJ, for some reason, would decline to take.

The Vermont winters were too brutal to get much work done in the river, so the fish ladder project was put on hold till the spring thaw. DJ spent the break back in Waltham. One day he drove to a construction site at a hospital in Boston. He stopped a guy in a hard hat and asked, "Can you get Timmy O'Leary for me?"

"Who's looking for him?" the guy replied, guardedly.

"Tell him it's DJ from Local Fifty-six."

DJ had deliberately worn a shirt and hat from his union. This visit might have made him uncomfortable, but he wanted to make it clear to everyone that he was a thriving twenty-four-year-old *man*. He had every right to be there.

For a long time, DJ had been wrestling with the idea of trying to track down his biological father, a mysterious figure he'd grown up resenting but had never met. The only time his mom ever invoked his name was when DJ showed his fastidious side—for instance, delicately spraying Armor All on his steering wheel. His mom would quip, almost without thinking, "That's just like Timmy O'Leary."

Finally, DJ had decided to act on his curiosity. He knew his mother had no contact with his father, but with his aunt's help, he tracked down the man's brother. The brother cautioned DJ, "Your father's his own man. He's very private." But he did cough up the fact that his brother was working on the hospital project. "You'll know he's there if you see his truck," the brother said. "And you'll know it's his truck because it will be immaculate." In a muddy parking lot, his would be the truck without any mud on it and with a towel draped protectively over the dashboard.

DJ knew his father had chosen not to be in his life, and for years he had kept that piece of resentment like lint in his pocket. Now, as he was forced to wait for the mystery man, he wondered if he should have just left it there. Tired of standing by, DJ entered the job site and found the first guy he had talked to now walking toward him with another guy at his side. The second man spoke first, and defensively. "You looking for me?" he asked.

"You Tim O'Leary?" DJ replied.

He nodded. He was shorter than DJ and had the compact build and crew cut of a drill sergeant.

"I think you'd appreciate it," DJ said, "if we can speak by ourselves."

O'Leary waved him off. "Whatever you've got to say, you can say it right here."

"My name is Donald James Gillis," DJ replied, enunciating as clearly as if he were identifying himself on the stand in a courtroom. "I think you'd prefer if we talk privately."

His father's tone softened, and he led DJ to a quiet area.

Sensing that the man was bracing for some bad news, DJ told him there was no emergency. "It's just that time's ticking, everybody's getting older, and I wanted to see who my father was."

O'Leary apologized. "I tried to contact you a long time ago," he told his son. "Your aunt said you were living out west. There was no phone number. After a while I just gave up."

There was no denying the awkwardness between them, but both sides seemed interested in trying to ease that. After talking for a few minutes, his father asked DJ for his number and said he'd call him sometime.

DJ never got that call.

Standing on a rocky ledge while submerged in the river, DJ clutched the sinker drill. The tool resembled one of those jackhammers that road crews use to tear up old stretches of highway, but it also had a drill rotation. It was early 1994, and DJ was back in Vermont working on the cofferdam. The divers had drilled holes in the river bottom and used them to anchor the base of the giant watertight "loaf pan." Now they needed to secure its sides to the riverbank by drilling anchor holes there. This was turning out to be a much tougher task, owing to the unevenness of the rocky surface. DJ had begun by standing on steadier ground, positioning the sinker drill above his head, but that seemed too precarious a way to hold such a powerful tool while standing underwater. So he'd sought more leverage by climbing up onto a small ledge. Still, in order to get the drill where he needed it, he had to stand on his tiptoes. He jammed the tool under his armpit, holding it out horizontally in front of him, as if it were a big pipe he was carrying. Then he started to drill. Suddenly a strong river current tugged him, pulling him off the ledge. The drill remained jammed under his armpit and kept rotating for a spell, torqueing his torso before tossing him onto the riverbank. He was left writhing in pain.

So much about this Vermont job had changed for DJ. Early on he had raced back to Waltham whenever he could to see his girlfriend. During one visit, DJ had had dinner with Dana and her doting grandmother, an immigrant from Italy. When the grandmother said something in Italian, Dana blushed. DJ asked for a translation. "She wants to know when you're going to put a ring on my finger," Dana said. DJ was a young guy who loved women and the nightlife too much to consider settling down. Still, the more he thought about it, the more he realized that Dana really could be the one. If getting engaged was what it took to keep her, he told himself, he would do it. Dana was unlike his previous girlfriends. Although those girls had been attracted to DJ's life-of-the-party charisma, they always ended up trying to change him, determined to get him to commit more fully to their relationship. Dana seemed so much more secure in herself. As the Vermont job wore on, though, DJ began to feel like he was the one wanting more of a commitment and suffering self-doubt when his calls went unanswered. Eventually his doubts were realized when Dana broke things off.

Meanwhile, his best friend on the Vermont job, Ron Kozlowski, had chosen not to return after the fall. Fortunately, Tap Taylor was still around, and DJ's friendship with him had deepened. Despite the still-frigid water temps, Tap did most of his dives using only a band mask. Much cheaper than a diving helmet, the band mask had the same kind of front and regulator system as DJ's SuperLite 17, but rather than being made out of fiberglass, it was essentially just a rubber hood. So it did little to protect a diver's ears and head from the cold. As they worked more shifts together, DJ began to let Tap borrow his helmet. For a diver, this was no small thing. For starters, there was the matter of hygiene. In serving as a barrier from the cold water, a dive hat became a reservoir of saliva and mucus. So sharing a helmet meant sharing a whole lot more. Beyond that, a diver's hat was an essential component of his professional identity, something he viewed as protectively as a woman might treat her diamond engagement ring. The fact that DJ was letting Tap take turns using his hat said a lot about how close they had become.

At one point, Tap introduced DJ to another diver, whose name was Billy Juse. Tap had met Billy working on the other Kiewit site, downriver in Vernon. A small guy with a black mustache, Billy was a hell of a diver, and DJ could tell how well he and Tap worked together. The bosses in Vernon had noticed the same thing.

For Kiewit, the twin Vermont jobs on the Connecticut River were beginning to wind down. At the end of most sizable projects, there is a punch list of small tasks that need to be done but don't require a full crew and are scarcely worth a big company's time. So the big company will often try to subcontract out that remaining work. One of the Kiewit supervisors approached Tap to see if he and Billy would like to be considered for this kind of assignment. Tap was ecstatic. It bolstered the line of reasoning he had taken in his disagreement with Kozlowski: if you work hard for your employer, rather than acting as though they're always trying to screw you, good things will happen.

To be considered for the work, the Kiewit manager told Tap, he and Billy would need to have a fully insured and incorporated company. The supervisor had heard that Tap had a side business, called Black Dog Divers. Tap had started it years earlier to handle very basic dive jobs like changing moorings, but he'd put it on hold as he pursued his commercial diving career. "Is Black Dog incorporated and insured?" the supervisor asked. Tap explained that it would take some paperwork, and that if he was going to all that trouble, he'd probably change the name. He'd originally chosen Black Dog because he had a black Labrador puppy at the time, but now he worried the name might not sound serious enough to win jobs from big companies like Kiewit. "Don't you dare change that name," the supervisor said, pointing out that it would stick in people's heads. So Tap and Billy teamed up and began working to transform Black Dog Divers into a respected subcontractor.

Meanwhile, DJ continued to struggle following his fight with the sinker drill. The pain, which he had initially thought could be handled with rest and lots of Advil, only intensified. DJ couldn't bear the discomfort when he tried to go back into the water. Visits to a chiropractor and specialists brought no improvement. He concluded that he had to go out on workman's comp.

His time in Vermont had begun with such promise. Fresh with a new love and a new career focus, he was now heading back home with uncertainty, heartbreak, and excruciating pain. After saying his goodbyes, he hopped into his Bronco and started it up. But then a thought crossed his mind, and he cut the engine. He grabbed his SuperLite 17 and walked back over to where Tap was standing. His new friend might own a company that could boast a great name and the prospect of becoming a favored subcontractor to one of the nation's biggest builders, but Tap Taylor still didn't own a diving helmet. "Here," DJ said, handing over his prized possession. "Put that band mask away and use this until I need it back."

2

Island Secrets, 1675–1995

The clear glass bottle on Doug MacDonald's desk looked like it should be holding salad dressing, but the true contents were a lot less appetizing. The label read "Boston Harbor Water: 100% Authentic, Undiluted, Polluted." Near the top was a band of cloudy water. Settled at the bottom was thick, oozing, disgusting mud. Someone had hatched this gag idea of bottling water scooped up at the Boston coastline as a way to highlight what a front-page newspaper headline had memorably called the "Harbor of Shame." MacDonald kept the bottle as a reminder that it was his job to bring redemption to one of the most famous bodies of water in American history. But by the summer of 1995, he had to confront the reality of just how elusive that goal remained.

Three years earlier, MacDonald had left his job as a partner in a big Boston law firm, where he had been pulling down as much as a quarter of a million dollars a year, so he could take a government agency post that paid less than half that and brought with it an unholy amount of scrutiny from the press and the public. There had been only two predecessors in his position. The first guy had been ousted after pushing through rate hikes while using taxpayer dollars to furnish his office with twelve-hundred-dollar leather chairs and hundred-dollar marble ashtrays. The second guy had more than once returned home to find death threats on his answering machine and his young daughters in tears at the sight of angry protesters camped in their front yard.

MacDonald's job was executive director of the innocuous-sounding Massachusetts Water Resources Authority (MWRA), the quasi-public agency that oversees sewer and water services for Boston and nearly fifty other cities and towns in the eastern half of the state. What made the position such a magnet for controversy was that the MWRA was in charge of the massive cleanup of Boston Harbor. The harbor into which colonists had famously dumped tea to protest British rule had, over the centuries, served as a different kind of dumping ground: it had become a giant, stinking cesspool for the region's human waste.

A century earlier, Boston had unveiled a sewage treatment system that *Scientific American* had trumpeted as perhaps the most sophisticated in the world. In reality, that system didn't treat the sewage so much as use a network of tunnels to move it away from where the people were. By the 1960s, there were two plants treating the sewage of eastern Massachusetts, one on Deer Island (a peninsula attached to the small town of Winthrop, north of Boston) and one on Nut Island (a peninsula attached to the small city of Quincy, to the south). But from the start, these plants were hopelessly inadequate. Treatment basically consisted of letting the wastewater stand long enough so it could be separated into sludge and sewer water. After light treatment, the sludge was dumped right into the harbor along with the sewer water, negating much of the benefit of the separation process. About 140,000 pounds of that toxin-dense sludge and about half a billion gallons of that treated sewer water were released into the harbor every day. Even worse, as the population grew and public officials let the facilities rot, the plants couldn't keep up, especially when the skies opened. For days after big rainstorms, millions of gallons of untreated sewage would never make it to the plant. Instead, the sewage was diverted into a system of release-valve pipes and eventually let out into the harbor. (On Christmas Day in 1980, the skeleton crew on duty allowed a staggering fifty million gallons of raw sewage to flow into the harbor.) How little did Massachusetts officials invest in their sewer system? At one point, the Smithsonian Institution contacted them in the hopes of acquiring one of their historic pumps. To the Smithsonian's astonishment, the officials explained they couldn't part with the relic of a pump because they still relied on it for everyday use.

Long after the harbor had ceased being the linchpin of the American maritime economy in the eighteenth century, it had continued to be the source of bountiful fisheries—the fabled cod as well as softshell clams, lobster, and flounder—and home to seals and dolphins. But by the 1980s, the years of flagrant disregard had left the harbor thoroughly degraded. Beaches were regularly closed for health reasons, water tests found alarming levels of PCBs and pesticides, and marine life had taken on the pall of death. In the harbor's Dorchester Bay, the EPA had classified 86 percent of the clam beds as "grossly polluted," finding levels of fecal coliform bacteria that were ten times what was acceptable. Winter flounder were riddled with cancer and "fin rot," a malady as bad as its vivid name suggests. The seals and dolphins had mostly departed.

Studies found that the organic material from all that dumped sewage had settled and decayed on the seafloor, creating an oozing mud known as "black mayonnaise." Rich in hydrogen sulfide and lacking dissolved oxygen, this mud was snuffing out fauna and enfeebling the food chain on which the harbor's ecosystem had long relied. Fatigued locals turned to gallows humor to distract themselves from their disgust. They rebranded the tampon applicators floating out of the sewer pipes and into the water as "beach whistles" and the spent condoms as "Charles River whitefish." The latter was a reference to the river that feeds into the harbor, whose filth had won it infamy in the 1960s thanks to the Standells' rock 'n' roll paean to pollution, "Dirty Water": *Well, I love that dirty water / Oh, Boston, you're my home.*

In 1982 a local lawyer named Bill Golden went out for a jog along the beach and returned with his sneakers covered in sewage debris. The lawsuit he filed triggered the Boston Harbor cleanup. The sorry state of affairs mushroomed into a national issue in the presidential campaign of 1988, when Republican nominee George Bush hopped on a ferry and led camera crews around what he called "the dirtiest harbor in America." Following up his ferry attack with a raft of national television ads, Bush successfully embarrassed his Democratic opponent, Massachusetts governor Michael Dukakis. By then, the cleanup was already under way, but officials had delayed so long that the federal money had dried up. That meant the cleanup's entire price tag, initially estimated to exceed $6 billion, would have to be shouldered by average citizens. These rate payers, who had already seen their water and sewer bills triple, were incensed by the steeper hikes on the horizon.

It was MacDonald's job to try to mollify them. All the while, the fifty-year-old had to preside over a leviathan construction project with multiple components. A state-of-the-art sewage treatment plant—the second largest in North America—was being erected on Deer Island, a two-hundred-acre smudge on the map of Boston Harbor. And two separate, massive tunnels were being bored under the sea. The first of these was the five-mile Inter-Island Tunnel, which would carry sewage from Nut Island to Deer Island for treatment. Once the sludge was separated out, that smelly guck would be sent back to Nut Island, ingeniously in a small tube installed inside the new tunnel, where it would then be converted into fertilizer pellets and sold to commercial clients nationwide. The second tunnel was the far more complicated 9.5-mile-long Deer Island Outfall Tunnel, which would carry

the treated sewer water away from the harbor and release it deep into
Massachusetts Bay, an inlet of the Atlantic Ocean. Angry activists on
Cape Cod were pushing to halt construction of this tunnel, fearful
that all the wastewater flowing out of it would harm marine life and
damage the ecologically fragile waters off the Atlantic coast. Since
the tunnel's endpoint was about sixteen miles from the national marine sanctuary known as Stellwagen Bank, some environmentalists
worried that all the treated sewer water would harm the important
population of marine mammals, especially the endangered northern
right whale. As if this weren't enough to contend with, MacDonald
had to answer to a bow-tied federal judge who served as the cleanup
project's robed referee, threatening stiff fines or worse if the deadlines
he imposed were not met.

MacDonald was a recently divorced father of two and a diabetic
whose doctor had implored him to reduce the stress in his life. During one medical exam, he took a health quiz that asked him to check
off sources of stress in his personal and professional life; each item
on the list was worth ten points. He scored 180 out of a possible 200.
Yet he had aggressively campaigned for his thankless post. He was a
workaholic who could no more avoid stress than a pack-a-day smoker
could resist lighting up after a nice meal. His pace often left the people around him feeling drained, but he tempered his relentlessness
with an affable personality. He'd regularly flash a smile that would
turn his eyes into slits behind his boxy glasses and light up his jowly
face. Despite graduating *magna cum laude* from Harvard Law School,
he brought a childlike enthusiasm to everything he did, especially to
the construction side of his job.

He was responsible for running an agency with seventeen hundred employees while also overseeing a harbor cleanup that, at its
peak, employed more than three thousand construction workers and
burned through $3 million a day. A finance industry journal called
his agency "the most prolific issuer of municipal revenue bonds" in

the country and quoted a sewer specialist saying the Boston project was one of the most challenging that anyone "in Western Civilization has ever been involved with." Its price tag put it ahead of some of the biggest megaprojects of the time, from Denver's new airport to San Francisco's expanded transit system to New York City's third water tunnel. At one point, there were 360 different companies working on the Deer Island project alone. The management challenges he faced would have been a stretch for even the most experienced CEO, never mind MacDonald, who was serving as a manager for the first time in his life. It was no surprise, then, that after the MWRA board unanimously elected MacDonald to run the agency, a politically savvy friend offered him this bit of advice: "Demand a recount."

For MacDonald, there had been a promising start to 1995, since construction of the new Deer Island plant was months ahead of schedule. But that progress couldn't mask the fact that his team was woefully behind on one of the project's other key milestones. The federal judge's court order had set the summer of 1995 as the deadline for the completion of the outfall tunnel. By now, MacDonald knew there was no way that was going to be met.

MacDonald was just the latest in a long string of Deer Island overlords. Since childhood, he had been almost clinically inquisitive, so it seemed natural for the new executive to spend a good chunk of the little free time he had reading up on the tiny island's sordid past.

Although the island drew its name from the deer that had been chased off the mainland by wolves, on a map it looks more like a dog. More specifically, it resembles a Labrador's head in profile, tipped down, as though drinking from a bowl. The top of the dog's head, to the northeast, was the location of the deep shaft leading to the outfall tunnel. The side of the dog's face represented the center of the island, home to the new treatment plant. The dog's mouth and snout, at the southeastern end, was where the dozen egg-shaped giant digester tanks stood guard. And the dog's neck, to the northwest, was the spit of land that had attached the island to the mainland ever since a 1938 hurricane had moved around enough sand to close up the little channel.

The more MacDonald read, the more he found himself riveted by the island's story. Its darkest chapter dated back more than three centuries. In the mid-1600s, a Puritan minister named John Eliot had used his mastery of the Algonquian tongue to translate the Bible and convert hundreds of Indians to Christianity. His missionary work led to the creation of fourteen "praying towns," villages ringing Boston that were populated by Christianized Indians. In 1675, when the Wampanoag leader King Philip led the charge to oust the colonists from the area they had rebranded New England, most Christianized Indians remained loyal to the colonists. But the Massachusetts Council, seeing enemies all around, ordered that the Christianized Indians be rounded up from their praying towns and forced into confinement on Deer Island. They were penalized by the colonists for the sin of having believed them.

The combination of the island's barren soil and its lack of protection from the pounding waves of the Atlantic led to the deaths of about half of the five hundred interned Indians, if not more, during the brutal winter. When the survivors were released in the spring, many returned to their villages to find that their land had been confiscated. The island was then converted into a prisoner camp for captured combatant Indians, many of whom were later sold into slavery in the West Indies and Africa.

In the centuries that followed, Deer Island continued to be the place where Boston quarantined its problems. Even after that hurricane turned the island into a peninsula, the place still somehow managed to stand apart. In the mid-1800s, when the potato famine brought desperate Irish to the shores of Boston, city officials sent many of them to a holding station and hospital on Deer Island. In just a couple of years, more than eight hundred died on the island and were buried in a makeshift cemetery. In later years, when the island was housing the poor, the criminal, and the insane, several thousand more people ended up making their final rest there.

In 1920 more than four hundred eastern European immigrants were rounded up and herded into a filthy, overcrowded prison on Deer Island. They had been caught up in the so-called Palmer Raids, part of the nation's first Red Scare, which took its name from Attorney General A. Mitchell Palmer. He had ordered the arrest of several thousand eastern European immigrants in about thirty cities on baseless charges that they were Communists and anarchists conspiring to bring down the government. In the chaos of the first few days of confinement on Deer Island, one prisoner hurled himself to his death, another was committed as insane, and several others attempted suicide.

MacDonald was struck by the common themes in these dark incidents. Leaders had let fear get the best of them, deeming people suspect simply on account of their otherness, then had used the remoteness of a barren island as cover for their misdeeds. Deer Island was still being used as the solution to a problem few people wanted to think about, but at least human beings were no longer the problem that needed solving, just their waste. That seemed like progress.

As he was lowered more than four hundred feet down the shaft leading to the tunnel, MacDonald could feel his boyish enthusiasm begin to bubble up. While he was headed down, "muck"—the chunks of rock that had been mined and excavated below—was headed in the other direction, up to topside, via a vertical conveyor belt. Stepping out at the base of the shaft, with sunlight snaking down from above, MacDonald breathed in stale air that smelled like a musty basement. He hopped aboard an open-air tram, which was pulled by a locomotive—or "lokey"—and traveled along railroad ties that ran the length of the tunnel. Eventually, the lokey deposited him behind a gargantuan rock-eating beast called a tunnel-boring machine. This TBM looked like a giant flat-nosed bullet but functioned like an enormous drill bit. On its rotating faceplate, called a cutter head, the TBM had fifty steel teeth. While the cutter head rotated, pressure from the teeth hitting the bedrock—Cambridge argillite, formed hundreds of millions of years earlier—was powerful enough to turn it into chunks of muck.

The $15 million TBM was carving out a tunnel with a diameter that matched the width of the machine itself, twenty-six feet. Weighing seven hundred tons with the horsepower of thirty cars, it was energized by a 13,800-volt electrical power line, essentially a fat extension cord bracketed to the tunnel wall. The TBM's main body was forty feet in length, but a series of trailing sections stretched it out as long as a football field.

As the TBM ground away, foot by foot, a ventilation "bag line" of rubberized fabric material would unspool out of a canister at the back of the machine, like a supersize party noisemaker. Workers would hang the bag line to the tunnel ceiling, after which a pair of 250-horsepower fans sitting on Deer Island would pump air down the shaft and into the bag line. Like the ducts of an office building's HVAC system, the bag line would take the fresh air from topside and distribute it along the full length of the tunnel, providing ample ambient air for everyone working down under.

The general contractor building the tunnel was a joint venture called Kiewit-Atkinson-Kenny. Although there were three names on the letterhead, the Kiewit Corporation was the dominant partner, and shorthand for the joint venture naturally became "Kiewit" (pronounced *KEE-wit*). Erecting everything from bridges to power plants, the Omaha-based construction giant had built more miles of the U.S. highway system than any other contractor. Kiewit had won the Deer

Island tunnel job after being the lowest bidder, at a lump sum of $202 million. If Kiewit ended up spending more to build the tunnel, it would have to eat the losses. In addition, the contract required Kiewit to deliver a completed tunnel by mid-1995, or pay a penalty of $30,000 for every day it was late. So Kiewit was taking on considerable risk. Still, the $202 million price tag was a big deal at the time, representing the largest public contract in Massachusetts history and the biggest line item of the overall Boston Harbor cleanup project. (It preceded the Big Dig, the multibillion-dollar project that would replace Boston's ugly elevated interstate with a highway carved under the city.) And the Deer Island tunnel deal had been made during a recession, when many contractors were submitting unusually lean bids just to keep their crews busy.

When someone builds a house, there are usually three major players: the architect who designs it, the general contractor who builds it, and the owner who pays for it. The same was basically true for the tunnel project. The designer was a world-renowned firm called Parsons Brinckerhoff. The general contractor was the Kiewit joint venture. And the owner was the agency MacDonald ran, the MWRA. To protect its interests, his agency hired a separate firm called ICF Kaiser Engineers to serve as construction manager. Essentially, Kaiser functioned as the MWRA's advocate, or something akin to a buyer's agent, for the entire cleanup project.

Considering his inexperience in both construction and management, MacDonald deferred heavily to his deputies and to Kaiser's managers. They told him that, by any measure, the tunnel had gotten off to a horrible start. The problems had begun with the digging of the main shaft and had continued to mount. Once the TBM had finally been assembled and began turning and burning, it became evident that the machine's toughest assignment wasn't the actual boring. After the TBM had carved out a section of the tunnel, the entire circumference had to be lined with four-ton concrete segments, which reduced the tunnel diameter to twenty-four feet. Crews used an erector arm on the back of the TBM to position each five-foot-long concrete segment and then grout it into place. The concrete lining had to be thick enough to maintain the integrity of the tunnel, and the segments had to be aligned perfectly to prevent anything from interfering with the flow of the wastewater that the tunnel would eventually carry out to sea.

But the concrete segments cracked and chipped. And the gripper arms that were used to brace the TBM against the tunnel wall broke. And the pumps that were used to remove groundwater from inside the tunnel failed. (The tunnel had been designed to be built on a slight incline, so that water seeping in through cracks in the bedrock during construction would naturally drain back to the base of the shaft, where it could be pumped up to the surface.) Twice the high-voltage panel box that controlled the TBM's supersize power cord exploded.

MacDonald could tell that Kiewit officials were determined to get the project back into the black and make up for all the lost time and money. Under the contract, their best ways to do that would be to submit change orders for relatively small dollars and to file claims for big money; in both cases Kiewit would argue that MacDonald's agency had to pay them more because his team had somehow failed to adequately explain what the job would involve. That meant it was a zero-sum game. For Kiewit to win, MacDonald's team had to lose.

MacDonald watched as the brawny guy with black hair grabbed the jackhammer-type tool and held it up on his shoulder, like a mujahideen fighter pointing a Stinger antiaircraft missile toward the sky. Once the tool was in position, he leaned against a support brace and started drilling. Chips of rock began flying around him. The tool was called a jackleg, and it's what tunnel workers used for the smaller mining tasks that the TBM couldn't handle. The big guy with a booming voice was a forty-two-year-old sandhog named Bobby Malkasian. But he was better known by his nickname, Mad Man, which testified to the short fuse under his friendly demeanor.

The term *sandhog* comes from the soft, sandy soil that miners often encounter underground. On this job, every sandhog had a nickname—the raunchier the better. One guy was called Big Daddy because he was reputed to be hung like a mule. Walking Small got his nickname because the tiny, feral-faced guy would bark at other hogs, "Open your fucking mouth one more time, and I'll drive a fucking spud wrench through your fucking eye." And Fingers got his handle after he suffered a nasty tunnel accident and came out with only nine.

As MacDonald made more forays into the tunnel, he got to know Mad Man and many of the other sandhogs. Kiewit managers blamed a lot of the tunnel delays on the unruly union laborers, most of whom had no experience working with an advanced tunnel-boring machine. In other parts of the country, the contractor had developed a reputation for something approaching militaristic precision. But in Boston, Kiewit managers had a tortured relationship with the sandhogs and the equally untamable leaders of the Laborers' International Union's Local 88. The local's recording secretary was forced out after a drug conviction, and the results of its officer elections were tossed out after a federal court found evidence that challengers had been roughed up. One challenger said that during an argument, the head of the local had pulled a .22-caliber semiautomatic handgun out of a drawer, cocked it, and pointed it at him.

Sandhogs had always been a special breed, attracting hardened men (and very few women), including more than a few ex-cons. Given the punishing conditions that they put their bodies through underground—working with explosives, dealing with the ever-present danger of a wall collapse, breathing God knows what—even those who lived to see retirement could expect to meet an earlier grave than those in other trades. At the very least, their hearing would be shot, a consequence of working around noisy machinery and clanking wrenches, whose din is intensified by the way sound reverberates in a tunnel. Before government safety standards required tunnel workers to wear protective ear muffs, following the creation of the Occupational Safety and Health Administration (OSHA) during the Nixon years, many sandhogs would try to reduce the noise by sticking cigarette butts in their ears. Not surprisingly, when they were above ground, most sandhogs found themselves to be the loudest talkers in the room.

As much as MacDonald shared Kiewit's frustration with the antics of Local 88, he found himself charmed by the sandhogs. These guys did tough work and followed their own code. When things between two sandhogs became heated, one would throw down his tool belt and head behind the muck pile, and the other guy would instinctively follow so they could settle the matter with their fists.

For MacDonald, the sandhogs' roguish charm was epitomized by Mad Man, whose path into the field had been far from typical. After graduating from college, Mad Man had begun teaching high school in his hometown, Cambridge, when he bumped into a few sandhogs working on the extension of Boston's subway. Outside the gates of rarefied Harvard Yard, one sandhog proudly wore a sweatshirt that read, in big block letters, MOTHERFUCKER. Their work sounded interesting to Mad Man—more interesting, he would explain later, than "sitting in a classroom dealing with rectal sphincters." So he signed on. By the time of the Deer Island job, he had become a brassy veteran who wouldn't take crap from anyone. Once when he caught a sandhog defecating near where the guys stored their sodas, he and a few pals poured cement into the back seat of the offender's car. He especially had no use for the few sandhogs who kissed up to the bosses. He accused one of trying to fellate his way into management's good graces, nicknaming him Jimmy Blew.

MacDonald was especially impressed with how sandhogs like

Mad Man wielded the jackleg. The MWRA boss knew firsthand how hard that task was, having once received permission to use the jackleg himself, just to see what it felt like. After drilling two feet into the tunnel wall, he was wiped out, thinking to himself, *That's the hardest thing I've ever done in my life. And this guy does it all day?*

MacDonald knew that one of the biggest challenges of the tunnel job was that it was really two separate and incredibly complex construction projects—handled by two unaffiliated joint ventures. While the sandhogs had been using the TBM to mine the outfall tunnel, a different crew of drillers and commercial divers, employed by a different contractor, had been working off a jack-up barge parked nine miles out in Massachusetts Bay. Their job had been to sink the fifty-five vertical pipes, called risers, that the tunnel would eventually connect to in its final mile.

The setup of the tunnel and those risers was something like a water main running underneath a neighborhood street, with a bunch of smaller pipes shooting out from it, each one delivering a portion of the water supply to a different house. When everything was completed, the horizontal tunnel would connect with those fifty-five vertical risers, which would each extend from tunnel depth, climb up through the bedrock, and finally punch through the ocean floor. At the point where each riser poked through the seafloor, it would be topped with a mushroom-shaped discharge port, a domed contraption known as a diffuser head. After exiting the Deer Island plant, the wastewater would flow along the main tunnel, up through one of those fifty-five riser pipes, out through sprinkler nozzles on one of those fifty-five diffuser heads, and finally out into the ocean. The intricate process was designed to slow down and spread out the flow of the wastewater, "diffusing" it in an effort to limit its impact on marine life and water quality. The entire system was an engineering masterpiece. Gravity, rather than engines or pumps, would propel the wastewater along its epic journey under, and then out into the sea.

But in order for this marvel to work, the crews from those two separate joint ventures had to do their jobs in perfect alignment. That was asking a lot, since one operation had to be done from high above the water, miles out in the open ocean, while the other was being done underground, deep below the sea.

Surprisingly, the crew of the IB-909 jack-up barge (which had included DJ's friend Ron Kozlowski) completed the installation of those fifty-five risers ahead of schedule and on budget. But MacDonald knew that achievement would matter little if the locations of either the tunnel or the risers turned out to be off target. And in a vast ocean, there was very little margin of error for that horizontal tunnel to make its subsea rendezvous with all those vertical risers.

The moment of truth came in the waning days of 1995. By then the TBM had mined and lined more than eight miles of the outfall tunnel. Three days after Christmas, Kiewit's chief surveyor in the tunnel, a man by the name of Bill Currier, headed down the shaft. Currier had been responsible for making sure the TBM moved in precisely the right direction, both horizontally and vertically. On a job defined by its stiff challenges, Currier's was one of the stiffest. This far down into the earth, GPS was of no use. And with only one entrance to the tunnel, there were no intermediate points he could use to confirm his calculations. His primary tool was something called a north-seeking gyroscope, which sat on a tripod. He pulled it out every five hundred feet or so, using his measurements and earlier calculations to adjust the laser-based guidance system on the TBM. This was all in the service of keeping the earth-eating beast on the correct path. Still, he would have no way of knowing for sure if his calculations were right until sandhogs began drilling probe holes from the side of the tunnel, in the hopes of connecting with a riser.

Currier was a forty-four-year-old career surveyor who followed

The World's Dead-end Tunnel
The World's Longest Dead-end Tunnel

The tunnel was designed to carry treated sewer water nearly 10 miles into Massachusetts Bay. The wastewater would travel through the tunnel, up 55 risers, and then exit through diffuser heads sitting on the seafloor.

Winthrop **Tunnel** *Diffusers*

Logan Airport

BOSTON **Deer Island** *MASSACHUSETTS BAY*

Boston Harbor

1 MILE

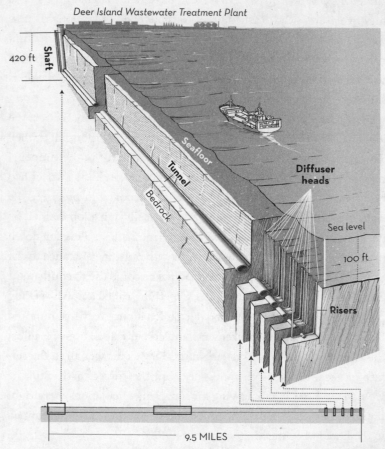

Deer Island Wastewater Treatment Plant

420 ft — **Shaft**

Seafloor

Tunnel

Bedrock

Diffuser heads

Sea level

100 ft

Risers

9.5 MILES

For more photos and illustrations, visit www.neilswidey.com

the adage favored by carpenters and tailors: measure twice, cut once. Considering the stakes, he took his caution to a higher level, taking his measurements closer to five times before making any adjustments to the direction of the boring machine.

At last the troubled TBM limped its way to the 8.15-mile mark. If Currier's numbers were correct, riser number 55—the one closest to Deer Island—would be nearly twenty feet due north of the tunnel. If his numbers were wrong, the project would be looking at more draining delays. Even worse would be if the survey had been so off-base that the riser wasn't north of the tunnel but instead was due east and therefore about to get plowed over by the TBM.

Before the jack-up barge crews installed the fifty-five risers, they had filled each one with eight thousand gallons of seawater, which had been dyed fluorescent green. So when the sandhogs in the tunnel drilled out to make the connection to each fiberglass riser, they'd quickly know if they'd hit it in the right spot. If they had, they'd see bright green liquid gushing back toward them.

Currier looked over his measurements one last time, grabbed a thick permanent marker, and marched over to the tunnel wall. He made a big X. That, he told the sandhogs, was where they should start coring into the bedrock. Then Currier, wearing his yellow rain slicker, stepped back, cracked open one of the Tom Clancy paperbacks he'd brought with him, and started to read. He knew it would take the sandhogs lots of careful work to get the probe hole all the way out to the riser, adding drill steel as they went along. Because they were drilling the probe hole into bedrock, they wouldn't see anything more than modest water seepage until they actually tapped into the eight thousand gallons of green sitting in the riser. This would be a tedious process.

Later that afternoon, when the swing-shift sandhogs arrived to relieve the first-shift guys, Currier kept right on reading. "Man, you're one cool cucumber," a sandhog told him. No use admitting how apprehensive he was feeling. After all, in the vast earth around them, their target was just thirty inches wide—the width of the riser. Later that night, when the graveyard-shift sandhogs showed up, Currier continued to keep his nose in his book. But as December 29 dawned, it became harder for him to concentrate. Partly, that was a function of knowing his moment of truth was drawing nearer. And partly it was a function of who was standing a few feet away from him. The graveyard-shift crew included none other than Mad Man Malkasian, perhaps the loudest character in a tunnel that was teeming with them.

By three in the morning, Currier had put aside his Clancy paperback and was watching closely. Finally the sandhogs had cored the correct length—nineteen feet, six inches—in precisely the right direction. For Currier, the seconds began to feel like hours. Then Mad Man shouted. In an instant, a torrent of water as green as Gatorade came pouring out, spraying and splashing Mad Man and Currier. "Hot damn!" Mad Man shouted, then pulled out his camera to capture the moment.

Their job was hardly done. But with all the setbacks the tunnel project had seen so far, it was a relief to be able to bask in the green glow of one step that had, against the odds, gone flawlessly.

For much of the twentieth century, the hazards of being a tunnel worker could be summed up by the rule of thumb that every civil engineer and sandhog could recite: "a man a mile." If you built a ten-mile-long tunnel, you could expect to lose ten men. Although Mac-Donald had come to his post with no experience in tunneling, it hadn't taken his staff long to share that bit of fatalistic shorthand with him.

Despite the introduction of worker safety measures over the years, tunneling remained extremely risky. The Channel Tunnel—the thirty-one-mile, two-entrance tube connecting England and France, which was winding down construction just as the Deer Island project was kicking into gear—had claimed ten worker lives. Even though the Boston Harbor cleanup involved the construction of some fifteen miles of tunnels, MacDonald had begun with the firm belief that the project would finish with no worker fatalities. With the right mind-set and precautions, he felt, that goal was attainable. But just five months into his tenure, he had been forced to bow to reality.

On the first day of July 1992, Kiewit had been completing its work on a small secondary tunnel that it had built to connect operations on Deer Island itself. Several sandhogs on that job were working their final shift. Their last task was to set a thirteen-ton concrete cylinder in place, completing a manhole that led to this small tunnel. Kiewit had already moved most of its equipment to a different job, so crews borrowed a crane from another company working on the island. Around lunchtime, the crane lifted the precast concrete cylinder, as Michael Lee and three other sandhogs worked to install a large rubber gasket that would sit underneath it. Lee, a forty-year-old welder with a big, swooping mustache, was an experienced sandhog popular for the cakes and sausage bread he regularly baked at home to share with the other guys. The workers were struggling to get the gasket into place, so Lee moved in closer to try to shift it. Just then, the crane's brake slipped, and the thirteen-ton load that was suspended in the air crashed down on Lee's head. He was pronounced dead at Massachusetts General Hospital.

As investigators began focusing on safety problems that had been overlooked on the borrowed crane, MacDonald came to understand a counterintuitive lesson about workplace safety: the most dangerous day on a job is often the last one. Workers and managers, feeling eager or pressured to get the job done and brimming with confidence, are more likely to let their guard down and take chances. A home-owner using an extension ladder to clean his gutters will likely begin the work by climbing down and moving the ladder every few feet, rather than risk a fall by stretching his arms too far. But by the time he's cleaning the last stretch of gutter, that same homeowner is more likely to chance an especially long reach rather than take the time to move the ladder yet again.

Just three weeks later, workers on the jack-up barge in Massachusetts Bay had successfully installed fifty-four of the fifty-five risers and had just begun working on the last one. Adding to the upbeat mood on the barge, Ricky Spears, a forty-seven-year-old, silver-haired foreman, was passing out novelty cigars whose label read "It's a boy." He was just back at work after his wife had delivered their baby son. Spears and his crew were standing on a drill platform that had been placed atop a two-hundred-foot pipe, which a crane had positioned perpendicular to the ocean floor, as straight up as a sequoia. The floor of the drill platform was actually a set of hydraulic doors that opened like a cellar bulkhead. Suddenly that sequoia-size pipe dropped deeper into the seabed, causing the bulkhead doors of the drill platform to fling open, catapulting Spears thirty feet into the air and costing him his life. Although he had lived to see his baby brought into the world, he had never had the chance to hold the boy in his arms.

Three years later, inside the tunnel, an inspector named Dick White had been attempting to board the railroad tram from the wrong side when the lokey that was pulling it had suddenly lurched forward. White was a crusty but well-liked character from Arizona who topped three hundred pounds and had piercing blue eyes. During this particular shift change, when the lokey had advanced without warning, White had been pinned between the railcar and a steel beam that was part of the TBM's trailing gear. His chest had been crushed. The sixty-four-year-old had been planning to retire in a few months.

Each of these deaths weighed heavily on MacDonald, a man whose tendency to tear up was nearly as well known around the agency as his workaholic ways. But he didn't feel he could afford much mourning. He was under federal court order to bring this stalled and stymied project to a close. As safely as humanly possible, he was determined to see it through.

3

Memo Wars

Just after two o'clock on a warm afternoon in mid-December 1996, thousands of people milled about a trendy New Orleans business district on the banks of the Mississippi. Shoppers settled on Christmas gifts at the boutiques in the upscale Riverwalk mall, diners lingered over lunch or munched on beignets at the cafés, and gamblers tried their luck at the docked riverboat Flamingo Casino.

At 2:07 p.m. an alarm began blaring in the distance, puncturing the holiday cheer. The complex of shops and restaurants was built on a wharf next to a parking garage and a Hilton Hotel. As pedestrians on walkways and shoppers on balconies craned their heads to look up the muddy Mississippi, it didn't take long for them to spot the source of the siren. A red and black beast of a freighter—stretching longer than two football fields, weighing nearly seventy thousand tons—was barreling toward them, seemingly out of control.

Three minutes after the pilot of the *Bright Field* sounded the alarm and notified the Coast Guard of an engine failure, the freighter barely missed colliding with two cruise ships, which were parked just upriver from the shops and loaded with passengers. Now, with its alarm still blaring and flares shooting up from its deck into the afternoon sky, the beast was headed for the mall. The shoppers and diners, who had initially worn puzzled looks, shifted into full panic, screaming and trampling each other in a desperate stampede.

The freighter slammed into the wharf at 2:11 p.m., collapsing huge chunks of the structures sitting above it. Amid the screams and the blare of the horn came the terrifying sounds of crumbling concrete and screeching metal. The freighter, loaded up with corn, then bounced off the wharf and headed toward the riverboat casino. A few of the Flamingo's eight hundred gamblers, after instantly calculating their odds, jumped overboard. Mercifully, the freighter came to a shrieking stop—seventy feet shy of the Flamingo. "It felt just like an earthquake," one survivor told the *Times-Picayune*. "It really just seemed like it was something that happened on TV."

The crash of the *Bright Field*, operated by a Chinese crew sailing under a Liberian flag, ended up injuring sixty-seven people and taking out fifteen stores and restaurants, as well as parts of the nearby parking garage and hotel. In all, it caused $20 million in damage. Investigators said it was a miracle no one had been killed. And they attributed that good fortune in part to the quick instincts of the freighter's American pilot, who had wasted no time in ordering that the *Bright Field*'s anchors be dropped to try to slow the beast's slide into the shopping mall.

Those were precisely the kinds of quick instincts that gave the designers of the Deer Island Outfall Tunnel nightmares.

There was more than ocean sitting above the Deer Island tunnel. The ribbon of Massachusetts Bay directly overhead just happened to be a busy shipping lane from the Atlantic into Boston Harbor. The decision to site the tunnel under that lane had been a fragile compromise following contentious negotiations, so there would be no going back on it. Yet the site path left the tunnel's designers fearful that a ship's anchor might dislodge one of those fifty-five domed diffuser heads sitting on the ocean floor. It was an extremely remote risk, since even if an anchor made contact, the domes were designed to withstand a direct hit. Still, the designers worried about a freak accident in which a captain, after losing control of his ship in a storm, began dragging his anchor in an emergency attempt to stay afloat. Although the tunnel drawings had been finalized long before that freighter took out the mall in New Orleans, the *Bright Field* accident represented the kind of worst-case scenario the designers had in mind.

Hydraulic studies predicted that if an anchor ever did succeed in ripping off a diffuser head during tunnel construction, the subterranean sandhogs would find themselves thrust into a disaster-movie scene even scarier than the one featuring all those Big Easy shoppers. Once the diffuser cap was breached, ocean water would gush down that riser and flood the empty tunnel at a crushing rate of 115,000 gallons per minute. It would resemble a flash flood, with a leading wall of water as wide as the tunnel and as high as five feet, moving at a rate of nearly four feet per second.

Designers shuddered at the prospect of an accident like this endangering both the expensive structure and the scores of workers who were building it. The picture of fifty or a hundred sandhogs racing for their lives against this violent wave of seawater was a terrifying one. The lokey-pulled tram would be of no use, of course, so the workers would be forced to try to either outrun or outswim the seawater. Even if they were able to access emergency rafts, many would likely not make it out alive. To guard against this disaster movie ever playing out in real life, the tunnel's design firm, Parsons Brinckerhoff (PB), had insisted that contractor Kiewit install a jumbo safety plug at the bottom of each of the fifty-five risers.

In the fall of 1996, just a couple of months before the New Orleans accident, the sandhogs had managed, finally, to get the TBM to its 9.5-

mile finish line. To reach the end, man and machine had excavated an astonishing 2.4 million tons of rock. It was impossible to drive the gargantuan TBM in reverse all the way back to Deer Island, because the tunnel's thick concrete lining had made it narrower than the machine itself. Instead, Kiewit instructed the sandhogs to use the TBM to keep boring an additional hundred feet to make a grave. Then, like something out of a mob hit, the $15 million bullet-shaped behemoth was buried in concrete.

This milestone came a full year after the federal judge's original deadline for the completion of the entire tunnel, and huge challenges remained. One of the most formidable involved turning the final stretch of the tunnel, which was the standard diameter of twenty-four feet, into something much narrower. Because gravity would be the only force powering the millions of gallons of sewer water along the miles of tunnel and up those fifty-five risers, the wastewater would have to be kept moving with enough velocity that any solids in it remained in suspension. Otherwise, the solids would settle to the bottom of the tunnel and eventually clog it. This same principle had guided the design of sewers going all the way back to Roman times. A good flow would also ensure that the sewer water climbed up the risers and exited out to sea with enough force to prevent an intrusion of ocean water from coming in from the other direction—a real risk since the salt in the ocean water made it slightly heavier than the treated sewer water. Those velocity demands meant the tunnel had to get progressively narrow along its last mile, as it connected with each of the fifty-five risers.

Kiewit chose a three-pronged approach to reduce the area inside the tunnel. First, workers built a long ramp, beginning near the start of the final mile, that would gradually reduce the tunnel's top-to-bottom space. Then, about halfway into that last mile, they installed a series of seven-foot-tall Jersey barriers—like the ones that line highway medians—to reduce the tunnel's side-to-side space. The farther east the tunnel went, the closer together those Jersey barriers were positioned. Finally, for the last six hundred feet or so of the tunnel, they laid in several precast pipes of decreasing diameters and then backfilled around them with concrete. The last of these precast pipes—which would house the connections to risers number 3, 2, and 1—was just five feet in diameter.

Meanwhile, sandhogs needed to finish hooking up the tunnel to all the risers. They had to repeat, fifty-four times, the same task that Mad Man and the others had completed at riser number 55, when thousands of gallons of Gatorade-green water had come gushing out. At each hookup, once the probe hole had been drilled and the riser had been drained of its green water, a new phase of excavation would begin. The sandhogs essentially had to dig fifty-five small side tunnels. Doing this allowed them to install concrete-encased elbow pieces that connected each of the thirty-inch-wide vertical risers with a horizontal riser pipe of the same width that would shoot off from the side of the main tunnel. Those elbows would create the clear path through which treated sewer water would eventually flow from the Deer Island plant, through the tunnel, up through a riser, and out to the ocean floor. The elbows had another important function. Preinstalled in each one was a sixty-five-pound domed safety plug that resembled a salad bowl from some industrial kitchen. Those plugs were designed to help protect the sandhogs from the ocean above.

By 1997 all fifty-five of these elbows and their safety plugs were in place. Only then did the moment of reckoning seem to begin for Kiewit: *When everything else is done, how in the world are we going to get these plugs out?* There was no good answer to this question because of an unusual provision buried in Kiewit's tunnel contract, a provision that would make the task of removing those plugs infinitely harder.

. . .

The roots of the safety plug decision could actually be traced back a decade, before PB had even become involved in the project. In 1989 the Massachusetts firm Metcalf & Eddy produced a conceptual design for the Deer Island tunnel and diffuser risers. To prevent ocean water from infiltrating the sprinkler nozzles on those diffuser heads during tunnel construction, M&E designers called for sturdy covers to be installed on each nozzle. But if an anchor ripped open the diffuser head, nozzle covers would obviously offer zero protection. So the designers decided a "secondary" safety plug should be placed lower in the riser pipe. They laid out three options for this secondary seal. Option 1 would put a plug inside the diffuser head, just out of reach of any anchor but still close enough to the ocean floor that the plug, like those nozzle covers, could ultimately be removed by divers swimming down to the sea bottom. Option 2 would put a plug inside each of the pipes connecting the risers to the tunnel, meaning the plug would ultimately have to be removed by crews working from inside the tunnel. Finally, option 3 would skip the secondary plug in favor of hiring a guard boat to patrol the waters above the diffuser heads, keeping that particular stretch of ocean free from boats and their anchors for as long as the sandhogs were in the tunnel.

M&E pointed out that the second option was problematic because of "the difficulty of re-entering the completed tunnel (with no track, lighting, or ventilation) to remove the plugs." The firm's report concluded with this assessment: "At first sight, solution 2 appears to be cheaper than 1, but it is believed that when the difficulties noted are taken into account, plugging from the top, which is the safest, will also turn out to be the cheapest."

But because M&E was responsible only for the conceptual design, its advice was not binding. The contract for the final design of the tunnel and diffusers went to Parsons Brinckerhoff, a bigger global infrastructure planning company whose letterhead boasted "Over a

Century of Engineering Excellence." PB settled on a different design for the diffuser head. This one was mushroom-shaped and resembled a domed Apollo 11 command module, a design that would improve its ability to withstand a hit from an errant anchor. Still, PB agreed that a secondary safety measure would be prudent. It settled on safety plugs—which it called bulkheads—that would sit in the elbows of the pipes connecting the tunnel to each of the fifty-five risers. PB decided that option 1, which M&E had identified as the safest route, would present too many problems when it came time to remove the plugs. So it went with an internal plug design that resembled option 2, the one M&E had characterized as appearing at first sight to be the cheapest route.

But no one—not tunnel owner MWRA, not tunnel designer PB, not construction manager Kaiser—spelled out how those plugs could be safely removed. Instead, they transferred the responsibility for figuring that out to the contractor, kicking the can down the road. But they did make one crucial stipulation. The contract mandated that the safety plugs could be removed only after the tunnel was completed, meaning after the sandhogs had cleared out, taking their extensive ventilation, transportation, and electrical systems with them.

The rationale behind this unusual stipulation appeared to involve worker safety. PB designers said the overriding goal should be to limit the amount of time when there would be lots of workers in the tunnel who were protected from the ocean by just one safety seal, namely the covers on the nozzle caps.

In 1990, after winning the bid to build the tunnel, Kiewit signed the contract that required it to come up with a plan for "providing lighting and ventilation (or breathing apparatus) for the personnel" who would remove the plugs. By suggesting equipment "such as battery powered vehicles [and] self-contained support services (i.e. scuba gear)," the contract clearly envisioned that the plugs would be removed only after the main ventilation system and other utilities had been dismantled. Although the contract was silent on exactly how such a complex operation could be safely accomplished, it clearly assigned Kiewit the responsibility for figuring it out. So when Kiewit would later complain about the grave dangers inherent in following these specs and sequence, the other project players could all respond, with some justification, by pointing to the contract and saying: *You bid it, you own it.*

In Kiewit's defense, though, the contract to build the world's longest single-entrance tunnel—one of the riskiest underground projects ever undertaken—sprawled over nearly a thousand pages, with a breathtaking inventory of challenges. Many of them seemed a whole lot more pressing than removing a few dozen backup safety plugs at the end of the job. To the extent that Kiewit executives thought about removing the plugs, they assumed they'd be able to yank them out at the same time they were pulling out the bag line, electricity, and rail tracks. In truth, during the early years of the project, they barely thought about the plugs at all.

As head of the agency that owned the tunnel, Doug MacDonald knew that as much as the thing was a one-of-a-kind technical marvel, for the contractor it was shaping up to be a financial disaster. Kiewit was already in the red tens of millions of dollars, since the project had become so much costlier and more complicated than it had anticipated in its original $202 million bid. On top of that, Kiewit was incurring late-fee penalties of $30,000 every day. Although these so-called liquidated damages are often negotiated downward at the end of a job, on paper at least the contractor had, by the spring of 1997, incurred a total penalty of more than $17 million—and counting. Complicating matters was the surprising way in which contractors were paid for big jobs structured like this one. Even though MacDonald's agency was considered the tunnel owner, Kiewit had to front pretty much all the construction costs and would get paid only when it "sold" completed sections of the tunnel to the MWRA, piece by piece. MacDonald was grateful at least that Kiewit was such a big operation, since a less capitalized company would likely have already gone into bankruptcy. What worried his team, though, was how aggressively Kiewit might use its deep pockets to try to crawl its way back into the black. To win the legal claims war, Kiewit would need to demonstrate that the MWRA or Kaiser or PB had made errors and should therefore be forced to pay for them. In other words, Kiewit would try to transfer huge chunks of that red ink to the others.

In fact, the MWRA had already settled several of Kiewit's claims, agreeing to pay the contractor an additional $32 million for unforeseen problems. Following Kiewit's corporate approach of matching power to power, those settlement negotiations had taken place directly between MacDonald and Ken Stinson, the chairman and CEO of Kiewit's construction group. MacDonald had to admit that he found himself a bit intimidated by Stinson, a tall, bald man in his midfifties who spoke in a crisp manner and exuded the aura of a natural-born leader. In his interactions with him, MacDonald felt as though Stinson were thinking, *Look, I'll talk to you. But you don't know what you're doing.*

Stinson had a long, successful track record. During the Vietnam War, he'd been a company commander for the Seabees, the navy's construction arm. In 1969, as a graduate student at Stanford, he'd taken a summer internship with Kiewit, which unexpectedly had turned into a lifelong career. He lived and breathed the Omaha company's corporate culture. It had been shaped by the late Peter Kiewit, who had turned his grandfather's bricklaying outfit into a construction powerhouse. With his motto "Know your costs," Peter Kiewit had insisted that all his managers keep close track of project expenses so there would be no nasty surprises at the end of a job. Because the company remained both privately held and intensely private, it had been able to maintain that corporate culture long after Kiewit's death in 1979. For Stinson, the Deer Island project presented a serious threat, given how far its costs had diverged from original estimates.

Meanwhile, MacDonald, despite his lack of management experience, felt he had proved his mettle as MWRA chief. Even though the tunnel piece of the harbor cleanup effort was late and over budget, he had managed to beat back his critics while keeping the overall project on firm financial footing. He told Stinson he wasn't going to let the cleanup become another Big Dig, the Boston megaproject whose price tag had already ballooned from $2.6 billion to $10.8 billion and continued to climb.

Somehow the two chiefs managed to forge a working relationship. After one session at MWRA headquarters, an agency staffer offered Stinson a ride to Logan Airport. As they approached the airport, the staffer asked the CEO which airline he was flying. Stinson coolly replied, "My *own*." Kiewit operated four corporate jets so it could easily get Stinson and other top executives to job sites where they were needed most.

In addition to the settled claims, Kiewit had either already filed, or was preparing to file, two far bigger claims. The first had to do with the rock that the TBM had encountered early on, which the contractor contended was far harder to penetrate than the MWRA's geotechnical estimates had predicted. The second claim involved the rock toward the end of the tunnel, which the contractor contended was so fractured that it allowed much more water to flow in than had been expected. Around MacDonald's office, people began referring to these two immense claims dismissively as "the rock was too hard" and "the water was too wet."

Kiewit signaled its aggressiveness by moving its Deer Island project manager to Washington, D.C., to work full time with company lawyers in assembling the rock and water claims. If those couldn't be settled between MacDonald and Stinson, they would go to a special review board and, if necessary, to the courts after that. Kiewit assigned a no-nonsense executive to be the tunnel's new "project sponsor," an overall Kiewit boss who would oversee the job from Omaha, and the company promoted a veteran engineer to run the tunnel project from Deer Island.

Against this tense backdrop in the spring of 1997, Kiewit fired the first shot in the safety plug battle. Here the conflict moved down a few rungs on the org chart, from MacDonald and Stinson to people closer to the tunnel's front lines. To make it easier for workers to yank out the safety plugs at the end of the job, Kiewit's new tunnel manager wrote a memo asking for permission to relocate the plugs now. The contractor wanted to remove them from their current position—inside the riser elbow, at the end of those connector pipes, which averaged twenty feet from the main tunnel—and place them just sixteen inches from the main tunnel wall. That would make the plugs a lot simpler to remove, sparing the workers from having to crawl through those narrow "off-take" connector pipes. Kiewit also asked the MWRA to pay the costs of relocating the plugs.

Kiewit sent its memo to Kaiser, the construction management firm that MacDonald's agency had hired to be its eyes and ears on the project and to ride herd on its many players. On tunnel matters, MacDonald and his deputies relied heavily on Kaiser's tunnel manager, Dave Corkum, a youthful-looking forty-three-year-old with a gap-toothed smile. A former geologist turned engineer, Corkum was now taking law school classes at night in preparation for yet another career change. With the tunnel action increasingly being driven by lawyers rather than miners, Corkum felt he had chosen wisely.

He firmly rejected Kiewit's request that the MWRA foot the relocation bill. He and his counterpart with designer PB said additional studies would be required to make sure moving the plugs wouldn't make them less effective. After all, the contract specified that "personnel safety is paramount."

Kiewit's construction manager replied with a sharply worded memo questioning Kaiser and PB's professed concern for worker safety. He disputed their contention that asking a worker to crawl "on his or her stomach inside a thirty-inch pipe under self-contained oxygen to drain approximately 565 gallons of water, removing a steel circlip, then finally removing the dome, which weighs approximately sixty-five pounds, to then crawl backward, all the while supporting the dome in the center of the off-take pipe (thereby allowing it to pass through) is the best [we] can do for paramount worker safety." Interpreting the refusal to have the MWRA pay the bill as an outright rejection, Kiewit decided to scrap the idea of relocating the plugs.

When spring turned to summer, Kiewit floated a new plan: replace the fifty-five plugs with one giant one. Near the tunnel's eight-mile mark was something called a venturi, a transition point between the end of the regular tunnel and the start of the much narrower section called the diffuser tunnel. Kiewit's idea was to remove all fifty-five plugs currently in the elbows and install one enormous bulkhead at the venturi. If that worst-case scenario of a ship anchor ripping off one of the diffuser heads actually came to pass, this new blockade at the venturi would hold back the waves, giving the sandhogs sufficient time to retreat to Deer Island. Once the tunnel was finished, the relatively quick job of removing this venturi bulkhead could be handled by a small crew relying on supplied air or even by remote-controlled explosive charges.

This proposal, like the other, produced sniping between Kaiser's Corkum and Kiewit over costs, responsibilities, and motivations. With the clock ticking, Corkum blasted Kiewit for having "only the vaguest idea as to how it intends to accomplish the task of removing the riser plugs. We would advise that this planning process should begin now!" Designer PB eventually quashed the idea of the venturi bulkhead, citing engineering concerns. Subsequent proposals and counterproposals went nowhere.

By that fall, things began to boil over. The Kiewit construction manager fired off a memo to Corkum saying the contractor was at wit's end. For seven months, he wrote, Kiewit had pursued various alternatives, only to have them shot down. Corkum, he complained, wouldn't tell him exactly how he and PB wanted the plugs removed. Rather, he would say only what was unacceptable, which appeared to be pretty much everything. It was an exhausting guessing game. Now, the Kiewit manager wrote, "time being of the essence, and with no acceptable alternative available, we feel the only solution is to proceed with pulling the safety plugs as soon as we are ready to begin backing out of the tunnel with utility removal and final cleanup operations." He noted that Kiewit would reach that point in just a couple of weeks.

The significance of this memo rested partly on the unusual piece-by-piece manner in which Kiewit was paid. Essentially, it would sell the tunnel to the MWRA in roughly one-thousand-foot sections, starting at its easternmost end and eventually concluding at the shaft. Before the MWRA would pay for a thousand-foot section, its construction manager Kaiser, in the person of Corkum, would inspect that section, creating a punch list of problems that needed to be fixed. When both sides agreed that the punch list had been dealt with satisfactorily, the "sale" of that tunnel section would go through, and the MWRA would cut a check in proportion to that section's size of the overall tunnel cost. Based on the contract, Kiewit wouldn't get paid for a section unless all the utilities had been stripped out and it was in what homebuyers and sellers refer to as "broom-clean condition." After all, the tunnel's ventilation, electrical, and transportation systems were so unwieldy that it would take months to dismantle them. It was hard to imagine how Kiewit could have delivered a broom-clean section if those utilities were still in place. But it was just as hard to imagine how crews could be safely sent in to remove the plugs once all those utilities were gone. The piecemeal payment arrangement, combined with the $30,000 in liquidated damages that Kiewit was incurring for every day it was late, made it all the more urgent for the contractor to move through this final cleanup stage as quickly as possible.

In his memo, the Kiewit manager pointed out that in the half-dozen years since the first risers were installed, not a single one of the diffuser heads sitting on the ocean floor had been damaged. What's more, he said, there hadn't even been a recorded close call with an errant anchor during this long stretch, which incidentally had included at least one "storm of the century." Also, the U.S. Coast Guard was prepared to institute additional safety measures aimed at keeping ship anchors away from the diffusers.

In his reply, Corkum said the Coast Guard's "no anchor" zone was all well and good, but the real risk would be from a ship losing its rudder in a storm. If a captain feared his boat was about to capsize, he wouldn't be worried about violating a "no anchor" zone marked on a map. Some ship captains coming into Boston Harbor likely wouldn't even be able to read English, he noted. Corkum invoked the freighter crash in New Orleans, in which language problems between the American pilot and the Chinese crew may have made matters worse. Even though the pilot had ordered that both of the freighter's anchors be lowered, the crew had dropped only one. Despite dragging that anchor, Corkum wrote, the freighter "still careened into a mall." Corkum rejected Kiewit's request to pull the plugs during cleanup, insisting that "all work in the tunnel be complete before the off-take plugs are removed."

The irony is that each side claimed worker safety was its primary concern. Corkum, writing on behalf of Kaiser and the MWRA, said it would be unwise to endanger the lives of up to a hundred sandhogs by leaving the tunnel vulnerable to a possible flood during the long cleanup period. Kiewit, meanwhile, said it would be insane to put a small number of workers at extreme risk by sending them into a tunnel that had no air or light, all in the name of protecting a larger group of workers from an exceedingly small risk. By waiting until the end to pull the plugs, the Kiewit manager wrote, *the risk of catastrophe would be exponentially higher!*

In frustration, Kiewit enlisted a former OSHA inspector named Fred Anderson as a consultant. In his report, Anderson stressed that the stakes were "enormous in terms of both money and political necessity." By insisting on installing backup plugs without a clear understanding of how they would be removed, the parties involved in the project had painted themselves into a corner, he wrote. But the tunnel would not be viable if they couldn't figure out a safe way to yank out the plugs. "They must come out!"

After reading the contract closely, Anderson noted, it was clear that the people who wrote the specs intended for the plugs to be removed by a crew "dependent on self-contained breathing apparatus in an unknown and uncontrollable environment." He stressed, "To me, this is a scary prospect." He warned that the hazardous assignment could cost lives. And if workers died, regulatory agencies would likely shut down the tunnel, adding further delays. Anderson strongly advised Kiewit to stand firm and insist on pulling the plugs before removing the ventilation, lighting, and rail systems. Asking workers to venture nearly ten miles into a dark, unventilated tunnel hundreds of feet below the ocean, he said, would be sentencing them to "an operation somewhat akin to a spacewalk."

In response to Corkum's rejection of the request to remove the plugs during cleanup, Kiewit in December declared an "emergency" in the contract. Company managers cited their consultant's advice and word from the Boston Fire Department that it would not provide rescue services for a plug-removal mission if the tunnel's utilities had already been removed. Corkum replied that there was no such emergency and directed Kiewit "to not remove or otherwise compromise the off-take plugs at this time."

At the start of 1998, after the yearlong memo war had produced little more than a mountain of paper, the MWRA tried for a fresh start by convening a brainstorming session with all the major players. A high-ranking Kiewit executive who had flown in from Omaha began the session with a blunt question to the MWRA: "What do you want us to do?" The MWRA manager demurred, saying he didn't have the extensive experience that the Kiewit folks had. The Kiewit executive countered that *no one* had experience in the kind of mission they were discussing.

As earnest as the MWRA manager was in trying to forge a consensus with all the bright minds around the table, the forces working against him had grown potent. Researchers in organizational behavior point out that as trust levels go down within a group, group members' creativity and willingness to seek new options also decrease. When intense time pressures are added to the mix, opposing sides tend to become even more fixed in their positions, relying more on cognitive shortcuts. They're unable to work collaboratively to solve a problem because they have become locked in an adversarial contest: *If you win, I lose.* But with both sides so hardened in their positions, all they were doing was ensuring that they'd have to spend more time together in the tunnel.

For years, Kiewit had kept a large sign hanging near the opening to the Deer Island shaft. It featured the joint venture's logo surrounded by the words GOOD AIR, GOOD LIGHT, GOOD HOUSEKEEPING, GOOD SAFETY. Now the Kiewit managers felt they were being asked to risk that last attribute by sending workers in without the benefit of the first two.

But Corkum saw something entirely different in Kiewit. He feared that financial concerns were driving the company's interest in pulling the plugs during cleanup. After all, as much as Kiewit framed its arguments in the context of worker safety, it was indisputable that removing the plugs while the contractor already had crews in that section of the tunnel, supported by plenty of air and light and transportation, would also be the cheapest route. "They're bleeding money like no tomorrow on this job," Corkum would explain later, "and all they want is out."

This, Corkum believed, was one of the realities of lump-sum, low-bid construction, the way most public works projects are bid across the country. To win the contract, a company has to bid the bare minimum. But then that contractor can make a profit on the job only if its costs turn out to be unusually low. Accordingly, if the contract called for X, Corkum expected Kiewit would try to do the very bare minimum of X, while he viewed his job as making sure the owner got the maximum. He suspected that Kiewit, in aggressively pushing to remove the plugs and utilities early, was engaging in a game of chicken. If it pulled down enough of the bag line and there was still no good option for removing the safety plugs without ventilation, maybe Corkum would simply relent and let Kiewit yank the plugs at the same time.

As Corkum progressed in his night classes for law school, he found new reasons to worry. If he read about some obscure legal maneuver, he'd immediately wonder if Kiewit's high-priced lawyers would employ it to try to extricate the contractor from its hole. Corkum's friends joked that he didn't have to attend class because he was getting his education in the law right on the job. His contracts class taught him the dangers of weighing in too much on a responsibility that was assigned to another party in a contract. The more you touched something, the more liability you assumed. That thinking informed his reluctance to come out and tell Kiewit exactly which approach to take in pulling the plugs.

Suspicions ran just as deep on the other side. Kiewit officials felt that Kaiser, PB, and the MWRA were unified in trying to push all of the risk onto them. These seasoned tunneling veterans couldn't understand why the MWRA seemed to be so unquestioning in its deference to Corkum. During the brainstorming session, one Kiewit executive argued that if Corkum and PB were truly concerned about worker safety, they would have accepted Kiewit's suggestion to remove all the plugs and install one giant bulkhead at the venturi. That would give workers plenty of time to make it safely back to the shaft. The fact that they remained opposed, he said, "indicates the tunnel is your priority—not personnel."

Despite Corkum's suspicions, Kiewit was moving aggressively on an entirely different legal front. The contractor had engaged one of Boston's more influential lawyers and power brokers, a man by the name of Fran Meaney, who had shifted his focus from the courtroom to the halls of government. Meaney, the former managing partner of Kiewit's law firm Mintz Levin, now ran its lobbying arm. He decided he needed to see the tunnel for himself. So he rode the lokey-pulled tram as far into the tunnel as it would take him, all the while snapping pictures and making mental notes.

The thought of sending workers into an endless, pitch-black tube and forcing them to bring in their own air struck Meaney as madness. He was convinced that if those in power were forced to confront this reality, they would come to the same conclusion. He went about turning his thoughts and photographs into a PowerPoint presentation, which he felt could be devastating. If he couldn't persuade the MWRA with his PowerPoint, he planned to take it to government regulators at OSHA. If that didn't work, he would take it to the news media.

While that was unfolding, Kiewit was following through on the one mildly promising idea to have emerged from the brainstorming session. Corkum had renewed his push for the contractor to consider installing a temporary bag line that would be much smaller and easier to remove than the big one that had brought air to the sandhogs for nearly a decade. But that idea was shaping up to be another dead end.

So that it could proceed with its schedule of installing the precast pipes at the very end of the tunnel and then "backing out," Kiewit had already removed the ventilation from the tunnel's final mile, running a temporary extension duct off the main line. At a meeting shortly after the brainstorming session, Kiewit had relayed its findings that a temporary bag line could be installed, but it would be "both difficult and expensive." To erect something that wouldn't kink and that would be strong enough to ventilate all nine and a half miles, Kiewit reported, crews would likely need to use electrical transformers and rail cars. If they needed all those services, how temporary and easily removable could this smaller bag line really be? Kiewit estimated that a temporary line would cost at least $1 million, if not much more. During that follow-up meeting, the Kiewit tunnel manager had stressed that if they went forward with this temporary-line route, they wanted the MWRA to foot the bill. In a memo summarizing the meeting, he wrote, "Mr. Corkum responded, presumably on behalf of MWRA, that if that was our position, we could forget such a plan. He stated we should go ahead and get out the bottled air, and he would be standing at the top of the shaft counting diffuser plugs as we brought them out." Although Corkum later disputed the suggestion that he had been so flippant, he freely admitted to having told Kiewit that the cost of installing a temporary bag line would be its alone.

Not long after this exchange, as Kiewit's lawyer-lobbyist was putting the finishing touches on his presentation, he received a startling call. Ken Stinson, the imposing Kiewit chief back in Omaha, had personally vetoed the plan for a public confrontation with the MWRA. As much of a brass-knuckle negotiator as the Vietnam veteran was, Stinson didn't want to see his company engage in a campaign of publicly shaming one of its clients. That just wasn't the Kiewit way. Meaney was told to shelve his devastating PowerPoint.

The Kiewit team had spent more than a year arguing vehemently against the idea of sending workers into the tunnel to pull plugs after the ventilation and lighting systems had been removed, given the "exponentially higher" risk of catastrophe. But now exhaustion led to an about-face. Kiewit called in a team of commercial divers with instructions to do exactly that.

4

Arranged Marriage

At the start of 1998, Tap Taylor got wind of an exciting job prospect. Once again Kiewit's headache could be his opportunity.

When Tap had been working in Vermont four years earlier, his Kiewit bosses had thrown him the small-bore work that the company couldn't be bothered with. He and diving buddy Billy Juse had used that work to launch Black Dog Divers as a legitimate business. In the years since, Black Dog had grown into a viable if still small company, thanks largely to the contracts and referrals Kiewit had continued to toss their way.

Now Tap was hearing that Kiewit managers were sniffing around for a dive company to help them finish their long-delayed Deer Island tunnel. Even if the Kiewit division running the tunnel job was different from the one that had been Tap's rainmaker, he figured his close association with the company might give Black Dog an edge. He made a few calls, and it wasn't long before he and Billy were invited out to Deer Island.

The Kiewit managers explained the challenge, stressing that it wasn't of their own making. Divers would have to journey to the end of the tunnel, crawl the length of each of the narrow pipes connecting the fifty-five risers to the tunnel, yank out each sixty-five-pound plug, and then transport them all back to Deer Island. The job sounded challenging but relatively straightforward—until they mentioned the key work condition, the reason Kiewit was casting about for divers in the first place: they'd be removing the plugs after the tunnel had been stripped of all its utilities. So the divers would need to bring in their own air.

To get a better feel for the assignment, Tap and Billy went down the 420-foot shaft and hopped on the lokey-pulled tram. The Black Dog partners were both in their early thirties. Physically, though, they were a study in contrasts, Tap with his thick, six-foot-two build, balding head of light brown hair, and vise-grip hands; Billy with his lean five-foot-nine frame, sweep of black hair, and dense, dark mustache.

Even though Tap was used to being at sea, and even though there was still plenty of ambient air in the tunnel, he found it eerie to be so far underground, so far *under* the sea. What was most striking was just how far the tunnel extended. Each concrete-lined section they passed on the tram looked like the one before it and the one after it, which made for one monotonous ride. Although it was noisy, even the noise was monotonous, dominated by the lokey's jet-engine roar and occasionally punctuated by the honk of its horn.

Still, it was cool to get this rare view of the underworld. Even more exciting was the prospect of Black Dog scoring a possible million-dollar job while solving a massive migraine for Kiewit. As usual, Tap was relieved to have Billy at his side. He had complete faith in Billy's abilities, going back to their first days working together in Vermont, when he'd be coming up from the water, looking for a certain tool, and Billy would always be holding exactly what he needed. Tap felt his buddy's mechanical skills bordered on wizardry. He would often compare Billy to MacGyver, the secret agent at the center of the 1980s

ABC drama of the same name who could fix a blown fuse with nothing but a gum wrapper and repair a leaking radiator by cracking eggs and dumping the whites inside it. As they left the tunnel, it was clear that for all of Kiewit's vast resources and experience, the company was stymied. The way Tap saw it, Kiewit needed them to come in and do what Billy had always done best: MacGyver them out of a jam.

Tap and Billy ran Black Dog out of a large A-frame garage they'd built in Tap's backyard in southern New Hampshire. Whenever they weren't out at a job site, Billy and Tap could usually be found there, working on equipment, preparing for the next assignment. For Billy, every piece of broken equipment was a personal challenge, so much so that he would not hesitate to spend four hours repairing something that could be replaced for eighteen dollars. "Billy," Tap would gripe, "your time is worth more than that." Like MacGyver, Billy never concerned himself with replacement costs. It was a pride thing.

Billy and Tap both knew their strengths. Tap would find and bid the jobs and handle client relations, while Billy would take care of equipment maintenance and supervise the job site, overseeing any divers brought on for a particular project. Billy often ate dinner with Tap and his wife and two kids, and he sometimes slept over.

Tap set a relentless schedule, usually fourteen-hour days, seven days a week, and Billy matched him hour for hour. Both of them had a lot invested in making Black Dog a success. For Billy, the company represented the answer to the question he had been wrestling with since high school: What should I do with my life?

His father, Bill, the son of Ukrainian-Russian immigrants, had managed to build a thriving trucking business despite having dropped out of school in the ninth grade. His mother, Olga, who had come to New York from Puerto Rico at age nine, had helped her husband grow their New Jersey company as office manager. She was a determined

soul, having even persuaded her husband to change the way they pronounced the family surname of Juse, from *Juice* to *Juss,* because she
preferred it that way. Olga was deeply spiritual and a serious worrier.
When Billy was a teenager, his daredevil ways on his dirt bike or
behind the wheel of a car gave her plenty to fret about. Before bed,
she would include a special prayer in her nightly list. *Dear God, please
let Billy live until he sees eighteen.* When he made it that far, she and her
husband hoped their son would build on his advantages by graduating from college and finding a good career. But Billy dropped out
after a couple of semesters, retreating to the family business to work
as a diesel mechanic. He grew a mustache to match his father's, but
other aspects of adulthood proved more elusive. Olga kept praying.

Billy then spent years bouncing around seasonal tourist areas in
the Caribbean and New England, scuba diving for fun and working
as a bartender for money, before settling in Wolfeboro, New Hampshire. In that picturesque village on the banks of Lake Winnipesau-
kee, he made some side money as a diver, retrieving the occasional
snowmobile that had fallen through the lake ice. Otherwise, bartend-
ing continued to pay the bills.

It was only at the prompting of his girlfriend, a go-getter marine biologist, that he had turned diving into something more than
a hobby. She made it clear that she didn't want to hitch herself to a
barkeep drifting from pub to pub. Billy followed her lead in looking
to the water to find a career. After graduating from dive school in
California, he returned with a new sense of purpose and, before long,
a new identity. Even though his relationship with the marine biologist
didn't last, he was immensely proud to be following his father into the
ranks of small-business owner.

For his first few years as a Black Dog partner, Billy gave the company everything he had. He was away from his apartment so much,
either at Tap's or working on jobs in rivers and lakes all over New
England, that it hardly made sense for him to keep paying rent. So

when his closest friends in Wolfeboro, a married couple he'd originally met while he was tending bar, offered him a room at their place,
he agreed. Even though the couple, Ken and Deb Jones, were not that
much older than Billy, they took a parental interest in him. Before
long Billy took to calling them "Ma and Pa Jones." Deb had recently
lost her brother to illness, so Billy filled a void for her. Billy only in-
frequently saw his own parents, who had moved to Florida and then
separated. So the arrangement worked well all around, even if Billy's
impulsiveness sometimes drove Deb nuts. She still gave him grief for
the time he had tried to MacGyver their clogged kitchen drain by
blasting it with a scuba tank full of compressed air, only to blow out
the whole plumbing line, leaving the walls in the bathroom covered
with excrement.

Despite the traction he and Tap were beginning to get with Black
Dog, Billy knew something important was missing in his life. That
changed at a Fourth of July party, when he met a brown-haired,
wholesome-looking pub manager named Michelle Rodrigue, and
they quickly became a couple. Still, Billy's punishing schedule gave
them only stolen moments together. Billy had always felt most comfortable in his tan Carhartt work overalls, and he'd think nothing of
wearing them out to eat with Michelle. She would jokingly ask if he'd
give her "just one day a year" when he wore something else. But beneath that joke, Billy could detect the frustration in her voice. In some
ways, the Carhartts were a reminder of how he always put his work
ahead of his personal life.

Billy told Tap something had to give. Though he shared Tap's commitment to their company, he needed more balance in his life. After
all, Tap was in a different place—he had a wife, two kids, and a house.
Was it too much for Billy to want that too? He told Tap he needed to
stop working on Sundays so he could golf with Ken in the morning
and spend the rest of the day with Michelle. Tap agreed, though Billy

could sense his reluctance. Tap clearly felt that Black Dog was on the verge, and this was no time to be slowing down.

By the time they toured the Deer Island tunnel, both Billy and Tap saw how the project could be a game changer. The mission was so high stakes and so high profile that it truly could bring Black Dog to the next level without consuming their every waking hour.

On the last Thursday in May 1998, Tap strode into a fifteen-story office building in Omaha's Kiewit Plaza. The Nebraskan city was home to a handful of industry-leading corporations, yet it retained such a small-town feel that they all seemed to be connected in one way or another. Insurance giant Mutual of Omaha sat just a block away, and legendary investor Warren Buffett kept his Berkshire Hathaway offices on the fourteenth floor of the Kiewit building. Despite all its advances and riches, Kiewit continued to cherish reminders of its past, notably the company's simple, black and yellow circular logo. While the exterior of the headquarters was spare and Brutalist-inspired, the interior revealed more of the company's prosperity, with a sleek lobby featuring mahogany wall panels and a graceful waterfall of black granite. After checking in at the security desk, Tap idled by the bronze bust of Peter Kiewit, the founder of the modern company whose "Know your costs" motto continued to serve as a bumper-sticker version of its corporate philosophy.

In the three months since the company had asked Tap to bid on the job to remove the tunnel safety plugs, he had thought about little else, refining his plan again and again. He estimated that winning the contract would generate at least three-quarters of a million dollars for Black Dog, perhaps closer to one million—making it by far the biggest contract in his small company's history. Moreover, it would establish Tap and Billy's dive operation as a prized go-to problem solver for one of the country's largest general contractors. Tap was already thinking of DJ and the other trusted divers he would recruit for the project. "This job," Tap would tell people, "is going to launch us into the big leagues."

Tom Corry, the Kiewit executive who oversaw the Deer Island project from Nebraska, had personally summoned Tap for further talks about his plan. But Tap was not the only one invited to Omaha.

After shaking Tap's hand, Corry introduced him to two other guests. They were both representatives of Norwesco Marine, a Washington State diving company similar to Black Dog though considerably larger. The first guy, Harald Grob, was a short and stocky thirty-eight-year-old Canadian engineer with a low voice and deep-set eyes on a round balding head. As soon as he started talking, he struck Tap as intelligent. The second Norwesco representative was a low-key forty-six-year-old named Roger Rouleau. He had a lantern jaw and thick, graying hair that he wore combed back, giving him a passing resemblance to the current occupant of the Oval Office, Bill Clinton. Roger was far more reserved than Harald, and Tap noticed that when he did speak, it was more likely to be about financial matters than about diving. But Tap was heartened to learn that Roger was in charge, since he owned Norwesco. Like Tap, Roger had started as a diver and had found his true calling as a businessman. Tap felt he could work with Roger and maybe even learn from him. Although Norwesco was only a few years older than Black Dog, it was doing $3 million a year in business, to Black Dog's quarter of a million.

The Kiewit executive explained that Black Dog and Norwesco had both submitted strong bids for the Deer Island job, and he now wanted them to work together to fashion the best possible approach for removing the plugs. Given all the energy and hope he had invested in crafting his plan, Tap was eager for it to get a full airing from Norwesco.

. . .

Roger Rouleau liked to joke that only a crazy man would start a diving business in landlocked Spokane, a small inland city sitting near the Idaho border and a four-and-a-half-hour drive from the Seattle coastline. In fact, his decision had previewed the savvy, pluck, and luck that he would ultimately rely on to make Norwesco successful.

He had gone to dive school, but after failing to find work on oil rigs in the Gulf of Mexico, he had returned to the Pacific Northwest and got by pumping gas and working as a farmhand. Things changed when, in his early thirties, he met a dynamic inventor named Phillip Nuytten, who ran a big-name company called Can-Dive Services in Vancouver. Nuytten's creations included the "Newtsuit," a bulky, puffy breathing apparatus that made a diver look like the Michelin Man but served as a sort of personal submarine, allowing deep dives without requiring decompression. Roger boldly asked Nuytten if he could open a U.S. affiliate office for Can-Dive in Spokane. At the time, Can-Dive's offshore work in the Arctic was beginning to slow. So Roger's offer to pound the pavement in search of small construction jobs on inland dams sounded worthwhile, especially since Roger was willing to do it for five hundred bucks a month.

Before long Roger, who'd always thought of himself as only a mediocre diver, stopped working in the water altogether. His real talent, he found, was bidding jobs and attaching the right divers to them. Success soon attracted some unwanted attention, though, when union workers picketed his office because he was using nonunion divers. Chastened, he switched to union labor. A bigger crisis emerged at the start of the 1990s, when Can-Dive sought legal protection from its creditors and shed assets in order to stay afloat. But Roger turned it into an opportunity, borrowing money from his parents to buy the equipment his office had been using, then incorporating his own business, Norwesco Marine.

Not long after that, Roger bid on a job for Kiewit to build a guide wall as part of an Oregon dam's new navigation lock system for barges. Given the complexity of the project, a few Can-Dive associates suggested to Roger that he bring on an engineer by the name of Harald Grob. Roger was immediately taken with Harald's brains and credentials, and the two men were soon collaborating on lots of projects. Here was a self-assured, worldly pro who had earned his civil engineering degree from the University of Waterloo, spoke both English and German, and was an innovator with experience supervising dives in zero visibility and at depths of up to 240 feet. Unlike other project managers, Harald seemed relatively uninterested in the size of his payday. When he came to Spokane to work on a bid with Roger, he would sleep in a room above Norwesco's office in an industrial park. What seemed to motivate Harald most was the challenge inherent in a job.

Their relationship wasn't without some bumps. Roger would sometimes hear complaints from divers who simply couldn't handle working with Harald. As Roger saw it, Harald's biggest deficiency as a manager was his lack of social skills. The same single-minded determination that fueled Harald's desire to push boundaries in his field worked against him when he interacted with others. "The guy's kinda like a nerd," Roger once said of Harald. "It's like talking to an eraser-head." If a problem arose on the job, Harald, rather than pulling together all the divers and encouraging them to toss around ideas for the best solution, would draw deeper into his own shell, determined to work things out himself. And if Harald didn't respect someone else's opinion, he couldn't even feign interest in hearing that person out. He would simply walk away. Things got so bad on one job in 1995 that Roger had to remove Harald as project manager.

Still, the gusto with which Harald pursued towering engineering challenges made him a huge asset to Roger's company. And for all of Harald's social awkwardness, Roger's wife, Tawnie, would point out how naturally Harald interacted with their young daughter, whose cerebral palsy made many other people uncomfortable. Tawnie would tell her husband that she felt Harald had a kind way about him.

In 1997 Harald helped Roger land a transformative job for Norwesco, working on a new international wastewater treatment project to curb pollution in the Tijuana–San Diego area. Norwesco divers helped install a 150-foot-long, nine-foot-wide riser pipe, which would connect to a new outfall tunnel. When the crew of the drill rig had trouble penetrating the seabed for the riser, Harald sent the divers down to investigate. Instead of the expected sand and silt on the sea bottom, the divers found about five thousand boulders, up to two feet in diameter, all of which had to be removed. Harald also designed a system for divers to do inspections with cameras mounted on their helmets, and for deeper areas, they relied on remote-operated vehicles, or ROVs.

For Roger, the San Diego job led to even higher-paying work. For Harald, it answered the desire he had noted on his résumé: "Of particular interest are special projects where difficult conditions exist or new technologies are developed and used." Harald went on to deliver a lecture about it during a major industry convention and write a feature-length story about it for *UnderWater* magazine.

Their alliance was feeding the needs of both men: Harald's desire to be seen as a bright light in the dive world, and Roger's desire to succeed in ways he could have scarcely imagined when he'd been pumping gas and working as a farmhand. It continued to flourish in 1998, when they landed their biggest project. This one took place in Lake Mead, Nevada, outside Las Vegas, and once again the employer was Kiewit. Lake Mead, the country's largest reservoir, had been created by damming the Colorado River to build the Hoover Dam in the 1930s. By the 1990s it was supplying water to a large swath of the rapidly growing Southwest. But now Lake Mead was struggling to quench the thirst of the Las Vegas Valley, which could claim the world's biggest supply of hotel rooms, even before the planned addition of another twenty thousand on top of the existing one hundred thousand rooms.

The $80 million Kiewit job involved sinking a four-hundred-foot shaft on a peninsula in Lake Mead, driving a tunnel out to the lake, and "tapping" the lake, which more or less meant installing a drain at the bottom of it so it could more efficiently supply drinking water to the region. Creating the water tap required excavating a twelve-foot-diameter, sixty-foot-deep shaft in 240 feet of water. That's where Norwesco came in. The dive job would keep crews from Roger's company busy for about a year and a half, and Kiewit would pay Norwesco close to $3 million. Because of the water depth, divers would be required to spend so much time decompressing that on an average day, they would each usually get less than half an hour of time working at the bottom of the lake. So Harald designed the job to get the most out of each dive. Among other steps, he employed ROVs and used prefabricated construction so divers could multitask during their limited bottom time.

Harald's innovative approaches had not gone unnoticed by Kiewit managers, including Tom Corry, the executive who had solicited the proposals for the Deer Island job in Boston. In fact, Roger knew that Harald was the main reason Norwesco had been asked to bid.

The Kiewit managers, in searching for a solution that would extricate themselves from their Deer Island quagmire, had first reached out to Black Dog and two larger companies in New England. However, after touring the tunnel and seeing what was involved, the two larger companies had declined even to bid on the job, citing safety concerns. "We do not possess the engineering prowess or the safety program/procedures/controls to conduct this work in the manner it requires to provide a successful and safe outcome," wrote the president of one of those companies.

So Kiewit was left with just one bid for this high-risk job, submitted by Tap Taylor and his tiny Black Dog outfit. Seeking reassurance on Tap's plan, Corry sent it to the hotshot Canadian engineer who was wowing him and other Kiewit managers at Lake Mead. At the time, Norwesco hadn't been asked to bid the job, so Kiewit was treating Harald like a consultant in requesting that he review Tap's proposal.

In late March 1998 Harald sent Kiewit his extensive comments about the Black Dog plan. Two weeks later Kiewit changed course, asking Harald to bid on the job, shifting his role from a consultant commenting on Tap's plan to his competitor. After Harald made a site visit to Deer Island, he and Roger submitted Norwesco's bid to Kiewit on May 21. Harald, who had previously been sent Tap's plan without Tap's knowledge, took no chances now with his own submission. Using his Canadian spelling, he included this warning: "Any information, concepts, and equipment contained in this document which was hitherto not known by the recipient, or in which there subsists a proprietary right must not be disclosed or used by any other organisation without the written permission of Norwesco Marine Inc."

Now, just one week after Harald submitted the bid, he and Roger were sitting in Omaha. Tap was also there, knowing Kiewit had asked them to work together, but he was beginning to realize that his company would be asked to play only a supporting part. By now, it was becoming clear to everyone that Harald's plan would carry the day.

Tap was used to dealing with engineers who always believed they were the smartest guys in the room. But Harald, whose CV distended over seven pages including attachments, struck Tap as someone who took this self-assuredness to an extreme. Tap found himself learning a lot about Harald—how he had written for respected dive magazines, how he had done mixed-gas dives with zero visibility, how he had experience with those "Michelin Man" personal submarines called Newtsuits, how his designs had solved one knotty dive challenge after another. Tap, who had never even gone to dive school, couldn't help but be impressed. He knew he couldn't begin to compete with Harald's credentials.

Early on in the meeting, the Kiewit executive excused himself so that Tap, Roger, and Harald could iron out more of the details on their own. To Tap's disappointment, they devoted no time to talking about the proposal he had spent months crafting. Tap's plan involved a diesel flatbed truck that would tow, deep into the tunnel, a twenty-foot-long tube trailer carrying seventy-five thousand cubic feet of high-pressure, or HP, compressed air. That, Tap estimated, would be enough air to sustain his ten-man dive crew for up to twenty-four hours. The tube trailer was not an exotic piece of equipment—it looked like the same trailers that regularly travel along interstates. Tap estimated the truck could make it as far as riser number 22, and then a small excursion crew could use a battery-operated vehicle to keep moving east.

Tap still didn't know that Kiewit had asked Harald to review his plan and that Harald hadn't been shy with his criticism, calling it "overly complex," questioning the maneuverability of that tube trailer, and faulting Tap for having only one main air supply. "Reliance on a single large source is unsafe," Harald noted. Now in Omaha, as Harald presented his own plan, Tap found it so outside-the-box that he struggled to comprehend it. The differences between the two plans were stark. For transportation in the tunnel, Harald proposed using a pair of diesel Humvees, one towed by the other. The off-road vehicles would be customized to work in the oxygen-deficient tunnel, with a scrubber attached to their exhausts, and an oxygen bleed and monitor added to their air intake systems. From the Humvee setup around riser 12, an excursion crew, attached to the Humvees by umbilicals, would slog on foot to the end of the tunnel.

By far the biggest difference between Harald and Tap's plans concerned the air supply. While Tap's had relied exclusively on the tractor-trailer of HP air, Harald had determined that the cramped confines and nearly ten-mile length of the tunnel would make it impractical to transport a sufficient supply of bottled air. Instead, he called for three separate sources of breathing air. There would be a small HP air supply, as well as backup contraptions known as rebreathers. But the most unusual source was a cutting-edge system by which liquid oxygen and liquid nitrogen would be mixed in the tunnel and converted into gaseous breathing air. Simple arithmetic explained why he was calling for liquid gas. He'd get the same amount of breathing gas from "vaporizing" four 180-liter liquid tanks as he could from using seventy standard "K" bottles of HP air, a conventional supply that would exceed four tons in weight. So the liquid option would spare the divers a whole lot of lugging.

In addition to being more nimble, Harald's plan even carried a lower estimated price tag. Whereas Tap's would cost about three-quarters of a million dollars, Harald's would come in at around six hundred thousand. His plan called for one week of training and two weeks to complete the plug-removal operation.

During the Omaha meeting, Tap asked a few general questions about Harald's plan, but stayed mostly silent about the liquid air breathing supply. He had absolutely no experience with that kind of work, and didn't want to give Harald the chance to lecture him if he asked a question that somehow revealed his ignorance.

By now, Tap knew one of the reasons Kiewit had decided against giving him the main contract. He had been unable to secure bonding—financial protection for the operation, in the event the job couldn't be completed. Unlike Norwesco, Black Dog was simply too small to obtain bonding for such a complex project. Given the hell that Kiewit had gone through on Deer Island, Tap could understand why the company was not willing to roll the dice with an unbonded subcontractor.

When Corry, the Kiewit executive, came back into the room and asked how the two subcontractors had made out, Tap smiled along with the others. It was now clear to him that he was being asked to play a junior role to Harald and Norwesco, most likely because Tap could provide some cover with the Boston labor unions for Kiewit's decision to bring in a West Coast company with a history of union problems. At the same time, he would offer logistical support for the operation, lining up local workers, equipment, and facilities. Even if the project no longer presented the game-changing, million-dollar payday that Tap had once envisioned, it would still allow his divers to earn some good wages while giving Black Dog a subcontract that would easily exceed $100,000. What's more, Tap saw it as preserving Black Dog's position as a trusted problem-solver for Kiewit.

In his hotel room that night and on the flight home, Tap repeatedly read over Harald's plan. When he returned to New Hampshire, he had Billy do the same. Even if Tap hadn't made a great connection with Harald, he had to admit that the Norwesco project manager's design, with its military feel, was dazzling.

The white Chrysler K car pulled up to the terminal at Logan Airport, and Roger squeezed in. Waiting in the car for him was Tom Corry, who'd flown in from Omaha, and two Kiewit managers from Deer Island. It was July 31, 1998, two months after the Omaha gathering. Roger had been asked to fly to Boston to attend a meeting that Kiewit viewed as critical.

Roger was grateful that the Deer Island job appeared to be shaping up well for Norwesco. He knew there were plans for him to sign a revised contract with Kiewit that would pay Norwesco $776,100, or nearly $200,000 more than Harald's original estimate, once additional costs and services had been factored in. That promised a tidy profit, even after Norwesco's considerable outlay of cash to buy the Humvees. Still, Roger wasn't happy to be in Boston. For him, the trip had come about unexpectedly, after Harald had told him, "I can't make the meeting. Can you go?"

"Jeez, I don't know much about this job, Harald," Roger had replied. "What am I going to be doing over there?"

"You're going to be meeting with OSHA," Harald explained. "They want to go over the plan."

"Well, that means I better know a whole lot more about this thing," Roger replied. He spent his long, expensive, last-minute flight to Boston via Minneapolis reading over Harald's revised plan. There were parts of the design he still didn't fully understand. But given his confidence in Harald's abilities, Roger reassured himself, *We're cool. This thing's good.*

Now, as they made their way in the K car to the OSHA regional office in suburban Boston, the Kiewit team briefed Roger on what to expect. They would be meeting with three representatives of the federal workplace safety agency, including its assistant area director. The Norwesco plan had already won plaudits from Dave Corkum, the Kaiser manager with whom Kiewit had been sparring for years, as well as from other key players. Roger knew that, in addition to Harald's creative solution to the plug problem, one of Norwesco's supplemental selling points to weary project officials was the company's status as fresh-faced outsider, an expert consultant arriving with new ideas and a can-do attitude.

At the start of the meeting in the OSHA conference room, Corry introduced Roger as the president of Norwesco. But mostly Roger kept quiet, letting the Kiewit executive lead the discussion. The three OSHA officials taking part in the meeting were quite familiar with the Deer Island tunnel, since, during its long construction, the agency had fined Kiewit for a series of safety violations. Now Corry handed them copies of Harald's Norwesco plan, and Roger watched as they flipped through it. Sitting at the table, Roger felt like a puppet. Of course, some of the important details of the plan had yet to be worked out, such as the precise manner in which the liquid oxygen and liquid nitrogen would be mixed and delivered to the divers. But the main components were all there. The plan presented to OSHA made it abundantly clear that by the time of the plug-pulling operation, the tunnel would be entirely stripped of its ventilation and other utilities, so dive crews would be bringing in their own air for life support. The meeting lasted about an hour.

Before he knew it, Roger was once again in the K car. Although OSHA did not have statutory authority to preapprove a plan, the Kiewit team talked about their relief in seeing how receptive the government officials had been. Despite the plan's unconventional use of supplied air, the OSHA reps had raised no red flags about it, and their only questions had struck Roger as rather mild.

Roger, though, left the meeting feeling annoyed. His command performance would cost him the better part of two workdays in travel time, not to mention significant travel expenses, all so he could sit in a room and watch a few government employees breezily flip through the plan Harald had crafted. Roger thought to himself, *What a waste of time.*

A month later Tap was on the move, inside the rapidly darkening tunnel. While Roger was back in Spokane, Tap and Billy and two other Black Dog divers joined Harald and a couple of Norwesco divers for a fact-finding mission into the Deer Island tunnel. By this time, in late August 1998, sandhogs had removed the bag line, rail tracks, and lighting from more than half of the tunnel. As they backed out, the sandhogs had used a special dam car to pump the standing water— the seepage from those cracks in the bedrock that Kiewit had complained about in its "water too wet" claims—gradually moving the water westward back toward the shaft.

The goal of this fact-finding trip was to get a more precise handle on conditions in the tunnel, particularly the presence of any explosive or toxic gases, so there would be no awful surprises during the actual mission. The divers' specific assignment was to measure the levels of oxygen and various other gases in the ambient air far from the shaft, so Harald could refine his plans for the ultimate operation. Consistent with the pattern that had been established in Omaha, Harald had taken the lead in designing and running this mission, with Tap playing a supporting role.

As the divers rode the tram along the remaining rail tracks, they occasionally spotted numbers spray-painted onto the tunnel wall. Those numbers represented "rings," the metric by which the sandhogs had long charted distance in the tunnel. Each ring represented the five-foot length of one of the concrete segments lining the tunnel wall. So if the spray paint on the wall read "1000," that meant the workers were at ring 1000, or roughly one mile in. Around ring 4000, or just shy of the four-mile mark, the tracks ended. There the divers stepped off the tram and began unloading gear. They put equipment for communications, air monitoring, and air supply into two flat-bottomed aluminum boats. Then they attached wheels on collapsible legs to the boats, allowing the vessels to float or be pushed like hospital stretchers, depending on how deep the standing water was in certain spots.

Even though the ventilation line no longer extended past the four-mile mark, the divers found that oxygen levels were sufficient to sustain human life well beyond that point. But they knew that by the time the tunnel cleanup had been completed and the actual plug-removal mission had begun, those oxygen levels would be lower, for two reasons. First, the cleanup wouldn't be considered complete until the entire bag line was yanked out. Second, the remaining oxygen at the end of the tunnel would essentially begin using itself up. In the dank, confined space of the tunnel, oxygen would be depleted by things like the growth of aerobic bacteria and the rusting of metals, such as bolts. There was also the very real possibility that oxygen would be displaced by highly toxic gases, such as carbon monoxide, methane, and hydrogen sulfide, which is produced when certain organisms decay.

Tap watched as Harald set up his communications command post, purposefully striding around the tunnel, wielding a radio antenna like an orchestra conductor's baton. Harald wore a yellow slicker and high rain boots over his tan Norwesco jumpsuit and gray hard hat. He would remain in contact with the divers via VHF radios, which were designed to work for up to four miles, though no one was sure how well they would function this deep into the earth.

At the command post, two teams of divers put on extra layers of rain gear over their coveralls, then left Harald and ventured deeper into the tunnel. They stopped at regular intervals to measure the levels of oxygen and flammable gases, logging their readings into black notebooks.

Once the oxygen levels dipped to a certain point, Tap and the other divers paused to don their main breathing apparatus for this mission: rebreathers. These Dräger-brand contraptions consisted of facemasks attached to bulky backpacks that looked like roller suitcases. Inside each rebreather was a small bottle of oxygen and a soda lime filter that absorbed the carbon dioxide from the exhaled air so it could be reused.

Tap was relieved that their readings showed no significant levels of explosive gases. Even though the divers weren't asked to venture all the way to the end of the tunnel, the partial journey gave Tap enough of a taste to conclude that his body was simply too big to be messing around in such a confined space. For the actual plug-pulling mission, he would keep himself out of the tunnel and stay topside with Harald.

5

Mine Rescue

One month after Neil Armstrong took his giant lunar leap for mankind in the summer of 1969, a NASA manager pulled Jim Lovell aside. Lovell had been aboard the Apollo 8 when it made the first manned orbit of the moon, and he was scheduled to conduct his own moonwalk in late 1970, as commander of Apollo 14. But now his boss asked him if he'd be willing to take command of an earlier mission instead. Lovell said yes. Even though the training period would be condensed, he and his two crewmates would still have nearly a year to prepare.

They would be taking a tougher path to the moon and back, since this time the landing spot would be a treacherous Appalachia-like range rather than the desolate lunar plains. So NASA toughened its already exacting standards for preparation, putting Lovell's crew through months of flight simulation. With the headset-wearing staff of Mission Control at their consoles and the astronauts in mockups of the Apollo command and lunar modules, a separate team would surprise them with a series of simulated problems. This endless wargaming had twin aims: to forge cohesion among the crew and with Mission Control, and to arm the entire team with the knowledge and confidence to tackle any crisis that might emerge during the actual space flight.

NASA had always taken great care in assembling its teams, using psychological tests to exclude those who didn't demonstrate the right qualities, such as the ability to work well with others. But this simulation allowed the astronauts to be tested on what's known as space flight resource management, notably their ability to work through a crisis by following proven protocols rather than by making panicked gut decisions.

After they had spent so many hours together, Lovell could detect every nuance and inflection in the voice of his crewmate Ken Mattingly. This was important, since four days into the actual mission, they would be separated, as Lovell and the third crewmate would land the tiny lunar module on the moon, while Mattingly hovered sixty miles above in the main command module. A few weeks before launch, NASA ran a simulation of precisely this part of the mission. Staring at his monitor, the electrical and environmental command officer from Mission Control, known as EECOM, noticed a drop in the cabin-pressure reading for Mattingly's command module. Because this blip disappeared after a second, he dismissed it. Forty minutes later Mattingly's voice crackled into everyone's headsets. "We had a sudden depressurization here, Houston. Cabin pressure is down to zero," he said, adding that he was now relying on the pressure from his space suit.

The EECOM froze. He had failed this pop quiz spectacularly— grounds enough for him to be fired. But the no-nonsense flight director in Houston, Gene Kranz, was feeling charitable. "All right," he said, "let's work through the problem." For several hours the EECOM and the entire team ran through an obscure survival simulation, blasting the lunar module back up to dock with the command module to serve as its cramped "lifeboat" for all three astronauts.

Just days before launch came a different, nonsimulated crisis. A member of the backup crew came down with the German measles. Lovell's crew underwent blood tests to see if they were susceptible. Mattingly was the only member who'd never had the measles, so NASA doctors decided to ground him.

Lovell went crazy. "Now?" he thundered. Crew cohesion that had developed over nearly a year of training should matter more, he argued, than the remote possibility that Mattingly might come down with the measles. Lovell's trust in Mattingly dated back to Apollo 8, when he had served as "Capcom," the astronaut stationed in Houston who is the primary conduit for information between Mission Control and the crew hurtling through space.

Although NASA gave its crews great latitude in critiquing mission plans, it gave its doctors greater deference in medical matters. Forty-eight hours before launch, Mattingly's replacement was certified to fly.

Fifty-five hours, fifty-four minutes, and fifty-three seconds into the mission, Lovell heard a dull bang. "Ah, Houston, we've had a problem," he said, reporting a low voltage reading from the instrument panel. As it turned out, a tank of liquid oxygen had exploded, leading to a crippling power failure in the command module.

Anyone who saw the 1995 Ron Howard–Tom Hanks movie *Apollo 13* will recall the dramatized version of events that followed that dull bang (which the movie depicted as a big blast). On the screen and in real life, Lovell's crew suffered an unholy number of setbacks, any one of which should have been a death sentence: the trail of liquid oxygen venting out the back of the module, the loss of cabin heat, the shortage of water, the buildup of carbon dioxide, and the need to fly blind during a power shutdown. Although Kranz never actually uttered the unforgettable line from the movie "Failure is not an option," the flight director did provide inspiring, sure-footed leadership.

"This crew is coming home," he told his team. "You have to believe it. And we must make it happen." (Kranz did like the "failure" line well enough to make it the title of his autobiography.)

Kranz in Houston and Lovell in space were both able to draw the best out of their crews, working collaboratively and leveraging all their long hours in training to will a miracle. Mattingly (who remained measles-free) was even pressed back into service, returning to the mock-up modules in Houston to test work-arounds for the power problem. And the EECOM's embarrassing failure during simulation turned out to have been a blessing. It had forced everyone to practice the obscure lifeboat maneuver that the trio of astronauts would ultimately use to save their lives.

In addition to being serious about diving and partying, DJ Gillis was a history buff and adventure nut who loved watching the History Channel and the Discovery Channel. When Tap called him in the spring of 1999 to offer him a slot on what some people were starting to call a moonwalk under the sea, DJ was intrigued.

Tap explained that the plug-pulling mission in the tunnel was expected to take a couple of weeks. But because of its unusual nature, the job would be preceded by several weeks of preparation, all to be hosted at Black Dog's place in New Hampshire. It would be their version of NASA training.

At the time, DJ was working as a pile driver on a Big Dig–related project, building soldier pile retaining walls. He was enjoying steady work at good pay. But as pile drivers often joked, divers were so addicted to the thrill of being in the water that they'd give up a lucrative yearlong pile job on land for the chance to dive in a puddle for a couple of weeks. Of course, Tap wasn't calling DJ to offer him traditional dive work. The details were even more enticing. He'd be going down a 420-foot shaft. Traveling in souped-up Humvees to the end of a dark tunnel running for miles under the ocean. Relying on a cutting-edge mixed-gas breathing system. "Tap, this sounds real cool," DJ said.

Tap told him it was a complicated job, the likes of which had never been done before, so they would have the opportunity to make history in their field. It would involve working with a crew from the West Coast. Tap said he'd feel more comfortable knowing that a talented diver like DJ, with Gulf experience, was there representing Black Dog. It didn't take DJ long to give notice at his pile driver job and sign on to Deer Island.

Tap told DJ that a representative from the Dräger equipment company would be coming to New Hampshire to train the divers on how to use the specialized breathing gear. A trainer from a mine rescue company would also be there to conduct sessions on working in confined spaces. For DJ, the fact that a company that trained coal miners would be instructing them was as good an indication as any of just how different this job would be. That training would be essential because the Boston Fire Department had already made it clear that if the dive team ran into trouble miles into an unventilated tunnel, firefighter crews would not be going in to save them. If the dive crews wanted the security of knowing a rescue crew was available, they'd have to form one themselves.

Tap explained to DJ that he wouldn't be in charge of this job. He'd be answering to a Canadian engineer named Harald Grob who worked for a company called Norwesco out of Washington. Tap leveled with DJ that he didn't much care for Harald. But he said the guy seemed to be razor sharp and had a ton of high-level experience, so at least he knew what he was doing.

On May 13, 1999, Harald phoned Kiewit executive Tom Corry to report some good news: the first two customized Humvees would be leaving Norwesco's Spokane office in a week, loaded, along with other equipment, onto a tractor-trailer bound for New England. In addition, Harald had lined up safety specialists to conduct separate training sessions for the Norwesco and Black Dog dive crews. Now that the sandhogs had removed the last of the tunnel's utilities, Harald seemed to be putting the final touches on his plug-removal plan. He informed Corry that the crews would be assembling in Tap's backyard in a month and would then begin their convoy to Deer Island. Nine months after the fact-finding mission, one year after the arranged-marriage meeting in Omaha, and nine years after Kiewit had won the big contract, the proverbial light at the end of the tunnel finally appeared to be flickering into view.

But if Harald was truly putting the final touches on his plan for the T-minus-one-month mission, his actions behind the scenes suggested something completely different. Four days after Harald's upbeat phone call to Corry, Harald's primary equipment supplier in Spokane sent a fax indicating how very fluid the plan still was. With the mission's start date looming, the key component of Harald's plan, the cutting-edge system he envisioned for supplying the divers with enough portable breathing air, was still something of a back-of-a-napkin wish.

Jerry Anderskow, Harald's contact at supplier A-L Compressed Gases, Inc., faxed his inquiry letter to the customer service department of a Minnesota company called MVE. He needed a company specializing in cryogenics because converting cold liquid gas into a gaseous state would be the crux of Harald's innovative breathing system. "Dear Sir," the May 17 letter to MVE began, "I need your help on a very important application. . . . Our customer has a project on the East Coast and needs all the appropriate equipment in place ASAP. The specifications are open for engineering and recommendation." After describing the circumstances of the Deer Island job, and Harald's desire for a system that would vaporize liquid oxygen, Anderskow's fax concluded, "Please quote me a turn-key system for new equipment. Any recommendation on how systems should be set-up would be appreciated. Time & Delivery is critical."

It wasn't clear what Harald had been doing for the nine months since the fact-finding mission and the finalization of Norwesco's contract with Kiewit. That he and his primary supplier were still desperately searching for a "turn-key" system just a month before the mission's scheduled start date painted a vastly different picture from the rosy one he had presented to Kiewit.

Two weeks later Anderskow faxed Harald an update. After consulting with MVE, the nation's largest manufacturer of cryogenic storage systems, he recommended a setup consisting of one liquid oxygen cylinder and either one or two liquid nitrogen cylinders, for a price tag of just under $5,000 per setup. But Anderskow included this word of caution. "*Problem*: We cannot find a breathing air approved mixer/blender," he wrote. "Do you have access to this equipment?" The mixer was essential since it would be responsible for blending the vapors from the liquid oxygen and liquid nitrogen to provide the divers with their primary source of breathing air. Without the mixer, the system would not only fail to be "turn-key." It would fail to work.

In early June, DJ climbed the stairs of an office building in Portsmouth, New Hampshire, to attend the East Coast sessions of the mine rescue training. Harald had arranged for the Norwesco divers to undergo the same kind of training at Lake Mead, Nevada, where they were finishing up work on that Kiewit job. Black Dog had two offices in southern New Hampshire. The A-frame garage in Tap's backyard in Greenland served as the company's storage shop, while a small office a few miles away in downtown Portsmouth was the place where Tap bid on jobs and met with clients. That office, in a tired building

with creaky wooden floors and drop ceiling tiles, was too tiny to ac-
commodate the training sessions, so he reserved a conference room
a few doors away.

The first day's session focused on the Dräger equipment, which
included the handheld analyzers that the divers would use to measure gas levels in the tunnel, the firefighter-type masks they would
wear, and the rebreathers they would use as a backup air supply. From
the fact-finding mission, Tap and Billy had already gained some experience with those rebreathers, the roller-suitcase-type backpacks
that scrubbed and recirculated breathing air. But the Dräger trainer
explained that for the actual plug-pulling mission, the divers would
be required to be clean-shaven, because beards and sideburns could
interfere with the seal on the facemasks and cause leaks. DJ let out
a laugh. He knew Billy would comply by making sure there was no
stubble on his beard, but there was absolutely no way his friend was
going to part with that mustache.

Although DJ had met Billy only in passing back in the early 1990s
when he'd worked with Tap in Vermont, they were now close friends.
Over the last two years, he and Billy had logged time together on a
bunch of Black Dog jobs, bonding with, and busting on, each other.
Every time DJ made a crack about Billy's 'stache, Billy would shoot
right back, needling his buddy about his curious inability to grow
more than a thin band of hair above his upper lip. "Hey, DJ," Billy
would quip, "some middle school kid was here looking for you. He
wants his mustache back." Billy was like no one else DJ knew. He had
once shown up at a dive job with blood on his coveralls and a wine-
colored substance wobbling around inside a Tupperware container.
"This is lunch!" he told DJ excitedly. On the drive to work that morning, a deer had leaped out in front of his truck, and Billy hadn't been
able to stop in time. So he pulled over, removed the animal's heart,
and packed it up. When DJ asked how he would cook it, Billy laughed
and said he'd use one of their hot-water hoses as his heat source.

DJ was relieved that the easy relationship between Tap and Billy seemed to be back on display. He'd heard that the tension between the two partners over how much the business should consume their lives had escalated to the point where Billy, feeling thoroughly burned out, had actually sold his ownership share back to Tap. Still, a few months before the Deer Island project, Billy had returned from a big dive job in Turkey feeling so rejuvenated that he had decided to return to Black Dog as an hourly employee. The transition had been surprisingly smooth. Billy was back working with Tap at the business they'd both poured their souls into, but he'd figured out how to do it without sacrificing everything else in his life.

DJ's life had finally improved, too. For three years after his fight with the sinker drill in Vermont, he'd been hobbled by a worsening pain that no doctor was able to treat, no matter how many he tried. They all told DJ it was a soft-tissue injury, and there was little they could do besides prescribe ever more powerful pain medication and anti-inflammatory steroids. He'd begun by swallowing fistfuls of Advil and before long was taking enough prednisone to treat a team of Clydesdales.

The only good thing to happen to DJ during this dark period was that he became a father, though the circumstances around his son's arrival in 1995 were hardly ideal. Still pining over Dana following their breakup, DJ had begun dating the ambitious ex-girlfriend of one of his buddies. Just three months into their relationship, she discovered she was pregnant. They tried making a go of it as a family, moving into a small apartment in Waltham. But DJ was still too wild to settle down, and they were both too young and strong-willed to make it work. Regardless, DJ was determined to maintain strong ties with his boy, Cody, and not become the kind of disappearing act his own father had been.

Just after Cody turned two, DJ had been driving to pick him up when he suddenly felt cold and worried he was about to pass out.

Instinctively, he maneuvered his truck into the middle of a busy intersection and threw it into park. There, he slumped his head over the steering wheel and began vomiting blood. Then he blacked out.

After DJ awoke in the intensive care unit of the hospital, he learned that his decision to use his last moment of consciousness to plant his truck in the middle of the intersection had been a wise one. The attention he attracted led to the speedy arrival of paramedics, who found DJ in shock.

At the hospital, doctors quickly pumped blood into him to replenish what he had lost, and they continued to pump in more—twenty-eight pints in all—as they searched for the cause of his internal bleeding. Finally a gastroenterologist inserted an endoscope into DJ to find the source. It was an ulcer. It turned out that much of the back pain he'd been enduring for the years since the Vermont job had actually been referred pain from a posterior duodenal ulcer, which itself had apparently been caused by the massive amounts of medication he had been taking.

By the fall of 1997, after two additional surgeries, DJ was back diving. And by the start of 1999, his love life had come full circle. It began when he bumped into Dana's sister, Lisa, in Waltham. She invited him to join her and Dana for a night out at one of their old haunts. Even though he hadn't spoken to Dana in years, the laughs began flowing as easily as the beers. At the end of the night, Dana drove DJ home. Idling in the car outside his mother's house, she asked him, "So what's going on?"

"What's going on with what?"

"With you and me."

"I didn't know there was anything going on between you and me," he said tentatively. He told Dana he was seeing a nice girl, a bartender in town, and she was very good to him. Dana told him how sorry she was to have broken his heart and how much she wanted him back. Then, as quickly as he had left his pile-driving work for the tantalizing tunnel job, DJ broke up with the nice bartender to reconnect with Dana.

For DJ, the Deer Island tunnel job felt like a chance to reset the clock back to that pre-injury period when his future had seemed limitless. As he sat in the Portsmouth training session, one more reunion was taking place. Joining him on this tunnel assignment would be his old friend and mentor, the ex–Navy SEAL he hadn't seen since Vermont. Ron Kozlowski may have had a few more strands of gray in his sandy ponytail, but for a guy who was about to turn fifty, he seemed as high-energy as ever.

Five years after Vermont, here DJ was, back working with Kozlowski, Tap, and Billy, back working for construction powerhouse Kiewit, back dating Dana. Back where he belonged.

You're a firefighter charging into a burning building. You and your partner split up in the hallway and head into different rooms. Through the raging flames and suffocating smoke, you find a boy, perhaps six years old, lying on the floor, unresponsive. You call for your partner but get no answer. Heading into the next room, you find he's also down on the floor now, unconscious. The roof above you is threatening to collapse. You're running out of oxygen in your tank. You realize you can carry only one person out, at most. Who do you take? The boy or your partner?

In painting that scenario, the trainer had captured DJ's attention. And it didn't take DJ long to make up his mind. *Of course, I'd take the kid. He's got his whole life ahead of him.*

The trainer stressed that you have to respond with logic, not emotion. Given how long the fire has been blazing, there's a high probability that the boy is already dead. Your partner, meanwhile, will have gone down only a minute or two earlier. He still has a chance.

All right, DJ thought, *I'd better pay attention to this guy.*

A representative from the Central Mine Rescue Company of Idaho led the divers through days two and three of the training, following day one's instruction by the Dräger rep. If the Dräger session had showed the divers how they would be relying on different equipment for the mission, this mine rescue portion showed them how they would be entering a different world entirely.

In fact, although that firefighter scenario had been an attention grabber, DJ and the other divers would face an environment more treacherous and uncertain than what firefighters or even coal miners encountered. After all, most buildings and mines have at least two entrances, often many more than that. When miners get trapped, the first goal of rescuers is to establish a "fresh air base" staging area as close to them as possible. But if a nightmare happened nine and a half miles into the Deer Island tunnel, with its single entrance, the rescue challenges would be staggering.

The aim of the mine rescue training was to teach the divers about the overall forces they would be up against, arming them with the kind of clear-eyed, fact-based knowledge they could use in a crisis to counteract all the emotion that would be coursing through them. The trainer explained how, when confronted with extreme danger and stress, a person's fight-or-flight response kicks in, producing a flood of stress hormones, notably cortisol and adrenaline. This produces short-term benefits, sending more oxygen to the brain to fight off the threat. But the price for this response is a huge spike in the consumption of oxygen. During the Dräger training, the divers had been tutored on how to use their backup air supplies. In addition to the rebreathers, there were self-rescuer devices called Ocencos, which were designed to give up to sixty minutes of air, and a Dräger "bail-out," a small bottle that the divers would wear on their hips, offering up to ten minutes of emergency air. However, the trainer explained that in a crisis, with the fight-or-flight response gobbling up all that additional oxygen, those backup supplies would run out of air in about half the usual time. So an apparatus promising sixty minutes of air could be counted on for just thirty.

All that adrenaline might give the divers the strength to do things they never thought possible. However, the trainer cautioned, the divers needed to respect their limits. "Don't try to bluff your way through," the training manual warned. "Feeling nauseous with your apparatus on is just not safe. If you are in unsafe air and you vomit into your facepiece, you will not be able to take off your facepiece. . . . If you pass out or go down, you become a detriment to your team."

For DJ, the most sobering lesson in the mine rescue training was the one that focused on how to handle a fallen comrade. Notwithstanding that scenario about having to choose between your partner and a six-year-old boy, the awful message from the trainer was that, in a confined space like a mine or an endless, empty sewer tunnel, the divers probably would have no choice at all. If disaster struck and took a team member down, the other divers' main task had to be to ensure their own safe exit, rather than risk becoming another casualty by trying to revive their coworker or retrieve the body. One handout put it this way: "If the victim has collapsed as a result of an oxygen deficient atmosphere and been there for any length of time, it is very likely that he is dead and the discoverer's life is risked in vain."

To DJ, this seemed to go against what he had learned as a diver, where the expectation was that you'd do everything in your power to try to save your buddy. And it certainly violated the *Saving Private Ryan* message of duty and honor that he had gleaned from watching hours of World War II movies and documentaries.

But the trainer stressed that the tunnel's irrespirable air made it more hazardous than a battlefield. In an environment where just one or two breaths could be fatal, the main tenet in an emergency situation had to be: "Do no more harm." A mine rescue association had compiled numbers from mine disasters over the last half century and found that, despite their good intentions, people attempting to save miners had ended up losing their own lives at a rate that increased the total number of casualties by about 50 percent. The message to the divers was clear. You must not add to the body count.

In mine disasters, a special rescue and recovery team is typically responsible for retrieving bodies, after safe passage has been established. DJ winced at the graphic descriptions in the training manual, such as the one warning recovery team members how careful they had to be in handling bodies that might have decomposed: "In some cases, the skin has actually been pulled off the hand and resembles a hand-like glove, complete with fingernails and creases at the knuckles."

While the tunnel mission would be different from a mine operation, there would be enough Black Dog and Norwesco divers hired and trained to allow for a separate backup crew. Those divers would remain on standby anytime the main team was working in the tunnel. But exactly how and under what circumstances that backup crew would be activated remained unclear.

Everything boils down to oxygen. Human life depends on it. Yet oxygen makes up just 21 percent of the air humans normally breathe. By far the biggest component is nitrogen, at 78 percent (with the remaining 1 percent made up of other gases). But that big dose of nitrogen is inert, meaning it's just along for the ride. It does nothing to support human breathing and becomes a factor only in the negative, if it somehow becomes an asphyxiant by displacing the amount of available oxygen. Although oxygen makes up just one-fifth of our breathing air, when the usual concentration slips even slightly, bad things begin to happen.

The mine rescue trainer had the divers flip to a printed table in their manual, which spelled out the dire consequences of falling oxygen concentration.

17 percent: Panting, decreased ability to perform tasks

15 percent: Tightness in forehead, headaches, dizziness, impaired
   judgment

9 percent: Unconsciousness

6 percent: Death

The trainer explained that the most insidious part of oxygen deficiency is that the sufferer is typically not aware of what's happening
until it's too late.

For DJ and the other divers, number-dense charts were nothing
new. One of their early lessons in dive school had dealt with how
to read the tables in the *U.S. Navy Diving Manual*. These spelled out
the appropriate number of decompression stops they'd have to make
on a dive, based on depth, and the total time they'd need to spend
breathing oxygen in a chamber to decompress. Deep-sea divers had to
understand and respect these numbers if they were to avoid complications or even death. The two biggest concerns were decompression
sickness, or "the bends," which can cause paralysis if divers ascend too
quickly; and nitrogen narcosis, which produces a feeling of intoxica-
tion and slows mental functioning. Divers often refer to the effects of
narcosis as Martini's Law, which holds that, for every fifty-foot depth,
a diver is hit with the anesthetic equivalent of drinking one martini.
Both of those were nitrogen-related dangers, however. Surprisingly,
when it came to oxygen, commercial divers on deep dives typically
had to be more concerned about getting too much oxygen rather than
too little. Astronauts have no problem breathing pure oxygen before
and during spacewalks, since that helps reduce the nitrogen levels in
their tissues. But for divers working underwater, oxygen is not necessarily their friend. That's because the deeper the divers descend, the
more they must contend with the increasing pressure from above.
For every thirty-three-foot descent in salt water, divers carry the pressure of one additional "atmosphere"—that is, the total weight of all

the air above ground. Deep-sea divers know that they must breathe higher concentrations of oxygen to counteract that increased partial pressure. But elevated oxygen levels can quickly become toxic, potentially leading to convulsions or seizures.

On this job, although the divers would be working hundreds of feet below the sea, the tunnel walls would protect them against pressure. So they wouldn't have to worry about too much oxygen, only too little.

The challenge was getting just the right amount. That glam rock group from the 1970s had it right with the hit song "Love Is Like Oxygen."

> *You get too much, you get too high*
> *Not enough and you're gonna die*

If DJ's initial impression of the Vermont dive crew reunion was pure joy, it didn't take long for the old tensions between Tap and Kozlowski to resurface. Here were two guys he respected and liked, yet something about their personalities just refused to mesh. Part of it seemed to be a perception problem. When they'd last worked together, Tap had been the untrained rookie and Kozlowski the seasoned pro. Kozlowski had faulted Tap for being insufficiently deferential to the veteran divers and yet too deferential to the Kiewit bosses. Tap, meanwhile, had found Kozlowski to be overly boastful and intent on chipping away at management for every possible perk. Now, five years later, Kozlowski found himself working for Tap, who was once again working for Kiewit. When DJ asked why Tap had hired Kozlowski, given their past sniping, Tap explained that the union hall had recommended him. Because Tap knew nothing about the kind of mixed-gas systems underpinning Harald's plan, he wanted to make sure there were experienced guys representing Black Dog who did.

The new friction between Kozlowski and Tap began over money.

Divers were typically given an extra stipend for something called penetration pay. That differential kicked in if they were required to enter an environment like a pipe or tunnel, where their overhead was obstructed in a way that prevented them from making a clear ascent to the surface of the water. On those jobs, they would typically be paid a few dollars on top of their normal wage for every foot of penetration. Since these divers were being asked to enter an underwater tunnel, Kozlowski naturally assumed they would be getting penetration pay.

But during this training session, Tap informed him that full penetration pay was not possible. After all, the tunnel was nearly fifty thousand feet long. Did Kozlowski seriously think each diver was going to be paid a bonus of $50,000 or more? Tap stressed that Black Dog was merely a subcontractor to Norwesco on this job, so his influence was limited. Still, he said, he had negotiated as forcefully as he could with Norwesco owner Roger Rouleau. He felt he'd secured the divers a good deal, with a minimum day rate of $742 and several hundred more on the days when they were actually working in the tunnel. They'd get time-and-a-half pay after eight hours, double-time after twelve hours, and even a small form of penetration pay. When a diver got to the point where he was crawling into one of the thirty-inch-wide "off-take" connector pipes to remove a safety plug, he would see his pay goosed, based on however long that pipe happened to be.

Kozlowski countered that the so-called penetration pay that Tap had negotiated was chump change, since those off-take pipes averaged just twenty feet in length, and there would be only one diver at a time crawling into them. Yet all the divers would be taking an enormous risk, traveling into this 9.5-mile, unventilated, unlit tunnel. Wet or dry, this was the very definition of penetration, and they should be paid accordingly. Kozlowski wanted to make sure that Tap, now that he was in management, hadn't forgotten about the guys like him and DJ who would be on the front lines.

Trying without success to shut Kozlowski down, Tap found his mind flashing back to that Vermont job, only now he was the employer that Kozlowski seemed determined to chip away at. Tap told him he'd done the best he could for the divers. If Kozlowski had further complaints about the job, he said, "take it up with the union hall."

Once again DJ felt trapped in the middle. He agreed with Kozlowski that the divers should be paid more for this unprecedented assignment. But he also knew Tap well enough to trust that he had done everything he could to advocate for his guys, and beating up on him would do little good. DJ was relieved when Kozlowski seemed to drop the matter.

Throughout the training sessions, Kozlowski repeatedly pressed for details on the mission. To Kozlowski, this was simply the prudent way to prepare for a job and try to minimize risk. But Tap seemed to see a different motivation at work. Given their history together, he had to wonder if Kozlowski was more interested in showing everyone how intelligent and experienced he was. When Tap hit his breaking point, he pulled Kozlowski aside. "What are you asking all these questions for?" Tap snapped. "It looks like we don't know what we're doing."

Kozlowski countered that he'd been a diver for thirty years now, in jobs all over the world, and he'd never once had to take a mine rescue class. "Tap, no one's ever done this before," he said. "We *don't* know what we're doing."

The divers had gained some knowledge during the training, consisting of a total of three days of classroom time, to be followed at some point by a week of equipment testing in Tap's backyard. But it was hardly the kind of rigorous, NASA-style preparation you might expect for such a high-risk, first-of-its-kind mission. After the Dräger and mine rescue sessions, a safety officer for Kiewit gave the divers a short presentation, offering more specifics about the tunnel. The of-

ficer explained how the tunnel choked down in size as it moved east,
and how, in its final mile, it connected with fifty-five vertical risers.
While most of this was new information to DJ and the other guys in
the room, no one had to tell Kozlowski about this setup. Back in the
early 1990s, back before he'd met DJ and Tap, Kozlowski had helped
install those riser pipes and the mushroom-shaped diffuser heads resting on the ocean floor. While he had been working off the jack-up
barge, nine miles out into Massachusetts Bay, he had seen lots of big
ships and tankers traveling along the shipping lane above the tunnel.
Now he mentioned the same nightmare scenario that the tunnel's
designer and construction manager had invoked when they insisted
that Kiewit keep the safety plugs in place for as long as the sandhogs
were down there.

Citing Murphy's Law, Kozlowski asked the safety officer, "What
happens if Captain Wrongway Peachfuzz comes flying into the harbor and hits the wrong lever and the anchor comes down and hits one
of these diffuser heads?"

"Well," the safety officer replied, "there's going to be a tidal wave
going through that tunnel."

The comment produced some nervous laughter in the room, but
not from Kozlowski.

At the end of the last day of training, he pulled DJ aside. With
nothing but seriousness in his blue eyes, Kozlowski told DJ that after
a lot of soul-searching, he had decided to back out of the job. He suggested DJ do the same. Despite all the concerns Kozlowski had been
raising, DJ was nonetheless stunned. He knew his friend was such a
pro that he would never walk off a job unless he had a damn good
reason. After all, the guy was an ex–Navy SEAL who had survived
firefights in Vietnam. Combat had also given him firsthand experience with how human nature takes over in life-or-death situations.
He told DJ the whole operation didn't feel right to him. The way he

saw it, the divers were being asked to remove the safety plugs that had protected the sandhogs from a tidal wave tearing through the tunnel. Just what would be there to protect the divers?

"The only way I'd go in is with a boogie board and a .45," Kozlowski said.

DJ flashed a perplexed look.

Kozlowski explained, "I'd use the boogie board to ride the wave out."

"What do you need the .45 for?" DJ asked.

"To keep you guys off my boogie board."

6

Mobilization

DJ pulled up to Tap's house in Greenland and navigated his pickup down the long driveway leading to the Black Dog shop. He was coming off a wild Fourth of July weekend spent with Dana at a friend's lakefront place in New Hampshire. Now, in the sober morning light, he was feeling a bit groggy. But he was relieved that the long-talked-about tunnel job, which had suffered yet another delay after the mine rescue training, was finally about to begin. First, though, there would be a week of equipment setup and walk-throughs in Tap's yard before the crews mobilized the Humvees, trailers, and breathing equipment and convoyed from southern New Hampshire down to Deer Island.

Because the Black Dog and Norwesco divers had done their mine rescue training separately, this setup period would give DJ his first opportunity to meet the West Coast divers he would be working alongside in the tunnel. Given the nomadic nature of commercial diving, a certain pride went along with being on your home turf. DJ was eager to take the lead and exercise all his local privileges. He knew the Norwesco team would probably be staying in some crummy motel and trying not to get lost as they drove unfamiliar roads in borrowed cars.

Meanwhile DJ lived just an hour away, he was cruising to the job site in his immaculate '96 Dodge Ram with the chrome wheels and tinted windows, and he knew the best bars up and down the coast.

Before the introductions began, though, Tap pulled DJ aside. Still fuming that Kozlowski had walked off the job a few weeks earlier, Tap had decided it was too late to try to find and train a replacement. They'd simply have to make do with one less man. The Black Dog contingent would consist of DJ, Billy, two young guys who were new to the company, and Tap himself. Tap would not be venturing into the tunnel but instead would be a permanent member of the standby team. Of all these New England guys, DJ had the closest relationship with the Local 56 union office in Boston, so he had been designated as the shop steward, meaning he'd be the union's voice on the job site. Now, standing in his yard, Tap told DJ he needed him to set a good tone, especially in his dealings with the engineer from Norwesco who would be the project manager. "I know you're not going to like this guy Harald," Tap told DJ. "But I need you to do me a favor. Keep your opinions to yourself. Don't tell him to fuck off."

DJ laughed. Tap knew him too well. But DJ could also tell how serious Tap was. After all the time and energy he had invested in this job, Tap had to come to terms with the fact that he had very little control over it. He had only recently signed his subcontract with Norwesco, and now all they had to do was get through a few weeks without any blowups. "It's for the good of the job," Tap said, "for the good of us."

DJ had complete trust in Tap and Billy. But when the unfamiliar Norwesco divers showed up, the two camps eyed each other like rival rappers from opposite coasts. Their first task was to unload the Humvees and all the other equipment that had been shipped cross-country by tractor-trailer. The cylinders of liquid oxygen and liquid nitrogen as well as the backup HP compressed air would be delivered shortly by the New Hampshire facility that Harald's supplier back in Spokane had lined up. DJ figured the morning called for an all-hands-on-deck approach, so he quickly began hauling the heavy gear around. After he and the others had unloaded the equipment and begun shifting to the setup process, Harald, the short boss Tap had warned him about, marched around the yard. DJ didn't care for the way the guy seemed to strut, with his chest pushed out and his arms curled, as if he had a watermelon under each one. And he didn't like how particular Harald got about who touched what. DJ sensed that Harald had been content to let him sweat through the heavy work, lugging around hundreds of pounds of gear, but then seemed reluctant to let anyone besides his Norwesco guys do the mentally taxing work of assembling the equipment in the correct manner. DJ was nobody's bellhop, and he started to seethe. Recalling the promise he'd made to Tap, though, he decided to adopt the can't-be-bothered attitude of the hired hand that Harald seemed to view him as. If Tap asked him to do something, DJ would jump right on it. Otherwise he hung back and watched.

His frustration increased when Norwesco diver Dave Riggs asked him to avoid using foul language around him. Riggs's accent and his slow speaking cadence made it clear that he'd grown up in the South, and it hadn't taken DJ long to conclude that he was a Bible Belt kind of guy. Even so, DJ couldn't get over the idea of a diver actually getting worked up over a little cursing. Although Riggs had grown up in Texas, he now lived with his family in Nevada. He had one of those last names that sounded like a nickname, so that's what everybody called him.

The combination of Riggs and Harald gave DJ pause. The way he saw it, he was being asked to work *with* a guy who seemed to be waiting for the Second Coming and *for* another guy who struck him as Napoleon reincarnated. DJ had been asked to be part of a risky, first-of-its-kind journey underground, and *this* was his team? All he could do was shake his head and mumble, "You guys are a bunch of strange fucking birds."

But for a guy who loved driving trucks and watching tanks on the

History Channel, DJ was beguiled by the setup. The opposite-facing Hummer-brand Humvees were both four-door models, one of them a wagon and the other a pickup. (A second identical pair of Hummers, to be used by the standby crew, had been shipped directly to Deer Island.) Each vehicle required considerable welding to retrofit it for the tunnel job, including the conversion of one of the rear windows into a receptacle for hose hookups. An inflatable Zodiac boat had to be attached to the roof and an outboard motor for the boat attached to the back of the Humvee, in case there was a flood in the tunnel. There were umbilical hoses and manifolds and rebreathers and a two-way trailer. Once the tanks of liquid gas and HP air showed up, they all had to be plumbed into the breathing system. DJ couldn't get over just how many pieces there were to this colossal puzzle.

As the setup operation continued into the next day, DJ came to appreciate that the Norwesco crew wasn't one undifferentiated team but rather a group of diverse characters. And he found himself enjoying the company of a couple of them. Tap and Billy had already talked up a guy known as Hoss, who had gone on the fact-finding mission with them. And now that DJ had worked with the tall, blond-haired, tobacco-chewing guy from Idaho, he saw what they meant. DJ felt an immediate connection to him when he heard Riggs make the same kind of pointed request to Hoss that he'd made to DJ, asking him not to take the Lord's name in vain. *All right,* DJ thought, *at least I'm not alone.* DJ and Hoss even shared the same first name of Donald—which neither of them ever used much.

DJ also got a good vibe off Hoss's close friend, a six-foot, 220-pound easygoing guy by the name of Tim Nordeen. After work, they grabbed a few beers and swapped war stories. Tim had thick eyebrows that arched down toward his nose, and a big, honest smile. Although he lived in Washington State, he was originally from Texas. He was no Bible-thumper, though, and cursed almost as much as DJ. They figured out they'd both graduated from the same Houston dive

school, though Tim had gone there a full decade earlier. Tim and
Hoss were both gun collectors, and they had already made plans to
do some gun shopping during their time in New England. For them,
no dive job was complete unless they could squeeze in some gun pur-
chases during their downtime. Tim's wife, Judy Milner, knew he liked
guns and that he kept a few of them in their home. But as Hoss explained with a knowing laugh, Judy had no idea just how much Tim
liked guns. After one shopping spree, Hoss asked Tim, "How many
guns do you have?"

"Between fifty and sixty," Tim said.

"How many does Judy think you got?"

"Probably four or five."

Tim was a conservative Republican, blue-collar gun lover, while
Judy was a liberal Democrat, a child psychiatrist, and a gun-control
advocate. So he'd found ways to hide his pieces in various corners of
their house, including a false box spring in the guest room, all so he
could keep up his hobby without causing the woman he loved any
grief. DJ thought to himself, *Now, these are guys I can get along with.*

Hoss placed a work boot at the mouth of the pipe that lay on the grass
in Tap's yard, then stepped back to snap a photo. It wasn't uncommon
for divers to do what Hoss was doing: take pictures to fill photo al-
bums from each of their assignments. The fiberglass pipe was about
twenty feet long and attached to a ninety-degree concrete elbow,
which housed one of the much-talked-about safety plugs. The idea
behind bringing this mock-up to Tap's yard was to give the divers
practice shimmying into the pipe, detaching the plug with a special
wrench, then shimmying back out with the plug—all without disturbing their breathing hoses. The guys had heard the pipes they'd
be crawling through would be just thirty inches wide. But there was
nothing like seeing how a man's work boot partially blocked the

mouth of the pipe to hammer home just how narrow their passage-
way would be.

Although he'd never served in the military, Hoss carried himself
like someone in the Special Forces: strong and fast, cocky and quick-
witted. He guessed that the camaraderie he found with the other
guys in the overwhelmingly male world of diving was similar to what
a soldier might find in a foxhole. After all, a diver, like a soldier, relied
on his comrades to keep him alive. He had to be able to trust that the
guy at the other end of his hose knew what he was doing. If there was
a lazy screwup in their midst, divers were never shy about calling him
out before his incompetence cost a life. Also like soldiers, divers spent
so much time together, and away from their families, that tight bonds
inevitably formed. The bonding was usually expressed in the macho
language of ball busting—Hoss's buddies mocked him for being a
supposedly tough guy who fastidiously ironed every last wrinkle out
of his shirts—but it was intimate just the same.

Sometimes Hoss wondered how he'd managed to stick with diving, given his early experiences. When he was a sixteen-year-old
growing up in the northern Idaho lakefront resort town of Coeur
d'Alene, he'd decided to get certified in scuba. During one of his first
lessons, as he was preparing to take the regulator from his trainer, he
accidentally exhaled all his air, then sucked in a bunch of water. In
a panic, he foolishly raced to the surface, which, if he'd been much
deeper, would have surely given him the bends. He made it there un-
harmed but shaken by the knowledge of how close he had come to
injury. After dive school in Seattle, he headed to the Gulf of Mexico
to work offshore. On his second day there, he saw a fellow worker
fall nearly two hundred feet, from a drill tower to the deck. As a helicopter whisked the guy away, Hoss asked himself, *What the hell am I
getting myself into?*

Because work in the Gulf was slow at the time, he was thrilled
when he got a call from Roger Rouleau offering him an entry-level

tender position with Norwesco. That dive company's office in Spokane was an easy drive from his home over the Idaho border, meaning Hoss could remain near his girlfriend, Heather, a hairstylist he'd begun dating right out of high school. At Norwesco, the older divers seemed to respect Hoss for being a quick study and eager to learn.

In the five years since he joined Roger's company, Hoss had come to be seen as a rising star. The twenty-four-year-old, who had recently married Heather, had just finished working on the Lake Mead project. Although that had been largely a positive experience for him, he'd seen what had happened the one time he had complained to a Norwesco supervisor about a safety concern: He'd been quickly moved to another part of the project. Hoss figured the fastest way to jeopardize his rising-star status would be to squawk too much.

After Roger offered him a spot on the Deer Island team, Hoss consulted one of his most experienced diver friends, who raised doubts based on what he'd heard of Harald's plan. This mentor suggested he take a pass. Hoss valued his advice, but he knew that, unlike his mentor, he didn't have more than two decades in the business, so he couldn't afford to pick and choose assignments. He was a young guy with a young wife who'd just bought a beautiful piece of land in Coeur d'Alene on which they planned to build their dream house. He needed to keep his climb at the company going and take good opportunities where they came.

For Hoss, a big selling point of the tunnel job was that one of his closest friends from Norwesco had already signed on. The fact that Tim, who had fifteen years on Hoss in both diving and in life, was comfortable with the mission said a lot. The two friends made for an unlikely pair. With his spiky blond hair and clean-shaven face, Hoss was lean and impatient. With gray threaded through his brown hair and bushy beard, Tim was burly and laid-back.

Tim's calm approach to life had served him well, even during the occasional crisis. On a salvage dive years earlier, Tim had been serving as tender to one of his buddies, when the guy relayed a panicked message on his comm wire. His umbilical had become dangerously entwined with some ship lines, to the point where he feared he wouldn't make it. Without hesitation or drama, Tim put on his dive hat and went down to rescue his friend. The umbilical had become so ensnared with old ropes that Tim had had to methodically untangle them, one by one, all while his buddy's panic level continued to rise. Somehow Tim managed to keep his nerves in check.

About the only time that Tim had found himself agitated was during his four-year stint living with Judy in New York City. They had met at a party when he was working in Galveston, Texas, and she was doing her residency there. When she earned a prestigious fellowship in child and adolescent psychiatry at Columbia University's medical school, they moved into an apartment in an upscale New York co-op building. Because he refused to depart from his usual uniform of overalls and work boots, Tim was occasionally mistaken for a handyman by his fellow co-op residents. He found relief from his fish-out-of-water discomfort by taking dive jobs that kept him away for long stretches. He once joined a team hunting for sunken treasure off the Florida Keys. They never found it, but he did come away with a great barstool story about how a rival treasure-hunting team had shot at his gang. An unhurried outdoorsman, Tim viewed Manhattan as a penitentiary, with its blocks darkened by skyscrapers and clogged with tightly wound, overdressed people trampling each other as they raced to their next appointment.

Their next move, to Snohomish in rural Washington, had been a compromise. Judy was happy that the bustle of Seattle was only an hour away. Tim was thrilled that their five-acre property was so secluded that it could have been its own nature preserve. He had painstakingly built a wooden walkway that wound its way for five hundred feet over the wetlands in their backyard. He also devised an elaborate series of pulleys to feed the birds, and he regularly hand-fed a mother raccoon and her babies. He built a small wooden coffin for one of his cats when she died, holding a graveside memorial service that the neighborhood dogs sat through silently in the rain. To Tim, animals were creatures that deserved loving care. That was also why, as much as he loved guns, he had little interest in hunting. Judy would often remind him of how he had won her heart during one of their first dates, at the beach, when she watched him pick up his vulnerable puppy and tenderly carry it to safety away from the waves. Tim had found that brown and white mutt a few years earlier as a stray and named her Shackles, the term for the horseshoe-shaped metal fasteners that divers use in underwater rigging. When he and Judy moved to New York, Tim had left Shackles in his parents' care in Texas. Just a few months before Tim left for Deer Island, his folks had broken the news to him that Shackles had died, at the age of fifteen.

During a break in Tap's yard, Tim called home to Washington to check in with Judy. He told his wife that they were getting the chance to work with what seemed to be millions of dollars' worth of equipment. "Well, is it the right equipment?" Judy asked. "Is it going to work?"

Tim said he thought it would, though he acknowledged that none of them had ever seen it before.

Not wanting to worry his wife further about the exotic tunnel job, Tim stressed how smart Harald was, and how Harald had assured the divers that his sophisticated system was fail-safe, with multiple backups. Tim knew that the only way to deal with the hazards associated with his line of work was to believe in the people he was working with. Still, as trusting as Tim was, even he had to concede that this Deer Island job was a radical departure from his usual assignments.

In a moment of candor, he admitted to Judy, "It's so far out that if we get all the way out there and something goes wrong, we are totally fucked."

With the Humvees and breathing equipment assembled, the final two days in Tap's yard were devoted to training. Harald gave the divers a primer on the breathing, transportation, and communications systems they'd use during their mission. He went over everything from quarter-turn valves and facemasks to battery-operated water pumps and handheld radios.

By the time the crew began doing run-throughs, DJ and Billy had established a good rapport with Hoss and Tim. DJ especially liked how Hoss talked with more than a little cowboy in him, tossing around lines like, "Don't try to baffle me with your bullshit."

Although he still didn't know quite what to make of Riggs and his curious aversion to profanity, DJ had been seeing evidence that at least the guy knew what he was doing as a diver. The questions and recommendations that Riggs brought up about Harald's system struck DJ as sensible ones. If the divers were going to have to be lugging around heavy equipment when they were in the tunnel pulling plugs, Riggs suggested, they should be doing that kind of strenuous work when they were testing their breathing equipment in Tap's yard, rather than doing easy, sweat-free simulations shimmying through the pipe. Yet as far as DJ could tell, Harald seemed as unresponsive to Riggs as he was to the others, like a taxi driver traveling with the cruising light on his roof turned off, advertising to the world his lack of interest in interacting. To DJ, the fact that he and Riggs were in the same predicament with Harald suggested that Riggs might be all right after all. Tim had also reassured DJ that Riggs, though a bit prickly, was deep down a good guy who knew his stuff.

As the week in Tap's yard progressed and the divers became more comfortable with one another, the distance between them and Harald became more pronounced. The divers began to bond over their shared frustration with how Harald, in the face of setbacks, seemed to be drawing inward rather than turning to them for advice. The divers were experienced and perceptive enough to sense how the burden of the calendar seemed to be affecting the job. DJ felt a distinct time-is-money pressure hovering in the air. And the guys began to wonder aloud why the centerpiece of Harald's breathing system, the gas mixer that would blend the vapors from the liquid oxygen and liquid nitrogen, was not on site during this training period. As Harald explained it, the MAP Mix 9000 was coming from Europe and appeared to have gotten held up in customs along the way.

Harald did not share the mixer's backstory with the divers. After his supplier in Spokane told him a month earlier he couldn't find an approved mixer, Harald had searched the Internet and found three possible models. Harald's supplier determined that only one of them was appropriate, the MAP Mix 9000, manufactured by the Danish company PBI Dansensor. And although the supplier had arranged with a Massachusetts distributor called Topac to have the mixer imported and drop-shipped to Harald in Tap's yard, the unit continued to be AWOL as their planned mobilization to Deer Island drew near.

Harald tried to compensate for the missing mixer by showing the divers diagrams. He'd also brought along a copy of the device's slim instruction manual, which some of the guys flipped through. Although the divers went ahead and donned their facemasks attached to their umbilicals, it seemed almost pointless since they had no working mixed-gas breathing air system to plug into. Instead, they breathed off the HP air that would serve as their backup supply during the actual mission. Riggs termed it "playing like we have a mixer." The divers had been called in to be the closers on the multibillion-dollar, decade-long Boston Harbor cleanup, yet this sure seemed like a strangely improvised way to approach the bottom of the ninth.

When DJ persisted in posing even minor questions to Harald, the supervisor put up his guard, almost indignantly. Harald stressed that everything had already been worked out. The implication was clear: DJ was the hired help, and he should simply do what he was told. He saw Harald treat Tap the same way. Eventually DJ got the message. From then on, if he spotted something that didn't seem right, he reminded himself to keep his mouth shut and just collect his pay.

Harald was upbeat when a visitor from Boston showed up at Tap's yard to look over the operation. Dan Kuhs was the business agent for Local 56. A no-nonsense guy with a lineman's build, he was an experienced diver who was relatively new to the union front office. Kuhs quickly recognized the concrete elbow that was attached to the mock-up pipe lying in Tap's yard. Nearly a decade earlier he helped fabricate the real elbows that now connected the Deer Island tunnel to those fifty-five risers, and he'd also been part of the dive team that installed the risers from the jack-up barge in Massachusetts Bay. Even if the Norwesco guys were strangers to him, Kuhs had worked on jobs with Billy and had seen Tap and DJ around the union hall enough to get to know them. Although he was bogged down with the demands of supplying workers for Boston's insatiable Big Dig megaproject, Kuhs had decided to drive to Tap's place to look over the unusual breathing equipment he'd been hearing about.

Now when Kuhs met Harald, the engineer struck him as bright and all business. When Harald explained his system, Kuhs was not put off by the use of liquid oxygen, knowing how arduous it would have been to transport enough bottles of HP air to the end of the tunnel. Although the full system wasn't in place by the time of this visit, Kuhs left reassured by Harald's explanations of the multiple backup systems that the engineer had marbled into the setup. If any of his

Local 56 divers had concerns about Harald's system, they didn't mention them to Kuhs.

As much as the divers had managed to break down some of the barriers between the East and West Coast contingents, a week of working together could do only so much. There was no getting around the fact that the crew had been cobbled together from two distinct groups, to the point where they'd even taken their mine rescue training in different parts of the country. Even the equipment reflected this divide, since about half of it had been assembled by Norwesco crews back in Spokane, while the other half was put together by the combined East-West crew in Tap's yard. Because DJ and the other divers still had trouble reading each other's signals, it was hard for them to question Harald in any kind of unified way. So instead of seeing a gradual improvement in communication and comfort between the divers and their equivalent of Mission Control, the training week was devolving into little more than a series of testy flare-ups. DJ found himself wondering: *Is this any way to prepare for the kind of job nobody's ever done before?*

There were even fractures within the long-standing alliances the divers had brought to the job. DJ blew off the last day of work in Tap's yard, deciding he didn't need to waste time driving north to New Hampshire just so he could accompany the other divers on the drive south to Deer Island. That unexcused absence, coming on the heels of his flagrant tardiness earlier in the week, torqued off Tap. Privately he laid into DJ, telling the party boy that he needed to grow up and take the job more seriously, rather than putting him in a bind with the Norwesco guys, who already suspected Tap of giving him preferential treatment. Tap went so far as to report DJ's absence to the union hall, telling them they needed to get their shop steward in line. Now, as Tap walked on eggshells around Harald, DJ found himself in the unusual position of walking on eggshells around Tap.

As Harald ran a hose along Tap's driveway, he struggled to get the system working right. Finally he muttered something in frustration to Tap and then hopped into his car and drove off. Tap took that moment, late in the training week, to call all the divers into the Black Dog shop. All week he had been trying to walk a fine line. He was conscious that this was clearly Harald's project—not his—but he was also aware that the training was taking place in his own yard and that he had a responsibility to the divers, especially the Black Dog guys. "Is everyone comfortable with this?" he asked the divers. "Are we all right?" They told him how frustrated they were that key pieces of the equipment hadn't arrived yet, especially the mixer, and that Harald was so lousy at communicating with them. But the divers also told Tap they thought things would eventually work out.

Hoss raised a specific question. As designed, the system called for the three divers doing the actual plug-pulling to remain attached to the main breathing supply through fifteen hundred feet of umbilical. As they had been setting up the system during the week, Hoss noticed that the umbilical hoses had a relatively narrow diameter of three-eighths of an inch. Now he asked, Would that width be able to provide enough volume for the three guys working at the end of the tunnel? Wouldn't they be better off with a half-inch-diameter hose? By the time he raised that question, Harald had returned to the site and was ready with a reassuring reply. Everything had been calculated properly and approved, Harald said. The size of the umbilical was more than adequate.

To Tap's ears, Harald's response sounded uncomfortably familiar. All week, when Tap had asked him one question or another, Harald had invariably responded, "The system's been approved" or "It's an approved design." As if to underscore that point, a Kiewit safety officer and other officials with the contractor had visited Tap's yard to observe the setup, while project sponsor Tom Corry had flown in from Kiewit headquarters to Deer Island to see this final operation through to its conclusion. Tap knew that if the contractor or the sewer agency had concerns with the Norwesco plan, they had ample opportunities to raise them. Still, by the end of the week, Tap felt he needed to escalate things somehow. He called Norwesco owner Roger Rouleau back in Spokane and relayed the divers' concerns about how rushed the whole setup process seemed to be. "Hey, Roger," Tap said, "I don't think Harald's got this squared away very well."

Roger told Tap he'd look into it. Later that day he reached Harald by phone. Harald admitted that they'd run into some problems, but he reassured Roger that things had been worked out. "Is everything tested?" Roger asked.

"Yes it is," Harald replied.

When Roger pressed further, Harald said, "Things are going fine. No worries." Even though Roger owned the company, he had only a basic understanding of the system that Harald, in consultation with their suppliers, had designed. In fact, no one besides Harald seemed to have the complete picture, and he seemed to be guarding that exclusivity. Over their years working together, Roger had developed such abiding confidence in Harald's capabilities that he saw no reason to question him now. Trusting him had proved very good for Norwesco's bottom line. He suspected the angst that had been surfacing in Tap's yard was simply the result of Harald's poor social skills. Roger wished Harald could have found it in him to huddle up the crew and work through a problem collaboratively. But he knew Harald was too much of a lone wolf to do that. Still, it was enough for Roger to know that Harald had worked out the kinks all on his own.

And Tap really didn't feel he could argue with that. When he had pressed DJ and Billy and the other divers on whether they felt everything was going to be all right, they had replied in the affirmative. All dive jobs—especially the most complicated ones—have a few bugs early on. Sometimes those problems can be addressed during the preparation period. Other times you just have to push ahead and figure things out once the job actually begins.

7

The Cavalry

Just past seven o'clock in the morning on Monday, July 12, the caravan of Humvees that had left Tap's yard arrived on Deer Island, navigating across the narrow spit of land that connected it to the mainland. The trail of muscular guys in muscular Hummers charging onto the island gave the divers the impression of an invading army. And for many of the island workers who had spent the last decade struggling to get the tunnel done, it was a welcome sight. They adopted the posture of long-aggrieved peasants showering a conquering army with rice and flowers, on the assumption that the invaders couldn't help but improve their desperate lot.

Dave Corkum was no peasant, but he was just about ready to toss flowers. As point man in the tunnel for construction manager Kaiser, he'd spent a draining ten-year sentence on Deer Island. He knew it would all have been for nothing if they couldn't get those fifty-five plugs out. When Kiewit officials had introduced him to Harald Grob, Corkum had been wowed by the engineer's intelligence and poise. His faith in Harald had only grown when he saw the cool competence with which the Canadian engineer had fielded the questions about his novel plan from Corkum and his counterparts with tunnel owner MWRA and designer PB. When Corkum asked the engineer if he'd ever used a similar kind of mixed-gas system before, Harald had assured him, "This is something we do all the time."

"Where do you use it?" Corkum had asked.

"In small submarines."

That impressed Corkum, as did Harald's comment that they were using mostly "off-the-shelf" gadgets that were simply being combined in a unique way. Corkum was comforted by the notion that as long as it was in expert hands, Harald's setup was a lot less exotic than it appeared to be.

Of course, conscious of what he'd learned in his law school classes, Corkum had been careful not to "touch" the Norwesco plan too much. He didn't want to fall into a Kiewit legal trap that ensnared his employer or the MWRA into somehow "owning" a problem that contractually lay entirely in the contractor's lap. That helped explain why he had to tread so carefully around this Norwesco deal. Kiewit had failed to supply him with a report he'd requested, detailing what Norwesco had found on its fact-finding mission the previous summer, when crews had tested for levels of oxygen as well as explosive gases like methane and carbon monoxide. But he didn't push it. If they had found anything worrisome, Corkum assumed, he would have heard about it. The plug-pulling plan that Harald had crafted with the help of his suppliers had said only that liquid oxygen and liquid nitrogen would be blended for the divers in some way, but offered no specifics on what type of mixing equipment would be used. "As far as I knew," Corkum would say later, "it was a black box, and it came in liquid one end and came out a breathable mixture at the other end." He saw the full vetting of the design as Kiewit's responsibility. After some modest back-and-forth, Kaiser and PB had signed off on the Norwesco plan, although they were careful not to offer any kind of formal approval. Instead, they had returned it to the contractor with "no exceptions taken." About as far as Corkum had been willing to probe was in asking Norwesco for a detailed schematic for its oxygen-supply plan to ensure that the Humvees' internal combustion engines would operate in the tunnel. He asked for no similar schematic of the divers' breathing supply.

Despite Corkum's concern for keeping his distance, his meetings with Harald had persuaded him that they had finally found a capable guy who could extricate the project from its hopeless jam. That was exactly how MWRA chief Doug MacDonald felt as well. More than anyone, MacDonald had to answer for the long delays in the project—to rate payers angry with hiked sewer bills, to his critics who had opposed the tunnel from the start, and to A. David Mazzone, the federal judge overseeing the harbor cleanup. On his periodic tours of Deer Island, Judge Mazzone would jocularly address MWRA workers with greetings like "How are you defendants doing this fine morning?" But MacDonald knew that it was no joke when it came to him. As the head of the agency, he truly was a defendant in the case. And after years of frustration, he saw in Harald someone who had a real handle on the problem and an absolute determination to steer everyone toward resolution. MacDonald had heard that the divers had been undergoing extensive training in New Hampshire. So when Harald and his team began mobilizing on the island, MacDonald flashed one of his classic smiles that narrowed his eyes and lit up his round face. He thought to himself: *Special ops have just arrived.*

To Corkum, the approaching caravan of Humvees, each jam-packed with gear, was a supersize shot of adrenaline for the stymied project. After the divers parked the vehicles on a dusty clearing, Corkum walked over, flashing his gap-toothed smile. "Looks like you guys are going camping!" he cracked. No one had to remind Corkum that the tunnel was supposed to have been finished four years earlier.

But he also knew that the judge's final deadline for everything to be up and running—the new treatment plant and the two big tunnels—had all along been September 1999. The five-mile-long Inter-Island Tunnel was already operational, and the rest of the plant had moved into the final stage, as evidenced by the scent of freshly laid asphalt expanding in the summer heat. That meant that as long as they could get the Deer Island Outfall Tunnel done, they might still be able to meet the court's ultimate deadline. All that stood in the way were those fifty-five plugs.

Dave Riggs suspected some of the younger divers on the crew relished being cast as members of a Navy SEAL–type elite military team, swooping in on a rescue mission. But as a seasoned diver as well as an actual navy veteran, Riggs knew that seeing yourself that way could only get you into trouble. Military role-playing was best left to elementary school boys skulking around playgrounds. The divers were professionals, there to do a difficult job, make a decent wage, and then be on their way.

Riggs could sense that the sandhogs weren't crazy about seeing the divers charge onto the island. Maybe it was simply a union turf issue. In any case, it wasn't surprising that members of Local 88, who aggressively guarded their primacy in the underground, would resent seeing a bunch of divers—some from the other side of the country—stream into a tunnel that the Boston sandhogs had ruled for a decade.

The first day on the island was a blur of unloading and unpacking, after the divers went through the drug testing that the MWRA mandated of all workers on the job. Electricians wired up Harald's "Conex box," a white wheelless trailer he had shipped from Spokane. It would serve as his command post for the operation. Its rooftop antenna would allow Harald to monitor by radar the marine activity in the Massachusetts Bay shipping channel sitting above the tunnel.

Harald kept the radar monitor on a plywood table, alongside his lap-top computer and below a map of the bay that he pinned to the wall. His white Conex box had been placed near the tunnel shaft, less than ten feet away from Tap's eighteen-foot black trailer, which would serve as a combination changing area and break room for the divers. When the doors to both trailers were open, someone sitting in Tap's could hear what was being said in Harald's.

During the convoy from New Hampshire to Deer Island, the dive team had suffered a setback. The hitch assembly on the custom-built two-way trailer that the Humvee would tow into the tunnel had been damaged in a roadway accident. Given how essential the trailer was to the setup, this was a serious holdup. As soon as the divers arrived on the island, Hoss pulled out his welding torch and made it his focus to try to fix the hitch.

Riggs liked Hoss's resourcefulness, although there were other characteristics he cared for less. When they'd first begun working together, back in Nevada on the Lake Mead job, Hoss, who zoomed around with the speed of a sprinter, had yelled, "Riggs, help me move this!" It was a twenty-five-foot-long steel beam. As Riggs bent down to grab his end, he realized it was way too heavy for them to be lift-ing, especially since there was a forklift close by. Hoss evidently had no time for the sensible way to handle the task. His build was ropy more than muscular. But like a wire cable, he was deceptively power-ful. Riggs wondered if Hoss's strength, combined with his cockiness, could make him a little too fast and loose.

Later in the day, the tanks of liquid nitrogen and liquid oxygen were delivered to the island in a steel rack that would sit on the trailer attached to the Humvee. The tanks were, of course, the building blocks for Harald's innovative breathing system. Despite the ques-tions that everyone had asked back in Tap's yard, Riggs worried that the system's exotic feel might lead Hoss and the other younger guys to embrace it more quickly than they should. Riggs was thirty-eight, not an old man by any means. And his boyish face, with dimpled cheeks and a side part of brown hair descending onto a high forehead, made him look even younger. Still, he had lived through a hell of a lot more than a twenty-four-year-old like Hoss or even a twenty-nine-year-old like DJ. In a world where too many people moved and talked quickly, Riggs had come to rely on his more deliberative approach. His preference for thinking before talking, combined with his Texas roots, caused him to speak slowly, turning a choppy contraction like *wouldn't* into the sluggish and soft-bellied *woonninn*. Yet he packed a lot of punch into what he said, if you knew to listen for it. His wife, Karen, certainly did. During one phone call, when she asked how the tunnel job was going, Riggs drawled, "They're doing some real *original* stuff here with equipment—and the young guys think it's cool." Picking up on his sarcasm, she just laughed.

Riggs had joined Norwesco only one year earlier, when he'd been hired for the Lake Mead crew. He'd wasted little time in moving his wife and young kids from their place in Colorado to Boulder City, Nevada. After having spent so many long stretches away from home, Riggs got to enjoy the unfamiliar experience of eating dinner with his family most nights.

He'd met Karen in Colorado six years earlier, when the compact athlete was visiting from Minnesota as a coach for the Junior Olympics ski racing team. Riggs was by then splitting his time between working as a diver in California and running a business in Colorado called The Cruisin' Cameraman, where he would ski or snowboard alongside families while holding his camera, capturing their mountainside fun on videotape. At the time, Karen had twin four-year-old sons, but her marriage to their father was over. She and Riggs had hit it off immediately and married quickly. When Karen gave birth to their first child, Riggs insisted on naming the boy Enzo, after the macho character in the movie *The Big Blue*. That character was a "free diver," someone who holds his breath for unimaginably long stretches underwater. From the start, Riggs was an engaged father to both his son and his twin stepsons. But feeling the need to make more money, he scrapped the ski slope business and began working as a diver full time, even though that required him to chase work in locations as distant as the Persian Gulf. Shortly after Karen gave birth to their daughter, Swasey, Riggs received the offer to join Norwesco. The Lake Mead job was expected to last at least another eighteen months. The timing was perfect, since it would give him both steady, well-paying work and the ability to stay close to his young family.

Even though the itinerant nature of the profession required divers to move to whichever company happened to have work at the time, Riggs saw an opportunity with Norwesco to settle down, both professionally and personally. But at Lake Mead, he found the Norwesco circle of divers harder to penetrate than he'd hoped. That was partially because while they all went out for beers after work, he went home to his family. The one time he did join the guys at a sports bar, he nursed his O'Doul's nonalcoholic beer while they tossed back the hard stuff. After a long period of commotion in his life, Riggs had been sober for five years, having found focus in his family and through the Lord. He had no desire to change that.

His biggest disappointment with Lake Mead was that the job progressed too quickly. To his surprise, it began to wind down after he'd been there for just about a year. When he was asked to be part of the crew going to Deer Island, he quickly signed on. Even if it kept him away from his family for a while, at least it would allow him to stay with the same company. One of the things he had come to detest about life as a diver was the uncertainty and indignity of constantly having to scrap for the next job.

As long as he had to be away from his family, Riggs felt he might as well be working as many hours and earning as much overtime as possible. That's why he had no patience for the youthful nonsense that other divers like DJ seemed so interested in. Riggs had done his share of carousing and had seen how it could only lead to woe after the buzz inevitably wore off. He was at a place in life where he could get a high from a simple phone conversation with his four-year-old son, hearing Enzo pepper him with questions about what kind of diving Daddy was doing and relaying stories about the black and gold "diver man" toy he played with in the bathtub. When Riggs wasn't able to be on the clock, racking up the OT, he'd cool his heels in his room at a Howard Johnson's motel north of Boston. He'd even brought his PlayStation console from home to fill his downtime responsibly. No one ever had to be bailed out of jail after a night playing Formula One.

If party boys like DJ thought that made Riggs an uptight bore, so be it. Riggs had already seen enough of DJ's antics—his tardiness, his raunchy mouth, his attempts to paper over his lapses by playing buddy-buddy with Tap—to conclude that he didn't trust the guy. Riggs considered himself to be, first and foremost, a professional. He knew what he was doing and why he was there.

On Tuesday, the day after the cavalry charged onto the island in Humvees, the divers arrived in a more understated manner. They came packed into the kind of small school bus used to transport the elderly. All through the tunnel construction project, the MWRA had required construction workers to get to the island either by bus or barge. This was done to appease neighbors by reducing congestion along the narrow residential streets leading to the island. For the divers, the morning commute for the duration of the job would be the same. The Norwesco guys would meet in the lobby of their HoJo's motel in nearby Revere, pack into a rental car, and arrive by six-thirty a.m. at the parking lot of an old racetrack called Suffolk Downs. There they'd meet up with the Black Dog guys, who drove their own pickups to the track, and then everyone would pile into the school bus, which would get them to Deer Island by seven.

On the island that morning, Riggs watched as workers attached cables to the handles on the hoods of the Humvees, which a crane then
lowered down the tunnel shaft. There were two sets of Humvees—
one primary pair and one backup—with each set consisting of a white
four-door hard-top pickup and a four-door wagon in either gray or
green. Following the Hummers down the hole was an endless supply train of hoses, regulators, rebreathers, vaporizers, VHF radios,
pumps, sandbags, fire extinguishers, hand tools, manifolds, inflatable boats with outboard motors, flat-bottomed boats with collapsible
wheels, tanks of liquid nitrogen, tanks of liquid oxygen, and tanks of
HP air.

A cluster of guys in hard hats stood around the rim of the shaft,
watching the gear descend into the earth. In addition to the divers,
there were sandhogs, equipment operators, and supervisors and
safety officers from Kiewit, Kaiser, and the MWRA. Riggs could
sense Tap's frustration in not knowing what his role in the whole operation should be. At one point, the owner of Black Dog Divers looked
like a dad at his daughter's dance recital, gamely videotaping a Hummer being lowered down the hole. Riggs noted how the duct that had
supplied ventilation for the sandhogs was still blowing air down the
length of the shaft but stopped abruptly at the start of the tunnel.

Despite all his effort, Hoss had been unable to fix the trailer hitch,
and they had to call a welding company out to Deer Island to make
the repair. All around them the divers were feeling pressure, even
if they didn't know that Tom Corry, the high-level Kiewit executive
who'd flown in from Omaha, was keeping track of their delays with
notations in his day planner: "2nd Hummer trailer requires major re-
pairs. Lost one day."

By Wednesday, the divers were working at the base of the shaft,
assembling the various components of their intricate system. Kiewit
had arranged for a crew of sandhogs to install a temporary deck, made
of aluminum slats. The deck was wide enough for two Hummer

Under the Ocean, at the End of the Line

To reach the safety plugs, the 5-man crew drove a Humvee. A trailer in tow carried a rack of liquid gas tanks as well as another Humvee that faced the opposite direction for the return trip. The tunnel was too narrow to allow a vehicle to turn around.

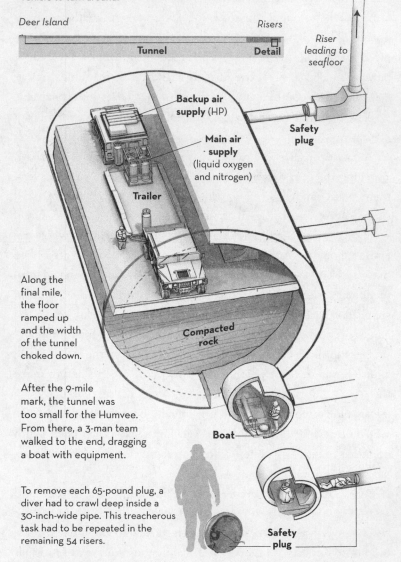

Deer Island

Risers

Tunnel

Detail

Riser leading to seafloor

Backup air supply (HP)

Main air supply (liquid oxygen and nitrogen)

Trailer

Safety plug

Along the final mile, the floor ramped up and the width of the tunnel choked down.

Compacted rock

After the 9-mile mark, the tunnel was too small for the Humvee. From there, a 3-man team walked to the end, dragging a boat with equipment.

Boat

To remove each 65-pound plug, a diver had to crawl deep inside a 30-inch-wide pipe. This treacherous task had to be repeated in the remaining 54 risers.

Safety plug

Humvees to be parked side by side, and high enough for the divers to stay dry from the three feet of standing water sloshing around the bottom of the tunnel and shimmering in the light through the slats. The divers spent the day essentially repeating the setup work they had done in Tap's yard, since most of the components had been disassembled in order to be transported from New Hampshire and then down the shaft.

Like the divers, each Hummer's internal combustion engine needed oxygen to function. A diesel engine uses a series of tiny explosions to convert the chemical energy available in fuel into mechanical energy, which in turn moves the pistons, which connect to a crankshaft that helps turn the vehicle's wheels. But instead of relying on spark plugs to ignite the explosion, as a gas-powered engine does, a diesel engine compresses the air before the fuel is injected. Once compressed by the pistons, the air gets hot. That heat is enough to light the fuel—provided, of course, there's enough oxygen around.

At the base of the shaft, and for the tunnel's first several miles, there would be sufficient oxygen in the ambient air for the Humvees to start. But as soon as the oxygen content in the tunnel air slipped to below 16 percent, the engines would likely stall. To get over that hurdle, Harald had turned to two different Spokane companies, Nex-Gen Manufacturing & Controls and K&N Electric Motors. They devised a sophisticated air system, which involved a liquid oxygen tank mounted on the back of each Humvee, along with a scrubber added to its exhaust pipe. Attached to the oxygen tank was a tall, sleek device of black metal that looked like an upright radiator in a trendy condo. This heat exchanger/vaporizer would augment the conversion of the liquid oxygen into a gaseous state and help ensure that the resulting gas remained at the correct temperatures and flow levels in order to feed the internal combustion engine of the Humvee. This heat exchange process was critical, since the liquid gas was extremely cold. Not for nothing do dermatologists turn to liquid nitrogen to freeze warts off their patients' skin. The extreme cold slows down the molecules of the nitrogen or oxygen, allowing them to be packed in more tightly.

Governing this whole vehicle-air-supply setup was a separate device called an oxygen-injection control. It would monitor the oxygen content of the air going into the engine, automatically making any necessary adjustments.

In contrast, the system that would provide the divers with breathing air looked fairly improvised. The liquid oxygen and nitrogen cylinders sat in their rack on the trailer and would connect via a half-inch-wide hose to the MAP Mix 9000, whenever that mysterious machine finally arrived from Denmark. The mixer would be placed in the back seat of one of the Humvees. Once it did its work, the blended gas would be sent out through a separate hose connected to a distribution manifold inside the vehicle. In reality, that manifold was little more than a piece of roughly cut, unpainted plywood. With hose inputs mounted onto it and fortified with lots of duct tape, the plywood manifold would be placed at an angle between the Humvee's bucket front seats. It would connect by hose to a second plywood manifold bolted to the outside of the Humvee, which would connect, through twelve hundred feet of narrower hose, to the flat-bottomed boat used by the three divers journeying on foot to the end of the tunnel. Those divers would each have an additional three-hundred-foot umbilical connecting them to the boat.

So all five divers would be relying on the same mixed-gas breathing system for the air they needed to survive. The only difference would be the length of their umbilicals. The three divers working at the end of the tunnel would be connected to the breathing system by fifteen hundred feet of hose. The two divers remaining in the Humvee would be connected by just ten feet.

Wednesday brought with it yet another problem involving the

trailer, even after the hitch assembly had been repaired. Of course, because the tunnel would eventually become too narrow for the divers to turn their vehicle around, they needed the trailer so that the lead Humvee could tow the second Hummer, which would be facing in the opposite direction for the drive home. But the crews noticed that once the bigger Humvee had been backed onto the trailer, the weight had pushed a couple of the trailer tires down into the gaps between the slats of the aluminum deck. And now those tires were flat. After changing them, the crew realized that they would likely get a repeat of the flats if they tried again to load the Humvee wagon onto the trailer while it was on the deck. So reluctantly they decided they would have to wait until both Humvees were off the deck before trying to back the bigger of the two onto the trailer. And that would require a particularly tricky maneuver, given all the standing water below the deck. At least they wouldn't have to worry about it until the next morning.

Despite this latest setback, these first few days in the tunnel gave Riggs a better feel for the Black Dog crew. Billy struck him as both energetic and knowledgeable, a real pro with a steady hand—and apparently a big appetite for a little guy, given the size of the cooler he packed his lunch into every day. Most surprising, Riggs had even found himself beginning to warm to DJ. Back in New Hampshire, he had seen DJ as a guy who tried to skate by and seemed most interested in extracurricular action. But once they began working in the tunnel, Riggs was pleased to see that DJ was all business.

Through the Humvee's rearview mirror, DJ could see Billy, standing in knee-deep water, giving him hand signals. *Go, but go slow.* DJ knew his assignment on this Thursday morning was precarious, especially given all the trouble they'd had with the first Humvee. Although the temporary deck they'd been working on at the base of the shaft was wide enough to fit two Hummers side by side, there was only a single-lane ramp leading from the deck down to the water-soaked tunnel floor. The ramp was scarcely an inch or two wider than the Humvee itself, and it was extremely steep. DJ had to back the gray Humvee—its cab and roof rack groaning with gear—down the sharp decline of the ramp. Hoss had already backed down the white Humvee, and it had gotten stuck along the way. It had remained stuck until Billy thought to wedge a two-by-six plank of wood underneath the Hum-vee's back wheels. All this drama, DJ thought, just to get the Hum-vees in position to *begin* the journey. What's more, the divers had to work in front of an audience that included sandhogs, Kiewit officials, and Dave Corkum. Still, as white-knuckle a task as it was, DJ got a jolt out of being behind the wheel of this powerful machine. He was, after all, a guy who could remember the dates of important events in his life according to which muscled pickup truck he was driving at the time.

Having learned from Hoss's experience, DJ managed to back his Humvee down the ramp on his first attempt. But that turned out to be the easy part. Now, with his vehicle idling in the standing water, he needed to back it up onto the two-way trailer that was already at-tached to the white Humvee.

The white, eastward-facing Humvee would be the divers' lead ve-hicle for the journey. Once they had traveled to the point in the nar-rowing tunnel where the Humvee could go no farther, they would unload the westward-facing gray Hummer off the trailer, positioning it for the drive back to the shaft. Because the trailer had mini ramps on either end of it, the divers could then back the white Humvee up onto it, converting that Hummer from transportation to cargo for the return trip. So this maneuver at the start of the tunnel was also a preview of the vehicular choreography to come.

DJ's task in backing the gray Humvee up onto the trailer was made more difficult by how crowded the trailer already was. At its far end, closest to the hitch with the white Hummer, stood the tall cage-like rack holding four 180-liter tanks—three silver ones containing the liquid nitrogen, and one white one containing the liquid oxygen. At the other end lay the greenish flat-bottomed boat, which was brimming with coils of yellow hose but just low enough to the ground to fit under the high-riding towed Humvee. In one corner of the boat, someone had left a sixteen-ounce bottle of Coke standing up. DJ would have to position his Hummer perfectly in order to begin backing it up onto the mini ramp at the end of the trailer. Then he'd have to keep moving in reverse so that the high-riding Humvee backed up over the boat. He had to be mindful of the big cylinder of liquid oxygen that was mounted to the back of his Humvee, supporting its diesel engine. Because space was at such a premium on the trailer, he had to back up the vehicle until it all but brushed up against the cagelike gas rack, but not so much that it actually collided with those four big cylinders inside it. After all, there was a reason "Mr. Science" used liquid nitrogen for his dramatic mist-filled demonstrations on TV, and liquid oxygen, under pressure and highly flammable, could explode if improperly handled. DJ had no interest in putting on a fireworks show.

Mustering all his concentration, he tapped lightly on the accelerator, moving the Hummer back in three stutter steps, then another longer one, before stopping the vehicle in perfect position. He could hear one sandhog yell "Yeah!" and another, "Wow!" DJ climbed out the driver's side, eager to check out his work. He was wearing slip-resistant boots and water-resistant coveralls over his faded jeans and long-sleeve collarless shirt, so he paid no mind to the water at his feet. As he walked around to the back of the Hummer, he was proud to see that the only impact he'd made had been with that plastic bottle of Coke, which now lay on its side in the boat.

Their shift had begun at seven a.m. After all the prep work, checklist review, and Humvee maneuvering, it was now after eleven. The morning was gone, and even though they wouldn't actually be removing safety plugs on this shift, they had more than a full day's work ahead of them.

The MAP Mix 9000 was still AWOL, but the calendar could wait no longer. So today the divers would go in. The continued absence of the breathing system's centerpiece had begun to weigh heavily on the guys. With Harald sitting topside in his trailer, and the divers preparing to climb into the Hummers, traveling east, DJ and Hoss talked it over. "Why is this thing even coming from overseas?" Hoss asked. "Why can't they get one from the U.S.?"

For this shift, they would be breathing off their backup system—the cylinders of HP air lying on the Humvee roof rack. A simple lever could change the source of air flowing into their umbilicals, from the supply of mixed gas to the HP supply. DJ knew that Tap, from the start, had proposed using HP air for this job. Of course, Tap's plan had lost out to Harald's. As Harald had explained it, crews would likely be able to carry only about eight hours' worth of HP gas with them, and considering that it would take about four hours of setup and travel to get to and from the end of the tunnel, that wouldn't leave much time for actual work. In contrast, Harald's mixed liquid gas system would allow the crews to work perhaps a twelve-hour shift, including travel and setup, and thus allow the job to proceed far more quickly. From his conversations with Tap, DJ had begun to wonder if Haraid's plan had won points simply for being more dazzling. It was way too late to try to replay that contest between Harald and Tap. Still, DJ found it funny that, for the first day of their journey into the tunnel under Harald's command, the divers would be relying on the kind of straightforward, workmanlike breathing system that Tap had originally pitched.

Harald had assigned Riggs and Tim the task of getting the standby-backup Hummers into shape, so that meant they would be doing their work from the base of the shaft. Meanwhile, DJ, Hoss, and Billy would form the main team heading into the tunnel for the day, joined by a strawberry-blond Norwesco veteran named Mike Mars and a young, stocky Black Dog guy named Tracy Markham. The Hummers began plodding eastward along the trying terrain of the tunnel, moving at a speed of just a few miles an hour. With Hoss driving the lead white Humvee, DJ and Billy walked ahead of the vehicles for the first stretch of the tunnel to try to clear it of any debris that could cause them problems. DJ noticed how quickly the light from the base of the shaft disappeared, leaving them in darkness except for the headlights on the Hummer and the headlamps on their hard hats. After walking a decent distance, he and Billy hopped onboard the Hummers. DJ watched as they passed one five-foot-long concrete tunnel "ring" after another, going by the markers the sandhogs had spray-painted on the wall like yard lines on a football field: ring 1000, ring 1800, ring 4000.

On the drive east, the divers stopped at regular intervals along the way, to take oxygen readings and to install "mine phones" that would connect them directly to Harald in his topside trailer. These phones, which were boxy yellow consoles with old-fashioned black receivers, could be spliced into a communications wire that had been temporarily strung along the side of the tunnel. The divers stopped roughly every two miles to patch a mine phone into that wire.

They used battery-powered VHF radios to remain in contact with the standby crew back at the base of the shaft. But they found that, for some reason, when the radio was on high, it interfered with the electronics of the vehicle's oxygen analyzer. As they ventured deeper into the tunnel, the reception of the VHF radios turned out to be weaker than expected. The crew also lost one of the connectors on the breathing system because it wasn't the locking type, requiring them to duct-tape the hose into place instead.

At each stop, they used their Dräger handheld gas monitors.

They were first testing to see when the oxygen content in the ambient air would drop below the standard 21 percent, to know when they'd have to don their facemasks. Considering how long they were going to be in the tunnel on Thursday, the divers wanted to delay turning to their life-support air supply until absolutely necessary. They knew they'd be fine with limited exposure to oxygen percentages in the upper teens. By ring 7500, the oxygen level had slipped to around 16 percent, and the divers quickly attached the masks that connected them, through umbilicals, to the HP air bottles on the Humvee roof.

Before long the divers passed through the venturi, the division between the main tunnel and the diffuser tunnel. The latter was the outfall tunnel's final mile, where it connected to the risers, starting with number 55 and descending to number 1 at its easternmost endpoint. Once the divers crossed into the diffuser tunnel, DJ noticed how the monotony of the view began to change. A concrete ramp on the floor and Jersey barriers coming in from the sides combined to gradually reduce the surface area. This made the Humvees' slow ride go even more slowly. At one point, Hoss's facemask started "free-flowing," meaning it was pumping out too much air. The "demand" regulator was designed to give air to a diver only when he needed it, stopping it automatically when he paused to exhale. After enough tinkering with his regulator, Hoss managed to fix the problem.

A couple of divers complained that the short hoses connecting their facemasks to the bailout bottle on their hips—the emergency supply offering up to ten minutes of air—wouldn't lock into place. They were told to duct-tape it to prevent an accidental disconnect. By the time the divers reached riser number 18, which was around the nine-mile mark, they were fast approaching the point in the tunnel where it would get too narrow for the vehicles to proceed. They couldn't afford to take off their masks, even momentarily, since that far into the tunnel the oxygen level in the stale ambient air had fallen to a consciousness-robbing 9 percent.

And it was 9:45 p.m. They'd arrived on site almost fifteen hours earlier, and they still had a two-hour drive back to the shaft. After doing some work at this stop, they talked to Harald by mine phone and decided to go no farther.

By the time they made it back to the base of the shaft and began driving up the steep ramp to the deck, it was after midnight. By the time they took the man cage up the shaft to topside, it was 1:15 a.m. DJ had been in the dark, dank tunnel for eighteen hours, and he was bone tired.

New concerns arose about the ability of the tires on the trailer to withstand the load. Rather than risk a flat deep in the tunnel, Harald decided to send out the tires to be filled with a special foam. This ran $400 a tire, but it made them much better suited to the dicey terrain. Those tires wouldn't be ready until the weekend, meaning the plug-pulling operation would now have to wait until Monday.

After their exhausting day, DJ was relieved to hear they'd be getting a rest. He detected that even Harald, who had always seemed so preoccupied with the demands of the calendar, appeared to be fine with the delay. That was probably because the day had ended with some rare good news. The elusive MAP Mix 9000 had not only arrived from Denmark, but it had cleared customs at Logan Airport and been delivered to the island. Two weeks after the training sessions had begun in Tap's yard, and one week after the divers had begun working on site, Harald's vaunted breathing system would finally be ready to go.

Riggs stood at the base of the shaft, hunched over, staring at the piece of equipment everyone had been waiting for. It was Friday morning, and Harald had selected him for the important job of helping to install the gas mixer that their breathing system had been built around. As he looked it over, Riggs mumbled, "Hey, this sure don't look like diving equipment."

It was, in a word, underwhelming. The stainless-steel unit looked more like an old-model toaster oven than a high-tech gas blender. It was small—just eight inches tall, nine inches wide, and sixteen inches deep. And it looked unsophisticated—with a gray faceplate featuring just two big dials and a small light sitting between them. The red dial on the left was labeled FLOW, and it went from 0 to 300. The blue dial on the right was MIX, and it went from 0 to 100 percent. The small red light in the middle was labeled ALARM. Across the top, the unit read PBI DANSENSOR, the name of its Danish manufacturer. In smaller letters across the bottom, it read MAP MIX 9000 GAS MIXER, to the right of a warning message in red letters: "!!Danger!! This mixer is mixing OXYGEN. Using oxygen may lead to mortal danger of EXPLOSION. Use only oxygen as specified by the rules of the specific country."

Even if the mixer seemed unimpressive, Riggs was pleased that Harald had picked him for this installation assignment. The only other diver he had chosen was Mike Mars, the thirty-six-year-old trusted veteran with Norwesco whom Harald had designated to be his crew foreman in the tunnel. Riggs took his selection as a good sign for his long-term goal of establishing a permanent home with Norwesco.

Unlike Mars, Riggs had had very limited experience working with Harald prior to Deer Island. Harald's role on the Lake Mead job had been largely at the front end, and by the time Riggs joined that project, day-to-day operations were in the hands of someone else. Riggs had interacted with Harald just twice during his year in Nevada, and he'd found the engineer to be quiet and low-key. So he hadn't been prepared for the imperious and impatient Harald he encountered in Tap's yard. He especially didn't care for the imperious part, the sug-

gestion that, as the smart engineer, he had done the hard thinking
for everyone else. Riggs's father had spent his career as an engineer,
and even though he never once told his son he wanted him to follow
in his footsteps or was anything but respectful of his choices, Riggs
could sense that his dad had higher aspirations for him than life as a
blue-collar laborer. Still, even if Riggs hadn't been a star student, he
knew he was smart. He didn't take kindly to anyone implying that
divers couldn't somehow measure up mentally, no matter how many
framed degrees that person might have.

But now, working alongside the stout engineer, Riggs saw a more
relaxed and collegial Harald, who wore a blue-gray jumpsuit with
NORWESCO printed on the back and his name written in cursive on the
front. Maybe, Riggs thought, Harald was finally settling into his role
and learning to deal with the pressure. Maybe, like DJ, he'd simply
made a bad first impression in New Hampshire.

When Riggs asked how this small machine could handle such
a big job, Harald explained that the elegant little proportional gas
mixer was fully up to the task. Just three hoses had to be hooked up
to the MAP Mix 9000: one input from the oxygen, one input from the
nitrogen, and one output for the blended gas once the MAP Mix had
worked its magic. Opening the regulators atop the liquid oxygen and
liquid nitrogen cylinders would send the liquid through a coil inside
the tank, vaporizing it so that it would be converted into a gaseous
state before traveling to the mixer. As for the two dials on the front
of the MAP Mix, Harald turned the FLOW up to its highest setting,
300 liters per minute. This would make it an "on demand" breathing
system, dispensing air to the divers as needed. He turned the MIX dial
to 22 percent, meaning that would be the desired oxygen content of
the blended gas supply—setting it a smidge over the usual 21 percent
would provide a small cushion. Once the blended air left the mixer,
which rested on the Hummer's back seat, it would travel by hose to
the plywood manifold wedged between the vehicle's front seats.

The MAP Mix 9000 might have been small, but it certainly wasn't cheap. Norwesco had paid more than $10,000 for the pair of mixers, buying two so that both the main pair of Humvees and the backup/ standby pair would have identical setups. Even though Topac, the mixer's Massachusetts distributor, had requested rush delivery for Harald, that hadn't been possible because the Danish manufacturer didn't normally stock models that mixed nitrogen and oxygen. Most of its customers used the mixers for industrial purposes and typically requested models that blended carbon dioxide with either oxygen or nitrogen. So Harald's mixers had to be specially assembled. During the long delay, Harald had warned the owner/salesman of Topac, Tony Drybanski, that every day lost was costing Norwesco $20,000, presumably given payroll, lodging, and other costs. Finally, after all the waiting, Drybanski had personally dropped off the mixers to Deer Island. Now, with relatively easy work on the part of Riggs, Mars, and Harald, both units were being installed.

Riggs asked Harald what the power supply would be for the mixer. He was surprised when Harald explained that the device was pneumatic, meaning it didn't require a power source. It was driven solely by pressure. Underneath the mixer's stainless-steel cover, there were multiple valves and diaphragms, including a large one that looked like a waffle maker resting on its side, with a rubber gasket inside it. As long as the pressure of the gas coming into the mixer—the vaporized oxygen and nitrogen—was greater than the pressure of the gas moving through and out of the unit, the machine would continue to hum along. That was because, on its own, gas flows from high pressure to low pressure, always attempting to achieve equilibrium. The same principle applies to something as basic as a balloon. When you blow one up and pinch the end of it, the air compressed into that stretched balloon is much higher in pressure than the air outside it. If you release your pinch, that high-pressure air will automatically flow out of the balloon, racing toward the low-pressure ambient air.

Like the miracle of the hydraulics in the outfall tunnel itself, where a billion gallons a day of treated water could flow along the 9.5-mile route, up one of the fifty-five risers, and finally out to sea—all without so much as a pump or a motor—so, too, could this mixer work without any electrical power.

Impressive, Riggs thought. But why then, he asked, was there a coiled-up black electrical power cord attached to the MAP Mix 9000? Riggs had already twisted off the factory tie and uncoiled the cord. Harald explained that they wouldn't need the power cord for their operation. "Don't worry about it," he said. "Just tie it back up."

8

Monday

Billy sat by the lake in New Hampshire, enjoying a lazy summer Sunday with his girlfriend, Michelle. R&R is exactly what he was looking for after such a stressful week on Deer Island. He knew the white waterfront bungalow where his close friends and surrogate parents lived was just the place where he could find it. He had started the day playing golf with Ken, and after Ken left with Deb for a wedding, Billy and Michelle had the place to themselves.

Before long, the relaxation gave way to worry—though not his own. Michelle rattled off all the things that were giving her pause about the tunnel project, cataloging the details he had shared with her as the job had progressed. She reminded him about the trailer getting damaged in a freak roadway accident on the way to the island, about the flats on the trailer tires, about the equipment that wasn't working right and the equipment that still hadn't shown up on site, and most of all, about how completely cut off from civilization the divers would be at the end of the tunnel. As silly as she knew it was, she threw in the fact that the divers were working on an Indian burial ground. "How many signs do you need, Billy?" she asked.

He decided to try to defuse the situation with a laugh. "You're supposed to be a good Catholic," he said, "and here you are doing this bad juju." No luck.

In their four years together, Billy and Michelle had built a strong relationship based on trust. He took her concerns seriously and never tried to snow her about the dangers of the jobs he was working on. Still, like his diving buddies, he was careful not to offer too many details. From the start with the tunnel job, he had accentuated the positive. "This has never been done before," he had beamed to her. "We could end up on the Discovery Channel." But as his involvement on Deer Island deepened, he had clearly let too many details slip.

Billy understood why Michelle was so anxious. He had his own troubles with the job, and they both knew that when he got worried about something, he tended to clam up. No doubt she had picked up on this and could sense how much his initial Discovery Channel–fueled excitement had cooled.

He tried changing the subject, focusing on the happy news in their life. He and Michelle had been renting a small house in Wakefield, a short drive from Deb and Ken's place, where he had previously kept a room. Now they were in the process of buying a beautiful old New England farmhouse in town. But when Billy tried to get Michelle to discuss that house instead of the tunnel, she wouldn't budge.

Rather than commute from New Hampshire to Deer Island every day, Billy had been crashing during the week at a friend's place outside Boston. He planned to do the same in week two of the job, heading south in the morning and not returning until the following weekend. He sensed that Michelle felt this was her last shot to get him to reconsider.

The somber news that had been dominating the airwaves all weekend hadn't helped ease her apprehension. That Friday night, John F. Kennedy, Jr., had been piloting his single-engine Piper Saratoga, with his wife and sister-in-law on board, headed to the Kennedy compound on Cape Cod for his cousin's wedding. By early Saturday morning, the Coast Guard had begun a massive search-and-rescue effort in the waters around Martha's Vineyard. By Sunday, the Coast Guard had reluctantly shifted its focus from rescue to recovery. No matter what he had done in life, JFK Jr. had always been remembered best as the little boy in short pants, poignantly saluting his father's casket during the funeral procession on November 25, 1963, which also happened to have been the little boy's third birthday. Now he was presumed dead at age thirty-eight. Bride-to-be Rory Kennedy, who had still been in her mother's womb the night her father, Bobby, was assassinated in California, postponed her wedding in grief.

Even after Michelle had to leave for work, Billy got no reprieve. By then Deb had returned, and in the living room, she told him just how distressed she was about Deer Island. He tried offering the same reassurances he had given Michelle. There was nothing to worry about. Everything would turn out fine. But it was clear he wasn't getting through.

Softly, Deb said, "Billy, please don't do this."

"Ma," he said, using his affectionate nickname for her. "I'm fine."

Feeling ganged up on, he walked out of the house and headed into the backyard. Near the lake, he found Ken sitting inside the tiny white shed he had turned into a watering hole, complete with wet bar and stocked refrigerator. They each grabbed a beer.

Billy had been having such a good run lately, now that he was working less and able to spend more weekend time relaxing with Michelle and golfing with Ken. Just a few weeks back, he had recorded the first eagle of his life, marking the accomplishment by jumping up and down on the course so fervently that Ken wondered if they'd be asked to leave. But this Deer Island job seemed to be casting a pall on them all.

Ken brought up the job, trying to explain Deb's concerns. "She's nervous," he told Billy.

"I know exactly what I'm doing," Billy replied. "You know that."

Back when they'd been training in Tap's yard, Ken had stopped by for his own look at the mock-up. "You've got to tell the girls that I know what I'm doing," Billy told him now. "Sure it's dangerous, but we've practiced. We've got all good guys."

That was enough for Ken. Later that night, he told his wife, "Look, he's going to be all right."

At four o'clock the next morning, Michelle awoke to hear Billy's truck running in the driveway. She was groggy but instantly became upset that he would leave without saying goodbye. They had an arrangement: before he took off for a job, especially one that would keep him away for a whole week, he had to jostle her awake for a goodbye kiss. But soon Billy walked through the door. He'd only stepped outside to pack up his truck. They kissed, and she rolled over and went back to sleep.

At seven-thirty a.m. on Monday, July 19, the divers gathered at the base of the shaft to begin going over the lengthy checklists Harald had handed them, ensuring that all their tanks were full, all their connections were tight, and all their equipment was operational. Only when they began reviewing the checklist for the Humvees did they discover a problem. The gray Hummer wagon wouldn't start. As it turned out, the battery was dead. Billy knew a dead battery at the base of the shaft was no big deal—they could simply jump-start it or swap it out. But it would be a different story if it happened again when they were deep inside the tunnel.

The divers were relieved that the mysterious MAP Mix 9000 was finally here and that their plug-pulling mission could begin in earnest. The day's assignment called for the five divers to drive the Humvees as far as they could into the tunnel, and then the three-man excursion crew would trudge on foot to the end, to begin yanking out the plugs. They would start with the plug in the elbow of riser number 1, then work their way back toward the shaft, with the goal of pulling several plugs by day's end.

As the divers maneuvered inside the Humvees while going over their checklists, their bodies rubbed against the thick plastic covering on the vehicle's upholstery. The official explanation for why the divers had to keep the clear protective plastic on the seats and interior door coverings, as if the menacing Hummer were some dowager's sofa, was that the plastic would be a barrier against the water in the tunnel. But by now the divers had figured out the real reason Norwesco wanted that barrier: after the job was done, the company planned to resell the two pairs of Humvees. Water stains on upholstery never helped with resale value.

At nine-thirty, with all the preliminaries done, the Humvees began their slow journey east, stopping at every mine phone so a diver could check in with Harald as he sat in his topside trailer. Forty-five minutes later, after reaching ring 6000, they stopped to fire up the oxygen-injection system that would feed the Humvee's diesel engine. Even though the oxygen level in the tunnel was dropping steadily, the divers decided to push it a bit more on the ambient air, before turning to their breathing supply. In addition to being more comfortable and natural, the ambient air allowed them to carry on conversations without having their voices distorted by their facemasks and the sound of hissing air.

On the drive out, Billy talked a lot with Tim. Although they'd known each other for only a couple of weeks now, the two found they got along well. They were both fun-loving guys who were good at what they did but low-key about it. They were both famous for their facial hair, Billy with his mustache and Tim with his beard, which he had reluctantly trimmed to a goatee for this job. And they each had a great woman at home who was growing increasingly worried about this assignment. The day before, around the same time that Michelle was sharing her fears with Billy, Judy was doing the same with Tim.

"I don't want you to go in the tunnel if you don't think it's safe," Judy told her husband on the phone. Even the usually upbeat Tim had been worn down by the frustrations and setbacks the divers had encountered during the first week. When Judy said, "Tell me again about the tunnel," he conceded, "It's cold, it's wet, and I'll be glad when this job is over." Now Tim and Billy were both looking forward to a productive, drama-free day so they'd have some calming news to report during their next call home.

When the divers arrived at the venturi at 11:15 a.m., they followed the protocol for activating the mixed-gas breathing system. Once the mixer and regulators and manifolds were all properly aligned, the divers put on their facemasks and started breathing the blend of the vaporized gas from the liquid oxygen and liquid nitrogen. Before long, the pair of Hummers began their trip into the diffuser tunnel.

Even after they had donned their facemasks, the divers would continue to make periodic stops to test the air. Harald had instructed them to use their Dräger Multiwarn handheld gas monitors to measure the ambient air in the tunnel, not just for its oxygen content but also for the presence of dangerous gases, such as methane, hydrogen sulfide, and carbon monoxide. Although no worrisome levels of those gases had turned up during the fact-finding mission a year earlier, back then the divers had ventured only partway into the tunnel. Now as they prepared to make their way to the very end, they needed to know everything they were up against.

Six weeks earlier during their Dräger training sessions, the divers had been schooled in how to use these Multiwarn analyzers to test ambient air. But just before the start of this Monday mission, they had been surprised when Harald told them they should use these same Multiwarn devices to measure the oxygen content of their own breathing supply—after that air had exited the MAP Mix 9000.

Because no one had made it out to the final stretch of the tunnel since the sandhogs had ripped out the ventilation, the environment there was shrouded in mystery, cordoned off like a crime scene. Still, the divers had good reason to believe that the air toward the far end of the tunnel had only gone from bad to worse, as rusting metals likely consumed much of the remaining oxygen. And thanks to all the water that had leaked in through cracks in the bedrock around it, the tunnel was now home to a host of microorganisms that thrive in damp oxygen-starved air.

These microbes are usually harmless. In fact, Deer Island's state-of-the-art sewage treatment plant had been cleverly designed to harness the power of microorganisms. Inside the 3 million-gallon egg-shaped digester tanks that defined the new island landscape, battalions of the single-cell bugs feasted like Romans at an orgy, except their menu consisted of thickened sludge from human waste. Playing the role of a giant collective stomach, those microbes broke down the sludge into water, carbon dioxide, and methane. That resulted in a vastly reduced amount of sludge, which could then be converted into fertilizer pellets. By far the biggest by-product of this digestion process—about 70 percent—was methane gas, which was sent across the island to boilers, where it was ingeniously converted into electricity to help power the plant.

In a controlled setting like the sophisticated sewer plant, methane wasn't a threat. But because the gas happens to be highly flammable, in a confined space like a coal mine or a dead-end tunnel, a simple spark could turn a pocket of methane into a calamitous explosion. That's why they had to be so vigilant in testing the ambient air.

Around two o'clock, the divers arrived at riser number 12. They spliced a mine phone into the wire running along the side of the tunnel and put a phone console inside the Humvee so there would be a direct line to Harald. They detached the towed Humvee wagon from the trailer and parked it a few feet away, facing back toward the shaft, so it would be in position for the return drive at the end of the shift. Billy and Tim climbed into the Humvee wagon, with Tim taking the driver's seat and Billy clambering around all the equipment packed inside the vehicle to get to his place in the back seat, right beside the MAP Mix 9000. Billy and Tim would remain in the Hummer to monitor the mixed-gas breathing system that all five divers would continue to rely on. Outside the Humvee, the oxygen level in the ambient air had already fallen to 8 percent. The three guys on the excursion crew would have to trek all the way to the end of the tunnel, where it choked down to five feet in diameter, alternately dragging and pushing the flat-bottomed boat that groaned with all their gear. DJ was supposed to have been part of the excursion crew, but Tap had benched him for being late, and Riggs was given a break from tunnel duty for the day. So Monday's excursion crew consisted of Hoss, Mike Mars, and Tracy Markham. Once the trio made it to their final destination, one diver would begin crawling into the thirty-inch off-take pipe leading to riser number 1.

After they removed each plug, the divers would be required to videotape their work so that Kiewit and the MWRA had verification of its removal. But even before the excursion crew left the Humvees, they tested the video camera and found it wasn't working. So much for the hope that Monday would be free of complications. The day was shaping up to be one in which even the easy stuff proved hard.

At ten minutes to five, the excursion crew began its schlep east. Hoss, Markham, and Mars were connected to the main breathing system via a twelve-hundred-foot umbilical—which was actually four separate three-hundred-foot hoses strung together. This umbilical stretched from the plywood manifold on the Humvee wagon to another plywood manifold in the boat. Attached to the manifold in the boat, each excursion diver also had his own umbilical that gave him an additional three hundred feet of running room. This was critical, since the boat could only comfortably go about as far as riser number 4.

That was because the final stretch of the tunnel was a series of three reinforced concrete pipes, each one narrower than the one before, with diameters shrinking from seven feet to six feet to five.

Almost from the start of their slog, Hoss realized that the mixed-gas breathing system was not producing enough air for him. He felt his mask getting sucked into his face, so much so that he occasionally had to pull it away for a second, a dangerous reflex since he risked inhaling the toxic tunnel air. In reality, Hoss had little choice in his reaction. To survive, the human body needs a constant supply of oxygen, moving through the lungs, into the bloodstream, and then out to nourish the organs and feed the limbs. The body is both hardwired to want oxygen and exquisitely sensitive in detecting even slight reductions in air volume. So when Hoss wasn't getting enough volume from his oxygen supply, his natural impulse was to suck in on his facemask, desperately seeking a more satisfying slug of air. The same dynamics are at play when a child, on a dare from a buddy, begins breathing into a paper lunch bag. Instinctively, he or she will begin taking deeper, more urgent breaths, until finally the child has no choice but to pull the bag away.

When the body fails to get enough oxygen—a condition known as hypoxia—the first sign is usually frustration. Then comes a fogginess that resembles the slight buzz you get after drinking a few beers, marked by impaired judgment and reduced motor skills. If nothing is done to restore the oxygen supply, the hypoxia will lead to unconsciousness and then inexorably to death. But along the way, the feeling often turns from frustration and fogginess into euphoria. Ernst Rodin, a neurologist who wrote an academic paper about his own near-death experience, called this altered state of consciousness "one of the most intense and happiest moments of my entire life." He attributed the euphoria to a kind of hallucinatory response from his oxygen-starved brain. Other scientists and anthropologists have found evidence of the quest for a similar euphoric, otherworldly experience in various ancient religious practices, such as some of the earliest baptisms, where the person being initiated was submerged under water to the point of near-drowning.

Fortunately, Hoss never got past the point of frustration. After a short period of struggle with his facemask, he persuaded Mars that they had to do something. Mars, who had been tapped to be foreman for that day's excursion crew, had no way of reaching Harald directly. But he did have a two-way comm wire that connected him directly with Tim back in the Humvee, and Tim had a mine phone that connected him directly to Harald. So Mars reached Tim, and Tim called Harald. A few minutes later Tim reached Mars and passed on Harald's instructions: they should switch to their backup supply of HP air. The way the setup was designed, the divers could continue to use their same facemasks and umbilicals. It was just that, after Billy stepped out of the Humvee and turned a valve on the outside manifold and Tim flipped a lever on the inside manifold, the source of their breathing air would no longer be the blended gas coming out of the MAP Mix 9000. Instead, it would be the four tanks of HP air strapped to the Humvee roof rack.

Hoss was relieved to find that the change immediately eased his facemask troubles. He and the other two guys resumed their march to the end of the tunnel. When they made it to riser number 4, Mars parked himself there, next to the boat, so he could monitor the manifold. Hoss—all six-foot-five of him—would then take the lead in duck-walking to the end of the tunnel and slithering into the thirty-inch-wide pipes to begin pulling the plugs. However, the divers had already consumed a lot of the HP air, given all the effort it had taken them to get to the end of the tunnel—which was as long as six Holland Tunnels laid end to end. By 6:25 p.m., Tim informed Mars that they had already burned through two of the four HP bottles on the Humvee roof.

They were finally in position to begin pulling plugs, but Hoss was worried that if they kept working at this pace, they would use up all their air and not have enough to get back to the shaft. After all, the HP air was supposed to be their backup. He told Mars they needed to abort. Once again Mars called Tim, and Tim called Harald, and word came back that they should continue working with the goal of getting at least the first two plugs removed. Harald reminded them that they had an additional HP bottle on the boat that the excursion crew could rely on if needed.

When Hoss heard that, he flipped. "Fuck that!" he yelled through his facemask. "That bottle on the boat is for *emergencies!*" Hoss may have been young, but he'd been a diver long enough to know that you never used your emergency air supply to conduct regular work. Even if Mars outranked him on the job, Hoss wasn't about to put his life on the line out of deference. After some heated back-and-forth, Mars agreed. They packed up the boat, turned around, and began their long walk back to the Humvee. Once there, Hoss spotted something unusual with the liquid oxygen tank sitting on the trailer. There was frost on the tank and its regulator.

When the Humvees finally returned to the ramp at the base of the shaft, there were representatives from contractor Kiewit and construction manager Kaiser waiting for them. A Kiewit representative made a final entry into that day's shift log. "Day's work unsuccessful: No plugs pulled."

Hoss was still angry as he made it up to topside just after nine-thirty p.m. Following the long buildup, he'd finally gotten to see Harald's mixed-gas breathing system in action. And as far as he was concerned, it just wasn't ready for prime time. Being sent nearly ten miles into a pitch-black, oxygen-starved tunnel under the ocean was nerve-racking enough. They shouldn't be forced to worry about whether their breathing system would be able to keep them alive.

When Hoss got back to his motel room, his lungs were hurting, probably because of the bad tunnel air he'd breathed when he'd been forced to pull away his facemask. He decided to go above Harald's head. In his five years with Norwesco, Hoss had come to view owner Roger Rouleau as a businessman focused mostly on dollars and cents. But he also knew that Roger had entered the business as a diver, so he hoped the man would take Hoss's concerns seriously. Roger had stayed back in Washington, but the novel nature of the Deer Island job had drawn him attention in Spokane's *Journal of Business* a few days earlier. In a feature story about Norwesco and its tunnel project, Roger had told the journal, "We've done a lot of wacky stuff. That's what makes this business fun. The weirder and wackier it is, the better it is for us."

After Hoss rattled off all the problems they'd encountered, Roger assured him he would look into it. While Roger considered Hoss an excellent worker, he also knew that he was young and felt he had a tendency to be a little mouthy. Maybe the kid was simply overreacting. Still, Hoss's grievances troubled Roger, since they came not long after Tap's concerned call from New Hampshire. After hanging up, Roger dialed Harald. Once again his project manager had reassuring answers. Harald admitted there were some glitches in the setup, but he stressed to Roger that they were working them out. In fact, he had just left a detailed message for the guy who had sold him the MAP Mix 9000, asking for guidance in addressing the pressure problems Hoss had experienced. Harald told Roger that he had a sound plan of attack. "This is going to work well," he said.

Even though Roger was the boss, he had always felt intellectually intimidated by Harald. That plus his boundless faith in Harald's abilities led Roger to accept these assurances, even if a small part of him wondered whether he should press a bit more.

9

Tuesday

When the school bus ferrying the divers arrived on the island at seven o'clock on Tuesday morning, Harald was on the phone with Tony Drybanski. The salesman-president of the mixer's local distributor was responding to Harald's message from the night before. He sensed Harald was agitated and impatient for an answer, but that was nothing new. "Every time Harald called, he spoke to you like a debt collector," Drybanksi would later say. "So he always had this sense of urgency in his voice. He was just all over me, all the time." To address the MAP Mix 9000's flow problems, Drybanski advised Harald to bump up the pressure on the machine. That couldn't be done simply by adjusting one of the two dials on the front of the device. Instead, he said, Harald would have to unscrew the mixer's stainless-steel cover and manually adjust the internal gauge that governed the pressure of the oxygen and nitrogen coming into the machine. The first page of the owner's manual warned that the "MAP Mix 9000 should be opened by authorised personnel only," and Harald clearly didn't fall into that category. Still, based on his previous discussions with people at the factory back in Denmark, Drybanksi explained to him that this inlet pressure gauge could be safely turned up. The mixer had left the factory with a default setting of 4.3 bar of pressure, but Drybanski said the gauge could be increased to 7 bar, or the equivalent of about 100 pounds per square inch (PSI). Rather than max it out, they settled on 6.6.

The goal was to get more juice coming out of the mixer. But there was apparently no discussion between Harald and Drybanski about how raising the pressure of the gas flowing into the mixer might affect the inner workings of the blending process. After all, the magic that powered the pneumatic MAP Mix 9000 was not electricity or gasoline but pressure—or more precisely, a differential in pressure, based on the way gas naturally moves from high pressure to low. To ensure that the machine's valves and diaphragms functioned properly, the inlet pressure coming from the oxygen and nitrogen tanks had to be at least 1 bar higher than the outlet pressure of the blended gas going out into the divers' umbilicals.

Harald took the man cage down to the base of the shaft. Enlisting DJ's help, they popped off the lid and made the adjustment. Harald then ran a test by turning on the liquid oxygen and liquid nitrogen and checking the flow meter on the manifold. The test showed that the increase in pressure created an increase in the flow rate—the gas came out faster and more forcefully. The adjustment had worked. Harald told the divers that the simple maneuver should address Hoss's complaints about the system.

Because his lungs were still hurting him from his facemask problems on Monday, Hoss would be staying out of the tunnel today. The Tuesday team consisted of Tim and DJ in the Humvee, and Billy, Mars, and Riggs on the excursion crew.

It was just after nine-thirty a.m., following the lengthy preparation and checklist period, when the divers piled into the Humvees and began their drive. About three hours later, they arrived at riser number 12, and Tim called Harald to tell them they were setting up.

After the excursion crew started moving east with the boat, DJ and Tim settled into their places in the Humvee wagon.

Because DJ had been out of the tunnel on Monday, this was his first experience working with the peculiar breathing system. With Tim behind the wheel, DJ sat in the back seat, next to the MAP Mix 9000. Like the trio on the excursion crew, DJ and Tim wore their facemasks, which supplied them with the blended air coming out of the mixer. It was dark inside the Humvee. The engine wasn't running, but they drew power from the battery to keep the vehicle's dim dome lights on. When Tim used the comm wire to talk with Mars, who was once again serving as the foreman of the excursion crew, DJ could hear both ends of the conversation. But when Tim was talking to Harald on the mine phone, DJ could hear only Tim.

As Billy had done the day before, DJ monitored the breathing system serving all five divers. When Harald had set the MIX dial on the front of the machine to 22, he was telling the mixer to blend oxygen and nitrogen in such a way that it produced breathing air containing 22 percent oxygen. But that was effectively just a request. There was no gauge or monitor on the mixer to confirm that the actual blended air was meeting the desired amount of oxygen. DJ's job was to use the Dräger Multiwarn monitor to ensure that it was actually happening. Now this battery-powered, handheld analyzer would be used not to test for methane and other gases in the dead ambient air of the tunnel, but to monitor the breathing air the divers needed to live.

Following Harald's protocol, DJ used a sampling port in the plywood manifold that was jammed between the Humvee seats. Once he opened the valve of that port, it siphoned off some of the breathing air—after the air had left the mixer but just before it entered the divers' umbilicals. DJ then held the probe from the Dräger analyzer in front of the air whooshing out of that open valve. When the digital readout told him the actual oxygen percentage of their breathing air, he would call it out to Tim, who would log it on his clipboard and then phone it in to Harald.

DJ did as he was told, but the whole setup seemed bizarre to him. During his time in the Gulf, he had done plenty of deep-water dives using some form of mixed gas, whether the helium-oxygen blend heliox or the nitrogen-oxygen blend nitrox. But that gas came premixed and therefore pretested. He'd never before been asked to mix his own gas on site and breathe it, right then and there. Regardless, he'd felt burned by his previous attempts to raise questions with Harald, so he wasn't about to do it again now.

The three divers on the excursion crew carried their own Dräger gas analyzer. With DJ in charge of monitoring everyone's breathing air back at the Humvee, the excursion divers used their handheld analyzer to measure the levels of oxygen and other gases in the ambient tunnel air. Even though their breathing supply was coming from their apparatus, prudence still required them to be mindful of everything in the air around them. If one of the divers encountered an emergency and felt the need to remove his facemask even for a moment, as Hoss nearly had on Monday, it would be critical to know the precise oxygen content of the ambient air. The divers had only to think back to the oxygen percentage chart they'd been given during their mine rescue training—decreased ability to perform tasks at 17 percent; dizziness and impaired judgment at 15 percent—to remember that not all oxygen-deficient environments are the same. While 9 percent would eventually lead to unconsciousness, just a few breaths at 6 percent would almost certainly mean death.

The Dräger handhelds had built-in alarms that were factory-set to detect when the oxygen level dropped below 19.5 percent. Back during setup in New Hampshire, Harald had told Tap that, because the tunnel would be so oxygen-deficient, they needed to adjust the alarm settings. If they didn't, the device's siren and flashing light would be

going off constantly, making it hard to get work done. He had instructed Tap to lower the alarm trigger point to 9 percent. But as the
divers ventured deeper into the tunnel, and the oxygen levels continued to drop, the alarm kept going off, even at the 9 percent setting.
So Harald had told Tap to ratchet it down even further, to 6 percent.
They might as well have set it to zero, since by keeping it at such a low
percentage, they were essentially disabling the alarm.

When DJ and Tim weren't taking their periodic measurements
and calling in with their scheduled updates, the Humvee assignment
could get mind-numbingly dull. Sitting in the dark, with little sound
except the constant hiss of the breathing system and the muffled
drips of water sliding down the tunnel walls outside the vehicle, DJ
struggled to keep from nodding off. He and Tim decided the best
way to stay alert was to keep talking, and they began swapping war
stories. After lots of laughs, Tim grew serious at one point, asking DJ
something that was gnawing at him now that he was back working
in the Northeast. Just before he and Judy had moved from New York
to Washington State, Tim had abandoned an old car on a Manhattan
street. "You think there could be a warrant out for me because of
that?" he asked DJ.

The question made DJ chuckle. Here was this big gun-collecting
Texan who was worried about being hunted down by cops from the
nation's largest city for the unpardonable sin of having ditched a ja-
lopy on some side street nearly a decade earlier? DJ knew the NYPD
was way too busy to be crossing state lines to investigate abandoned-
clunker cold cases.

"Nah, Tim," DJ said comfortingly, "I think you'll be fine."

Besides the boredom, there was hunger. During their mine rescue training, the divers had been told that, for obvious reasons, they
weren't supposed to eat while they were wearing their facemasks and
breathing supplied air. Harald had explained that there was an extra

port on the manifold, which was connected to the HP supply on the roof. Opening the valve to that port would fill the Humvee cabin with enough good air to allow the guys inside it to breathe without the use of their facemasks. So DJ and Tim tried that. They rolled up the Hummer windows nice and tight, cracked open the valve, and waited as the cabin filled with air. DJ stared at the gas analyzer as the oxygen level of the ambient air inside the vehicle began to climb. But it seemed to take forever to get the needle to move much, and the noise of the air was so loud that he and Tim worried they might miss a call from the excursion crew or from Harald. Also, they didn't like the idea of using up a full bottle of their backup air for a temporary upgrade in the comfort level of their dark cabin. So they scrapped the plan and kept their facemasks on.

The other divers' strategy for dealing with the long stretches without food had been to wolf down their snacks and drinks at the last possible moment before donning their masks. But with a voracious appetite and tall 225-pound frame to feed, DJ eventually decided that wasn't enough. Sitting in the back seat, he cracked open the bag of peanut butter crackers he had packed into his cooler, along with a bottle of Sprite and a corned beef sandwich. He inhaled deeply before lifting his facemask and popping a couple of crackers into his mouth. Pulling his mask away from his face for a few seconds while sitting inside the Hummer was much less risky than it would have been standing outside it, especially given that partial shot of HP air they'd blown into the cabin. The problem was, DJ tossed the crackers in too quickly and one got lodged in his throat. He started coughing and choking, smacking his hand on the seat to try to force the cracker loose.

Hearing this, Tim turned around in a panic. "What's wrong? What's the matter?" Because it was so dark, especially in the back seat, it was hard for him to make out exactly what was going on. Yet because Tim was illuminated by the dome lights, DJ could see the look of terror on his face. As DJ worked to dislodge the cracker bit from his throat, he held up the opened package to help Tim understand the source of the problem.

After DJ stopped hacking, Tim was ticked off. "What are you doing *eating*?" he asked. "You're not supposed to be doing that."

"Dude," DJ said. "I'm *always* hungry."

Crouched over, with his hands on his knees as though he were in a huddle from his junior high school football days, Riggs set himself up outside the off-take pipe leading to riser number 1. He began preparing himself, both physically and mentally, to crawl inside and yank out the plug. No one had to remind him that the horizontal pipe was just thirty inches in diameter, any more than he needed to be reminded that he was at the end point of an unventilated, unlit sewer tunnel with an entire ocean sitting above him. He knew most people would sooner sever a finger than subject themselves to this kind of remoteness and confinement, and he could certainly understand that view. After all, he suffered from mild claustrophobia himself.

Riggs could trace that back to a couple of childhood experiences that remained seared in his brain. When he was six, growing up in East Texas, one of his friends had a mattress box in his garage. Another friend had the idea one day to pile into the dark garage and seal one of the boys in their group inside that box. Then all the other boys would move the box around, kicking and punching it, while taking pleasure in the shrieks coming from the boy trapped inside. Riggs could still remember the holy terror he felt at being confined in that box, praying, *Please God, let me out of here.* What was less than a minute of pure fear had felt like an eternity.

The next unforgettable incident had come courtesy of his siblings. He and his sisters were goofing around and decided to try rolling each other up in a big carpet. When it was his turn inside the rug, he found the experience even worse than his confinement in the mattress box. He was totally restrained, with no hope of coming out until his sisters unrolled him. He had felt like an alligator whose mouth had been sealed shut with duct tape. No matter how strong the alligator might be, once constrained, he was utterly helpless. Rolled up in that rug, Riggs felt his blood curdling and his blood pressure spiking.

Through his many years diving, he had managed to tame, though never quite vanquish, his claustrophobia. He did this by thinking of it as a demon that he needed, through mental effort, to keep locked in a closet. As long as that demon stayed there, it would be as harmless as the alligator with its snout taped shut. On any serious dive job, instead of buckling under the weight of the uncontrollable forces, Riggs focused his mind squarely on the areas where he still had sway. Can I move my eyelids? My fingers? My hands and wrists? Simply being able to check off those items mentally went a long way toward easing his mind, allowing him to keep the demon locked in the closet. For Riggs, there was nothing contrived about the way he got himself into "the zone" before a hazardous assignment. It was how he coped, and coping was how he managed to excel.

He had found less success with other coping mechanisms. He had enlisted in the navy after high school, training in aviation electronics and developing a desire to become a SEAL. But when he thrived as a radar tech, the navy opted to keep him on that track, which effectively closed down his SEAL plans. In 1981 he was a twenty-year-old navy crewman aboard the nuclear-powered USS *Nimitz* as the world's largest aircraft carrier crawled sixty miles off the coast of Florida. Just before midnight, a navy pilot attempted to land his jet on the flight deck of the *Nimitz*. At the last minute, the plane veered, colliding with a row of other aircraft lined up on deck. The fiery crash claimed the lives of fourteen crew members, burned by jet fuel or struck by shrapnel. Another forty-eight were injured. Among the casualties were a couple of Riggs's friends.

Before long, he sought relief from his grief in the bottle. He continued drinking even after he left the navy and began his career as a commercial diver. After nearly a decade, though, he realized the booze was a crutch that was keeping him from real living. He put his trust in sobriety and turned to religion. And all the good in his life— especially his young family—had flowed from that decision.

Outside the connector pipe, Riggs and Billy crouched down near the tunnel floor. Among these off-take pipes that connected to the fifty-five risers, number 1 stood apart. It was the farthest from Deer Island. It shot out of the tunnel's south side, while the fifty-four others were connected on the north side. It was almost directly across the tunnel from the off-take pipe for riser number 2, while the others were spaced about 125 feet apart. Most important, it was the longest. The diver entering the thirty-inch pipe would have to crawl thirty-five feet to get to the safety plug—double the length of other off-takes. Before Riggs could crawl in, he and Billy needed to pump out the water that had accumulated in the pipe.

There had been some level of water seepage all along the tunnel, especially in its last mile. But here in the cramped five-foot-wide final stretch of the tunnel, and especially in the unimaginably tight thirty-inch-wide off-take pipes, even modest water buildup could become a big problem. Accordingly, the divers had to be smart in how they got around it. They would use battery-operated pumps to dewater the pipes, but even before they turned the pumps on, they needed to build a barrier of sandbags outside the pipe to prevent the exiting water from flowing right back in. So they made the slippery, football-field-length walk back to riser number 4, where Mars stood running the manifold. They grabbed sandbags and pumps from the boat and then lugged them all the way back to number 1, clambering with their heads bowed over.

The fact that Riggs was only five foot eight gave him an advantage in making his journey into the narrow horizontal pipe. Because he knew there would be both water and slimy algae inside it, he had chosen to put on a dry suit, a more advanced, thermally insulated version of the wet suits worn by scuba divers. Unlike the coveralls the other guys were wearing, Riggs's specially fabricated rubber outfit would keep him from getting drenched no matter how much water he encountered.

Based on the agreement that Tap had worked out with Norwesco, Riggs would be getting penetration pay, the small stipend for every foot he crawled into the pipe. But that hardly compensated for the unnerving conditions he would face. He decided to approach the assignment as though he were going for a swim in an ice-cold pool or eating a raw oyster for the first time. No toe dipping, no sipping around the edge—just bold, decisive action. When he crawled in, the array of tasks he needed to accomplish required so much coordination and concentration that it allowed him to avoid focusing on how ridiculously confined he was. He had to bring with him a special over-size hydraulic tool to remove the circular metal fastener—called a circlip—that held the sixty-five-pound giant-salad-bowl-shaped safety plug in place. So Riggs pushed that tool in ahead of him, holding it above his head as he wormed his way along the horizontal pipe. He lay on his belly, using his elbows and knees to push himself forward. When he reached the plug, he first needed to open a valve in the center of the salad bowl to release any water that might have collected behind it. Then he was supposed to loosen a pair of bolts and give a few pumps of the hydraulic tool to relieve the inward pressure of the circlip. But those bolts turned out to be extremely tight, requiring much more effort and ingenuity to pry them open than expected. Once he managed to remove the bolts, he dislodged an O-ring and a gasket, which finally allowed him to pull the plug out. All along, his only communication consisted of the occasional comment shouted at

Billy, who was crouched over just outside the pipe. And his only relief from the complete blackness of his captivity was the narrow beam from the lamp on his miner's helmet.

Then, of course, Riggs needed to crawl backward thirty-five feet out of the thirty-inch-wide pipe, dragging the plug and its components with him. He decided to slide the circlip and gasket down beside his body so he could push them out with his feet. He placed his tool and the O-ring inside the salad bowl dome, holding it above his head as he began crawling out of the horizontal pipe. He had to do all this while being extremely careful not to dislodge his facemask or umbilical so the air that was keeping him alive could continue to flow.

More than an hour after he had first crawled into the pipe, Riggs was once again outside it, crouched over with his hands on his knees, this time with the first safety plug successfully removed. It rested at his feet like the scalp of an enemy from some savage war. Only fifty-four more to go.

It was just past four o'clock. After using the repaired video camera to document their work in riser 1, he and Billy moved on to riser 2. About forty-five minutes later, they succeeded in yanking out the second plug.

Then Mars, who was still standing by the boat outside riser number 4, began having serious problems with his regulator. His mask started free-flowing, meaning too much air from his umbilical was rushing to his mouth and nose, regardless of his demand for it. Although it seemed like the exact opposite of the facemask-sucking problem that Hoss had endured a day earlier, it appeared to spring from the same defect—the flow was not being properly regulated. Mars and Billy conferred. Then Billy came over to where Riggs was standing outside riser 2. "We need to move, Riggs," he said. Rather than drag the two yanked-out safety plugs and the pumps back to the boat, as they had planned, he told Riggs they should leave them right where they were. They would worry about retrieving them tomorrow.

With that, the three men began trudging, boat in tow, back to the Humvees.

It wouldn't start. The gray Hummer wagon was positioned correctly to get the crew back to the shaft, but the goddamn engine wouldn't turn over.

It was six o'clock, and Tim and Billy took the lead in trying to get it going. Tim stepped forward because he'd been the Humvee driver two days in a row, and Billy because he'd been Tim's partner on the Humvee crew the day before and because, well, he was Billy and he could MacGyver anything. As confident as Billy was in his mechanical abilities, and as much as he'd always welcomed the toughest man-versus-machine challenges, even he had to admit how unwelcome this one was. After all, they weren't tinkering in the Black Dog shop behind Tap's house, or topside on Deer Island, or even at the base of the shaft, where there was still plenty of light and decent air. They were standing nine miles into a pitch-black tunnel buried under the ocean. The dive crew had aborted that day's plug-pulling operation because of concerns with their increasingly problematic breathing system, and now they had an even bigger problem on their hands. They did have a second Humvee, the pickup, of course, but it was of no use to them since it was facing in the wrong direction and blocked by the Hummer wagon. With the Jersey barriers coming out from its side walls, the tunnel was so narrow that they could barely open the vehicle doors all the way, never mind try to maneuver the two Hummers around each other to switch their positions.

Billy and Tim quickly concluded that there had to be something wrong with the oxygen-injector control. This was the key component allowing the retrofitted diesel Humvee to use its liquid oxygen supply and operate in the oxygen-depleted tunnel. They tried everything they could think of. Knowing there were already enough mechanics involved, Riggs tried to stay out of the way. But from his perch in the front seat, he watched everything closely.

After some tinkering by Billy, Tim cranked the engine, and it chugged a bit, ejecting some black smoke from the exhaust. But then the engine sputtered out. More tinkering, and the same result. They just couldn't keep it running. Billy knew from his days working as a diesel mechanic for his dad's trucking business that they had to find a way to manipulate the setup into getting more oxygen to the engine. He tried putting an independent power supply on the oxygen controller, but still no luck. He was frustrated that none of his usual go-to moves were working.

Again and again he and Tim would make an adjustment and then crank the engine. After more than an hour of this, their troubles worsened. When Tim went to turn the engine on, it wouldn't even click. Nothing. All that cranking had drained the Humvee's battery, and now it was dead. On top of everything else.

In that moment of exasperation, Billy's instincts kicked in. He headed over to the white Humvee pickup, the one they had driven out in, and popped the hood. He removed the battery from that vehicle and used it to replace the dead one in the Humvee wagon. That worked. But they still had to get the engine running.

More tinkering, more frustration. Finally they tried holding the override switch on the oxygen injector for an unusually long time. Whether it was that, or something else they did, or just sheer luck, the engine burped and, for the first time, didn't conk out. It was 7:23 p.m. The guys rushed to collect their tools that were scattered all around, then climbed into their seats in either the Hummer wagon or the towed pickup, and Tim threw the wagon into drive.

Hoss had spent much of the day sitting beside Harald topside in his trailer, so he'd heard all the calls about the air problems and the Hum-

vee not starting and the battery dying. He was still in pain because of
the problems from Monday's shift, but Tuesday had turned out to be
just as bad. He had felt himself getting more and more upset as the
day wore on. *This is bullshit,* he told himself. By ten o'clock that night,
when the divers had ridden the man cage up the shaft and were shuf-
fling over to Harald's trailer, Hoss was fuming. And so were the rest
of the guys.

They confronted Harald about the recurring problems, which
seemed only to be getting worse. Harald acknowledged there were
still a few kinks that needed to be worked out, but he assured them
things would go better tomorrow. To Hoss, though, the problems had
escalated to the point where they felt less like kinks than like fundamental flaws that were now compromising the divers' safety. And he
had no interest in hearing the same explanations and excuses.

Tap had been sitting in his black trailer with the door open, so he'd
heard enough of the divers' complaints to detect their seriousness.
After the divers cleared out and boarded their school bus, Tap walked
over to Harald's white trailer and asked, "What's going on?"

Harald talked about how they had to adjust a few of the valves to
improve the flow rates, and then everything should be fine.

"Okay," Tap said, tentatively. But as he walked back to his trailer,
he felt his stomach tightening. He knew the guys were upset, and he
wondered if he should raise a ruckus with the people above Harald.
But he didn't feel that he had enough concrete information to take
to the managers at Kiewit or the MWRA and ask for a halt to the
operation. If Harald gave them the same reassuring assessment he'd
just delivered to Tap, they'd probably defer to the distinguished Canadian engineer over him. *I'm dealing with these big, big companies that
can squash me,* Tap told himself. *I need to know what's going on before I
go to the powers that be.* Still, he promised himself that if things didn't
improve tomorrow, he'd step up his complaints.

As Hoss and the other guys rode the school bus off the island and

back to the racetrack parking lot, the tense meeting that had begun in
Harald's trailer erupted into an all-out bitch session. The consensus
was that the breathing system was lousy and the equipment unreliable. Even a normally easygoing guy like Tim, whose wife had made
him promise during their phone call Sunday night not to go back into
the tunnel if he felt it wasn't safe, had grown more concerned.

"This is fucked," Hoss told his wife when he called her from his
HoJo's motel room. Heather was used to her husband being gung-ho
about his work and relatively close-mouthed about any problems he
encountered on site. So when he opened up to her about how badly
the tunnel job was going, she was alarmed.

"Come home," she told him. "Just get out of there!"

Hoss said he couldn't do that. He couldn't abandon the other guys,
who after all were in the same predicament as he was. He also worried about the impact that walking off the job would have on his career at Norwesco, the only real employer he'd ever known as a diver.
Roger had flown him out to Boston twice, counting the fact-finding
mission from a year earlier, and had spent money on his training and
lodging. The owner was counting on him to help Norwesco finish
this big job. Hoss figured that if he walked now, it would be a surefire
way to extinguish his rising-star reputation. After he hung up with
his wife, he called his parents. They had the same reaction, encouraging him to come home. He told them what he had told Heather: he
couldn't let his team down. But he also assured them that if any more
problems arose, he would keep himself out of the tunnel and work as
part of the standby team.

As pissed off as Hoss was, a part of him still wondered if he might
be overreacting. He'd been on plenty of jobs before that saw setbacks
and tension, and somehow things had always worked out in the end.
Maybe that was all this was. But he sure didn't think so.

. . .

As they fanned out for the night, the other divers were feeling similar pangs of doubt. When Riggs talked with his wife about some of his concerns, Karen too grew worried, especially since he was usually so circumspect. But past experience told her that nothing she could say was going to influence his decision. So she tried to offer a few comforting words.

When Billy arrived at his buddy's house late that night and prepared to crash, there was a message from Michelle. He knew how uneasy she was with the job already, so he had no interest in adding to her agitation by detailing all the day's disturbing developments. He didn't have the energy for another lecture. Surprisingly, though, in her message Michelle didn't mention any of her qualms about the job. Instead, she told Billy she had an important question for him about what kind of homeowner's insurance they needed to line up in preparation for the closing on their house, which would take place in two weeks.

After years of scrimping, Billy and Michelle were finally going to get the beautiful old farmhouse they'd been longing for. Sitting on eleven acres, it had a gracious porch and a garage out back where Billy could keep his tools and do his tinkering. They hoped eventually to open a small business together, so in preparation Michelle planned to return to college to study accounting. Because neither of them had ever bought property before, Billy had been soliciting advice from his mother, Olga, who worked as an assistant to a real estate attorney in Florida. Olga had even delayed a planned summer visit to see Billy until after he and Michelle had moved into their new place. In several house-related phone calls, Billy had shared all the major details with his mother except one. He hadn't told her where his biggest source of funds for the house had come from. It had been the cash Tap paid him when he bought out Billy's partnership in Black Dog. Billy worried that his parents, especially his father, would be disappointed in him for giving up on being a small-business owner. He remembered how much his dad had beamed with pride when he'd come to New Hampshire to see Billy and Tap's Black Dog operation. So Billy had mentioned nothing of the sale to his family and had made Michelle promise to keep her lips sealed as well.

Billy didn't know the answer to Michelle's insurance question, but he knew his mother would. Still, knowing Olga was an early-to-bed, early-to-rise person, there was no way he was going to call her as the clock approached midnight. He decided to wait until morning to phone her, then call Michelle after that, once he had the information.

Meanwhile, before DJ made his way home to Waltham, he stopped off to see his girlfriend. Dana had been out for drinks with her sister, so DJ had to wait for a while outside her house before she showed up. Like the other guys, DJ was rattled by what he had experienced in the tunnel, especially given the steady accretion of problems on the job. Also like the other guys, he was looking to talk honestly with the woman he loved about what was on his mind, but without signaling so much concern that she became overly worried and he created a new headache for himself. When Dana finally showed up and he began offering a few details from the day, she didn't seem particularly concerned. Either that, or she failed to grasp the gravity of what he was trying to say. Frustrated, he told her he had to get going on account of his early start the next morning.

When he arrived home at his duplex apartment, his mother was waiting up for him. With her squinted-eye smile and a throaty laugh that testified to her longtime Benson & Hedges habit, Lorraine was an interesting mix: Despite the endearing brogue and almost courtly manner she maintained from her days growing up on a farm in Nova Scotia, adulthood had toughened her into a wary survivor. DJ knew how well his mother could read him, after all they'd been through. It would be no use trying to mask the misery he'd seen in the tunnel that day or the apprehension he felt about tomorrow. He didn't want to get her riled up, though, so he tried to spin his anxiousness in a positive way.

He knew how much strength his mother drew from her Catholic faith. Many a Sunday night he would walk her the few blocks from their house to St. Charles Church for the seven o'clock mass. And he certainly knew how much she'd adored her late father, the strapping carpenter who was the central character in most of Lorraine's favorite childhood stories. While driving home that night, DJ had recalled one particular story his mother had told him more than once. It involved his grandfather's time as a carpenter helping to erect Boston's fifty-two-story Prudential Tower, which when it opened in 1964 was the tallest building in North America outside New York. No matter how high above the ground he was, DJ's grandfather had said, he'd never been afraid because he kept a Virgin Mary medal in his pocket. Lorraine had come into possession of that Miraculous Medal sort of accidentally. She had always loved the smell of pipe smoke, because it reminded her of her dad. After he died, her mother had given Lorraine her dad's tobacco pouch as a memento. And nestled deep inside that aromatic pouch, Lorraine had found the Miraculous Medal.

Now, standing in their living room, DJ asked his mother, "You know Grandpa's medal? Is that still around?"

Lorraine felt a knot in her stomach as soon as the question came out of her son's mouth. With her thin eyebrows arched, she asked him why he wanted it. When DJ replied, "I'm a little concerned about the job," she could tell he was trying to backpedal a bit. He could downplay things all he wanted, but she had already gotten the message.

Lorraine said nothing as she climbed the stairs and fished the medal out of the drawer in a bedroom nightstand. She held it in her hand, gently buffing it with her thumb. It may have been called the Miraculous Medal, but it didn't look like much. Silver in color and oval in shape, it wasn't much longer than a large safety pin. There was a loop on the top of it, so it could be worn on a necklace or chain. In the center was a picture in relief of the Blessed Mother, arms out, halo around her head. Around that image was the message "O Mary, conceived without sin, pray for us who have recourse to thee."

As anxious as DJ's request made her, Lorraine wasn't about to add to her son's burden by forcing him to worry about her fears. She had faith that the Virgin Mary would wrap her protective arms around DJ in the same way she'd always watched over his grandfather. Lorraine descended the stairs and headed back into the living room.

"DJ dear," she said, handing her son the medal, "be careful not to lose this."

10

Wednesday

At five minutes to six on Wednesday morning, Billy made the call.

"Hello?"

"Ma," Billy said, "I'm sorry to call you so early."

"It's no problem, Billy," Olga Juse told her son soothingly, as she glanced over at her alarm clock. "You know I get up at six. My alarm would have gone off in five minutes anyway."

Olga, who was now sitting up on the edge of her bed at her home in Florida, was taken aback by the call. She never wanted to bother Billy when he was on a job, so she always waited for him to get in touch. When he did, it was almost always on a weekend, from home.

Billy quickly got to the point. Michelle had left him a message with a question about what type of homeowner's insurance policy they should get. Normally he would have waited to call, but with the closing date on the house just a couple of weeks away, they needed to get the matter settled. He was about to head into the tunnel, and that would keep him out of cell phone range until late that night, so this early-morning call was the best option. "I need to know what to tell her," he said.

Olga reassured Billy that it really was no trouble. And from her years working in a law office, she was ready with the answer. "Make sure Michelle gets the policy that says 'replacement value,' Billy," she told him.

Billy thanked his mother. Olga sensed an unusual heaviness in his voice. She knew his time was limited, but she didn't want to get off the phone too abruptly. So she was relieved when he brought up the job, expecting that he would perk up and deliver one of his standard upbeat reports. He didn't. "I'm working in this god-awful hole," he said.

Olga could feel the hairs on the back of her neck standing up. She hesitated. *Should she ask him to leave the job?* But before she could weigh the pros and cons of intervention, Billy moved to wrap things up. "Okay, thanks, I've got the information," he said. "I'll call Michelle." Then he hung up.

It was six o'clock now. Billy had to drive from his buddy's house and make it to the racetrack parking lot by six-thirty so he could meet the other divers and board the school bus for Deer Island. He dialed Michelle.

Given the tension hanging over them since Sunday, when Michelle had implored Billy to walk off the Deer Island job, he was bracing for round two. But Michelle restrained herself. She didn't pick up the argument where they'd left off. Nor did she tell Billy about how, driving home after her shift at the pub the night before, she had started bawling, out of the blue. She could predict his response: *We're not going to start this again, are we?* That's why she had chosen to restrict her Tuesday-night voice mail message to the homeowners' insurance question, making it sound a bit more urgent than it actually was. Her main goal was to make sure Billy returned her call. Short of getting him to change his mind, she knew the only thing that would make her feel better would be hearing his voice.

When Michelle picked up the phone, Billy relayed his mother's advice on insurance. Then they talked about a couple of other house matters for a minute. With some of the tension lifted, Michelle decided to risk pressing the issue once more. She asked him to reconsider going into the tunnel again.

Billy had no interest in reigniting their standoff, so he tried to laugh it off. "I've got the easy job today," he told Michelle. "I'm just going to be sitting in the Humvee."

Michelle didn't quarrel. Billy said he had to get going.

"Okay, I love you," she told him.

"Love you, too."

Riggs walked over to the lunch truck and bought a ham and cheese sandwich and a twenty-ounce bottle of Mountain Dew. He shoved both of them into his backpack and ambled over to the school bus that was idling in the parking lot. He kept his work boots unlaced because he knew he wouldn't be in them for long.

Just before seven in the morning, the bus dropped them off on Deer Island, and Riggs walked over to the Black Dog trailer, along with Hoss, DJ, Tim, and Billy. Riggs grabbed the dry suit he had worn the day before and began getting his gear ready. He noticed some slop on the edge of his facemask, extra material that he needed to trim so the mask would fit better. He borrowed a pair of scissors from Hoss, made the cut, and then wiped down the mask with alcohol.

As he began mentally preparing himself for the day ahead, he mulled over an idea that he felt could go a long way toward improving their breathing system. So he walked into Harald's trailer to float it by him.

The Humvee team had been following Harald's protocol for measuring the oxygen content in the breathing air that was being blended by the MAP Mix 9000 and then shot out to all five divers through their umbilicals. The diver sitting in the back seat next to the mixer would open the sampling-port valve to release some air, and then hold the Dräger Multiwarn analyzer in front of it to measure its oxygen level. DJ had handled that task yesterday, and during the bitch session following the shift, that protocol had come in for lots of complaints. After all, how accurate could it be to stick a probe in front of a whoosh of air that was inevitably mixing with the ambient air inside the Humvee?

Only after Riggs had been lying in his motel bed did a simple solution pop into his head. They needed an in-line analyzer. This relatively inexpensive gauge could be plumbed into the system. Rather than having to remove a sample of the breathing supply to measure it, they could plug this monitor right into the airflow, providing a continuous reading of the oxygen content in the breathing supply, like the temperature-control device in a refrigerator. The readings would be much easier for the guys to obtain—and trust. The idea made so much sense to Riggs that he was surprised he hadn't thought of it earlier. Then again, the equipment, environment, and procedures associated with this Deer Island job were so foreign to the divers that it was taking a while for everything to settle in.

"Hey, Harald," Riggs said. "Do we have any in-line analyzers back in Spokane, like you see on a gas rack or a chamber?"

Harald looked up. "Yeah. Why?"

"Could we get one of those hot-shotted out here?" Riggs asked. "I think it would be good to have an in-line reading. It's a lot more accurate, don't you think?"

"Well," Harald began. But then he got distracted by something else and walked away.

Riggs figured he'd press the issue again when he had a chance. But now he had to get ready. He slipped out of his boots and his black

Carhartts and put on his dry suit. Because it would be a while before he'd have to worry about getting wet, he kept the suit folded down at the waist, like the ski pro he'd been back in Colorado, walking around the lodge with his suspenders fashionably hanging down.

"Hey, Riggs!" yelled Hoss, who was standing over by the man cage, waiting to be lowered down the shaft. A little earlier Riggs had heard Mars asking Hoss if he'd take his place for the day. Mars had explained that he was feeling worn out from his back-to-back days working at the end of the tunnel. Hoss had agreed to the change, and now he was hollering for Riggs. "You ready?"

"Yup," Riggs called back, then strode over.

Riggs found Tap standing right outside the man cage, looking at his watch. Then Tap yelled back over to the trailer. "C'mon, DJ!"

Tap remained steamed even after DJ made it over to the man cage, lightening up only when DJ flipped over his hard hat to show the Virgin Mary medal dangling on a piece of twine.

At the base of the shaft, the three divers found Billy and Tim already there, flying through the premission checklists. With his hair pulled back into a short ponytail, and his gray sweatshirt and red sweatpants peeking through his green coveralls, Tim stood clutching a clipboard. He smiled as he checked off each item. A little while later, Harald came down the shaft and placed an additional handheld gas monitor in the gray Humvee. Even though he didn't have the in-line analyzer Riggs had asked for, Harald told the divers that at least they'd have this extra Multiwarn monitor to devote exclusively to measuring the blended air coming from the mixer. They could use the Humvee's other handheld to measure the content of oxygen and other gases in the ambient air.

Billy pulled DJ aside. He told him his back was hurting and asked if they could switch assignments for the day. Billy knew it would be a lot easier on his back if all he had to do was sit in the Humvee and monitor the breathing system. Although he had cleared the change with Tap the night before, he wasn't sure if DJ had been notified. When Billy broached the topic now, DJ acted as though it was news to him, calling Billy a pussy.

Billy was in no mood for it, but then he saw a smile forming at the corner of DJ's mouth. That told him Tap had already passed on the word. DJ, being DJ, was just trying to get some mileage out of it. So Billy played along.

"All right, I'll go," DJ told him. "But when I bring those plugs back, you're loading them in the Hummer."

Billy laughed.

The amended assignments were settled. Billy and Tim would remain in the Humvee, while Hoss, Riggs, and DJ journeyed to the end to pull plugs. Billy would serve as foreman of the Humvee team, while Hoss did the same for the excursion crew.

With their head start in getting to the base of the shaft, Billy and Tim had already managed to complete the bulk of the morning prep work. Despite all the setbacks the divers had suffered during the first few days in the tunnel, there was now a real rhythm to their work. They were more familiar with the system, and they also had made a few important refinements. The small team of remaining sandhogs had done a great job of improving the angle and length of the ramp off the aluminum deck, solving a problem that had given the divers considerable grief during their early attempts to get the Humvees into the tunnel. And the divers themselves had engineered other work-arounds in hopes of avoiding a repeat of the earlier troubles. More than anything, Billy was determined not to see the batteries die again in the Hummers. After discussing it with Harald, he disabled the interior dome lights in the Humvee wagon. While those lights had provided some relief from the dark for the two divers monitoring the breathing system, they decided that the dim light wasn't worth the added risk.

Billy climbed into the driver's seat of the lead Hummer, the white pickup, with Hoss riding shotgun, while Tim got behind the wheel of the towed gray wagon, with Riggs in the passenger's seat and DJ in the back. The divers' tweaks to the system and their improved muscle memory combined to give them a start to the Wednesday mission that was by far their best. While it had taken them until after 9:30 a.m. on both Monday and Tuesday to get the Humvees down the ramp, they now found themselves cruising into the tunnel at 8:20 a.m.

Before he settled in the back seat, DJ slipped a CD into the dashboard player. As they began their slow, roughly two-hour ride into the darkness, the Rolling Stones played "Sympathy for the Devil." DJ quietly began singing along: *I was round when Jesus Christ / had his moment of doubt and pain.*

After an hour of driving, Billy stopped the lead Humvee at ring 7500, just before the venturi, and all the guys climbed out. Before putting his dry suit on fully, Riggs walked far enough from the vehicles to secure himself some measure of privacy. Then he squatted near the tunnel wall and took a dump, letting his excrement drop into the ankle-deep water sitting stagnant in the concave bottom of the tunnel. His bowel movement offered a raw preview of the hundreds of millions of gallons of treated sewage that would soon begin coursing through the tunnel every day—that is, provided the divers were able to do their job and yank out all the plugs.

Riggs pulled up his suit and returned to the Humvee. He checked to make sure all his seals were tight and used some electric tape to shore up the mine lamp's attachment to his hard hat. After taking a final swig of his Mountain Dew, he put on his facemask. The other guys did the same. They were all now breathing the mix coming from the liquid oxygen and liquid nitrogen tanks and through the MAP Mix 9000. They climbed back into the Hummers, and Billy shifted the lead vehicle into drive.

Riggs said very little on this leg of the ride, using the time to concentrate his mind on the tasks ahead of him. Once again he would have the toughest job of the day, crawling into the off-take pipes to pull the plugs. He hoped to end this shift with a lot more than the two plugs they had managed to remove the day before. Tim turned to him and said, "You're the man in the barrel." Riggs told himself, *I can do this as well as anyone, if not better.* All he had to do was keep his focus, and keep that demon locked in the closet.

At ten-thirty they arrived at riser number 12, the Humvee parking spot near the nine-and-a-quarter-mile mark. The divers set up the yellow console of the mine phone inside the Humvee wagon, and Tim phoned in an update to Harald. The guys drove the gray Hummer wagon off the two-way trailer, positioning it for the westward drive back to the shaft. They then unhitched the trailer from the white Humvee pickup and attached it to the gray one. They pulled the flat-bottomed boat off the trailer. Then they slid the rack of liquid gas tanks to the opposite end of the trailer, closest to the gray Hummer. That way, at the end of their shift, they could back the white Hummer up the ramp at the other end of the trailer for the drive back to the shaft. They loaded up the boat with the hundreds of pounds of gear they needed for the excursion. In addition to the emergency bottle of HP air, the sandbags, and the water pumps, they had packed several hand tools, including a battery-operated Sawzall reciprocating saw. And, of course, there was the manifold that connected their umbilicals to the main breathing supply in the Humvee. A little after eleven o'clock, with all their setup work done, Billy joined Tim in the Hummer wagon while Hoss, Riggs, and DJ began trudging east.

Around eleven-thirty the excursion crew arrived at riser number 4. Riggs knew the pulled plugs from numbers 1 and 2 were still near the very end of the tunnel, so he walked over to them with DJ following behind. They lugged around lots of heavy gear, then each grabbed a plug and dragged it 125 feet. They set up sandbags outside the offtake pipe leading to riser number 3, so it could be pumped out. At that point, they resumed the process of dragging the pair of sixty-five-pound plugs all the way back to the boat, which was parked around number 4, where the tunnel's diameter was slightly less cramped.

Meanwhile, Hoss had cleared enough room in the boat to hold the plugs, since they'd eventually need to transport them back to the Humvees. He then started sandbagging outside number 4, so it would be ready to be pumped out after Riggs and DJ finished number 3.

As Riggs lifted one of the pulled plugs into the boat, he saw Hoss staring at the gauge on the manifold. He looked troubled. "The pressure's down," Hoss said. "It's down to 20 PSI." He gestured for Riggs and DJ to take a breather. Then he used his comm wire to call Tim and ask what was going on.

Tim said he and Billy hadn't done anything to change the pressure on their end, and their gauge looked fine, reading 120 PSI. Still, he said he'd confer with Harald.

A minute later, at 11:45, Tim called back, relaying Harald's instructions: Hoss, Riggs, and DJ should avoid all working hard at the same time, so as not to overbreathe the system. Instead, they should take turns.

Tap Taylor had spent the morning rushing around, making sure the guys had what they needed and were where they needed to be. But that instantly changed as soon as those Humvees started their drive into the tunnel. That's when the frenzy slowed to a crawl as he sat in his topside trailer. All of a sudden, like a snowed-in traveler staring at a screen full of cancellations at O'Hare, there was nothing for him to do except wait.

The entire Deer Island job had been a humbling one for Tap. He was used to being the guy in charge on his work sites, but on this one he felt like little more than Harald's lackey. He was still smarting over an incident in his backyard during mobilization. When he had posed a question to Harald about the system, the Canadian engineer had embarrassed him in front of the other divers, reminding him in no uncertain terms that his job was to provide labor and logistical support—nothing else. Even the equipment Tap had provided for the job now had Norwesco stickers covering the Black Dog logos. Once the divers were down under, Tap felt even more marginalized. Their sole communication with topside went directly to Harald, who sat at a table in his trailer, next to his mine phone, radar monitor, and laptop computer. From his own trailer ten feet away, Tap could overhear some of Harald's communications. But at best he could make out just one end of the conversation, and Harald wasn't particularly forthcoming when pressed for details.

Still, Tap had made a habit of popping his head into Harald's trailer periodically, asking, "How's everything going?" Invariably, the answer had been, "Fine."

A little before noon, Tap stuck his head in again. Tim had just phoned in an update to Harald, reporting that Riggs and DJ had dragged plugs 1 and 2 down to the boat, and Riggs was not far off from crawling into the pipe for riser number 3. Satisfied that everything appeared to be on schedule, Tap walked away from the trailers and headed to a rocky area of the island overlooking the water. This radiant July morning was more than just an important day on the plug-pulling calendar. It also happened to be Tap's thirty-sixth birthday.

If the morning's best out-of-the-gate performance carried through the day, Tap was hoping he might be able to get home to New Hampshire in time to share a late supper and a birthday cake with his wife and kids. His daughter was seven now, and his son was four. Tap was so focused on building his business into something bigger that he couldn't conceive of slowing down, even if that meant family time largely consisted of his kids wandering into the Black Dog shop in their backyard and watching their dad work. Tap knew Billy had a point about finding balance, but in his own life, that would have to wait.

In addition to being his birthday, this Wednesday offered another bright spot for Tap. A Boston TV station had recruited him to offer his expert advice in an on-air phone interview. Not only would the interview break up the tedium of his long day of waiting, but it provided a welcome ego boost, reminding him—and everyone else around— that despite all the indignities of this job, Tap Taylor was much more than somebody's gofer.

The TV station's interest in Tap actually had nothing to do with his involvement in the high-stakes operation to remove the tunnel plugs. As central as the divers' mission was to the rescue of the massive harbor cleanup project, they were doing their work completely off the media's radar. Instead, the anchorperson would be interviewing Tap about the news story that had been dominating the national airwaves for five days: the plane crash involving the son of Camelot, John F. Kennedy, Jr.

The night before, capping several days of searching, a remote-operated vehicle from the navy's USS *Grasp* had located the fuselage from Kennedy's plane, 110 feet below the surface of the water off the coast of Martha's Vineyard. That morning, as the Norwesco and Black Dog divers were at work in the Deer Island tunnel, a team of their diving counterparts from the U.S. Navy and the state police began searching the plane-wreckage area, looking for the bodies of JFK Jr. and his wife and sister-in-law.

Shortly before the noon newscast began, the dive team located the bodies. Senator Ted Kennedy, the family patriarch, boarded the USS *Grasp* to be briefed on recovery operations. Then, during Tap's live interview, he was asked for his insights on what the divers might have faced in trying to locate the bodies and how they would go about recovering them. Sitting on the rocks, looking out at the harbor as the warm sun beat down on him, Tap fielded the questions with ease.

At quarter past noon, with the water pumped out of the pipe to riser number 3, Riggs began his crawl in. He followed the same steps he had the day before in getting himself and all his tools into the slimy, pitch-black, constricted horizontal pipe. Once he removed the bolts from the safety plug, he opened its release valve. Then he yelled for DJ to pump up the hose leading to his hydraulic tool, which he used to release the circlip. After he dislodged the oversize-salad-bowl plug, he began backing out of the pipe, dragging the plug with him as he pushed the circlip out with his feet. The whole process to remove plug number 3 took about twenty minutes—considerably faster than yesterday—giving Riggs confidence they would be able to get a lot more done during this shift.

Once Riggs was out of the pipe, crouched over, DJ told him to take a break. As much as he was determined to keep the pace of the day's mission brisk, Riggs remembered Harald's most recent instructions, for the guys to take turns doing their heavy work. So he took a breather, watching as DJ dragged plug number 3 over to Hoss and the boat 125 feet away. After a couple of minutes, Riggs headed toward them with his umbilical dragging behind him. He joined Hoss at the boat, which now was almost completely full, given the three giant plugs that were sharing space inside it with all the other gear. Along the walk, Riggs passed DJ, going in the other direction, video camera in hand.

With DJ heading into the pipe to document the removal of plug number 3, Riggs headed toward number 4 with the hydraulic tool to begin pumping out the water left in the pipe.

"Hey, Riggs," Hoss said. "We should wait till DJ videos number 3 before we pull number 4."

Riggs nodded. "I'll just go in and get started on the bolts."

As Hoss called in a status update to Tim around 12:40 p.m., Riggs began making his way into pipe number 4. When he reached that plug, he realized there was something different about it. Unlike the other salad bowls, this one had not one but two ball valves at the center of it. And that second valve blocked off access to the circlip. Riggs tried moving it, but it was rusted in place. There was no way he was going to be able to get that plug out.

Then a work-around dawned on him. He crawled out of the pipe and headed straight for the boat. There he grabbed the Sawzall. If there would be no getting around the valve, he would go through it.

The valve would not go down without a fight. After Riggs had cut halfway into it, he noticed that he had burned out the blade of the saw. All its teeth were gone. So he maneuvered his way out of the pipe yet again. With Hoss standing beside him, Riggs leaned into the boat to search for a new blade. Meanwhile, DJ had finished videotaping and was now walking toward them.

With the blade in hand, Riggs began to head back into the pipe, but Hoss called him off. Riggs and DJ had spent the last couple of hours walking back and forth, back and forth, inside the cramped final section of the tunnel, working strenuously as they lugged around plugs and pumps and sandbags. In the process, they had turned their personal three-hundred-foot-long umbilicals into six hundred feet of tangled mess, like a big batch of angel hair pasta left on the stove overnight. "Let's take a few minutes and straighten these hoses out," Hoss said.

Riggs agreed. Even though Hoss was more than a decade his junior, and even though he was serving as foreman only because he was filling in for Mars, Riggs knew the guy was smart and capable.

Besides, as much as Riggs wanted to keep the momentum going, Hoss's suggestion to take some time and straighten things out made a lot of sense.

To disentangle their hoses, DJ walked back into the five-foot-diameter section, heading toward the very end of the tunnel. At the same time, Riggs walked several paces behind him, coiling up the hose that trailed behind DJ like some sort of oversize kite string. After several minutes of this, Riggs found himself feeling weird. *This is a simple task,* he told himself, *and I'm confused.* He looked to the tunnel wall, checking for spots. If he saw them, he'd know he was about to black out. But he didn't see any spots. So he kept walking, kept coiling. Then, without any warning, DJ just stopped walking. Riggs caught up to him so quickly that he nearly bumped into him, as if they were in a Keystone Kops skit. Suddenly Riggs could sense his hearing start to change, as if somebody had placed cones over his ears. His vision grew blurry. He wondered: *Is this what people mean when they talk about tunnel vision?*

He looked over to see DJ helplessly slide down onto the tunnel floor. As he heard Hoss yell, Riggs felt powerless to stop from falling down himself. He landed on one knee, settling on the tunnel floor across from DJ. "Hoss," Riggs called out, not quite sure where he was or if anyone was hearing him. "What's going on with the gas?"

Hoss had never been in such a strange situation in his life. Here he was, nearly ten miles into a dead-end tunnel under the sea, and he'd just seen the two other guys on his crew collapse. And yet in some ways, he'd never felt so good in his life. What he was seeing was danger and doom in the darkness around him. What he was feeling was freedom and lightness, as if he were being transported to a beautiful place. He had moved past the frustration stage he had felt two days earlier, when his facemask had sucked to his face, and he seemed to be moving briskly through fogginess, headed for euphoria. Somewhere at the base of this purple haze, he sensed that he had to fight it.

"I need to call Tim," he muttered.

Hoss reached Tim on the comm wire and told him he needed to know what the oxygen level was in their breathing supply. Tim assured him that he and Billy would check. A few seconds later, Tim called back. "Shit, it's 8.9!" he shouted.

"8.9!" Hoss repeated, alarmed even though his comforting narcoticlike haze had yet to lift.

When the line went dead, Hoss lunged toward the manifold in the boat and flicked the lever. That switched off the mixed gas that was flowing into their umbilicals and facemasks from the MAP Mix 9000 back in the Humvee, replacing it with air coming directly from the emergency HP cylinder they'd brought with them on the boat. The source of air for the breathing system supplying all five divers could also be changed, from the mixed-gas supply to the HP bottles on the roof of the Humvee. But making that change involved turning a couple of valves back at the Hummer. Hoss had changed the source solely for his three-man team.

He looked over to see DJ and Riggs, who were both sitting on the tunnel floor, begin to come around. They were obviously still disoriented, as though startled to wake up from an unexpected flash of sleep while behind the wheel of a car.

Hoss was still foggy himself, though that narcotic feeling quickly wore off. He felt lucky that he'd had enough presence of mind to lunge for that lever when he did. At the precise moment when the divers' lives depended on their ability to make fast, smart decisions, Hoss had every reason to believe they had all been suffering some form of oxygen deprivation from breathing that 8.9 percent air.

Even though humans cannot function without oxygen, we basically can't conserve it either. Brain cells, which have a high metabolic rate, are always hungry for oxygen and will generally consume whatever's around. As stress hormones kick in during high-pressure situations, the oxygen demand spikes. When the supply of oxygen to neurons starts to dwindle, human brains can't function like ship-wreck victims rationing their few remaining bottles of water and cans of tuna fish. The brain cells continue to devour oxygen until the supply begins to dry up. When that happens, the synapses start to malfunction. As the cells' oxygen supply gets exhausted, the execu-tive functions governed by the brain's prefrontal area weaken, mak-ing it more difficult to sense the consequences of dangerous actions. After enough oxygen deprivation, those brain cells simply die. With-out oxygen, the human brain resembles a lit candle with a glass placed over it. The flame will dim and flicker before it finally fades out for good.

Hoss tried calling Tim but couldn't get through. He figured Tim and Billy had stepped outside the Humvee to turn the valve that would switch everyone to the main backup HP supply. He spotted some water on the microphone attached to the comm wire and won-dered if that had somehow fried the system, making it impossible for Tim to reach him.

Still, Hoss was taking no chances. He decided the excursion crew should pack up and head back to the Humvee. Problem was, they needed to take the boat with them, because their emergency air and manifold were inside it. And Hoss had done such a good job organiz-ing all the gear in the boat that its total weight approached six hun-dred pounds—entirely too much for the guys to have to drag back with them when time was of the essence. Just one of those sixty-five-pound salad bowls took up nearly a third of the surface area in the well of the narrow boat, so Hoss had creatively jammed more pliable gear into the spaces around and on top of the plugs. Looking at the packed vessel now, Riggs told Hoss, "If we're bailing out, we need to get this stuff *out* of the boat."

"You're right," Hoss said.

. . .

As Hoss unloaded, Riggs resumed the task he had started before he'd blacked out, coiling up his and DJ's tangled hoses so they wouldn't trip over each other. He was relieved to see how much more success he was having now that his head was clear. He sent DJ back into the five-foot-diameter final section of the tunnel, coiling hose behind him. Once DJ's umbilical was straightened out, Riggs focused on his own, as DJ shifted his attention to helping Hoss unload the boat.

Everything was going well. Then, out of nowhere, Riggs's face-mask started free-flowing, leaving him unable to slow the hissing air that was rushing into his mouth. He knew this was a surefire way to burn through the one HP tank that all three divers were now relying on to stay alive. Riggs tried adjusting his regulator, to no avail. He tried kinking the hose in the hopes that it would reset the system and stabilize the flow. Still no luck.

In desperation, he decided his only option was to kink off his umbilical and hold it, releasing it just a little bit every time he needed to inhale, and then kinking it back up every time he exhaled. Handling this task demanded almost all of his attention, so he couldn't help with the boat. Instead, he stayed about ten feet behind Hoss and DJ as they began pushing the boat back toward the Humvees. In addition to handling his own airflow manually, Riggs focused on keeping everyone's umbilicals from getting tangled up.

They had a lot of ground to cover, about eleven hundred feet of eerie, misty darkness. And they needed to make sure the boat they were pushing didn't roll along the tunnel's Jersey barriers and flip over. They realized they could move only so quickly.

As they got about halfway, Hoss, who had been keeping a nervous eye on the HP bottle's pressure gauge, told everyone to stop. "Don't panic," he told Riggs and DJ. "Let's take our time so we don't over-breathe the system."

After a nearly twenty-minute trek, they had made it to within 150 feet of the Humvees. Riggs whistled as loud as he could. No response. Hoss was the first to spot the white Humvee pickup in the distance. He had expected to find a hive of activity, with Billy and Tim standing beside the vehicles, shining their miner's lamps, waiting to chew out the other guys for somehow having screwed up the communications system. Instead, there appeared to be no movement at all. Riggs whistled again.

Suddenly DJ yelled out, "Man down!" and sprinted ahead, his three-hundred-foot umbilical rapidly unspooling behind him, like a fire hose at the scene of a blaze.

DJ raced up to the white Humvee pickup and kept going, hustling even though he felt petrified by the stillness of the scene. He maneuvered himself along the tight space between the first vehicle and the Jersey-barrier sidewall to get past the trailer and finally make it to the second Humvee, the gray wagon, which was pointed west, back toward the shaft. There, on the tunnel floor next to the passenger's door, he found Billy, in his tan coveralls, with his facemask on. His legs were underneath the Hummer, as though he had stepped out of the vehicle, turned back to face it, and then accidentally slid right under it. His torso was at a forty-five-degree angle, pinned against the curved bottom of the Jersey barrier. The light on his hardhat was shining toward the tunnel ceiling.

As DJ knelt down beside Billy, he could see that his friend's eyes were open. But there was nothing in them. He tried shaking Billy to get a reaction, but there was no response. He needed to tilt Billy's head back to open his airway, but the space between the Humvee and the sidewall was too tight to give him enough leverage. He grabbed Billy by the waist and slid his legs farther under the Hummer until he was lying flat. Then he put his arm under Billy's neck and shifted his head back.

DJ wanted to rip off his facemask and start giving Billy mouth-to-mouth. But he knew from his training that if he did that in the oxygen-starved tunnel, after a few breaths he'd be lying unconscious next to his friend. That would do neither of them any good. It was like what the flight attendants always said during their safety announcements before takeoff—reminding passengers to affix their own oxygen mask before helping anyone else with theirs. Logically, it made perfect sense, but in the moment of crisis, was a father really going to feel morally right if he took time to put on his own mask while his terrified four-year-old daughter stared at him, waiting for help?

With mouth-to-mouth out of the question, DJ turned to the next-best thing. There was a purge button on the side of Billy's facemask. When DJ pressed it, it forced a rush of air into Billy's mouth and nose—sort of like mouth-to-mouth by remote control. Then he turned to the jaw joints on the side of Billy's mask. Jamming his fingers into these joints gave DJ the ability to pry Billy's mouth open. With that, he started doing chest compressions to try to force air into Billy's lungs. When none of this produced a response, DJ furiously repeated the steps, desperate for a reaction.

Hoss, who'd been attending to Tim in the Humvee, approached DJ and took over for him. He put his finger up to Billy's neck to check for a pulse but couldn't find one. He purged Billy's mask yet again.

At the same time, DJ climbed into the Humvee to get to the mine phone and call in the emergency. He found Tim sitting behind the wheel, his neck turned toward the breathing-system controls behind him, his head slumped over. Although DJ couldn't be sure why Billy and Tim had gone down, he instantly knew, based on their positions, what they had been doing at the moment it happened. They must have been trying to switch the source of air that supplied the main system for all the divers, from mixed gas to HP air. Making that change required turning a valve near the roof rack of the Humvee and another on the manifold inside the cabin. After Hoss's distress call had come through, Billy and Tim must have decided to do whatever it took to save the lives of their fellow divers eleven hundred feet away. Billy must have stepped outside the vehicle to turn on the HP air while Tim, with his right hand cradling the quarter-turn valve on the manifold, waited for the sign from him to make the switch.

If this scenario had become clear in DJ's mind, his head throbbed with countless other questions that seemed unanswerable. If he and Riggs and Hoss had been relying on the same system for breathing air as Billy and Tim, why were they still alert while the Humvee divers were lying as lifeless as mannequins? Why would the guys who had what Billy called the "easy job"—sitting in the Humvee—be the ones knocked down, rather than the guys doing the exhausting work in the scariest stretch of the tunnel? And if Billy and Tim had been in real trouble, why hadn't they blacked out earlier, when DJ and Riggs had gone down and Hoss had called to check in?

DJ picked up the receiver to the yellow mine phone and shouted to Harald, "We have two men down! Have medical assistance on standby." Then he hung up, having no time to wait for a response. He glanced down at his watch. It was 1:45 p.m. He approached the roof rack, which held four cylinders of HP air. DJ, Riggs, and Hoss had been expecting to turn to that backup supply before their single—and fast-dwindling—bottle of HP on the boat ran out. DJ noticed that the regulator to the first HP tank on the roof was open, so he moved to open the second one. All four tanks on the roof rack were connected as part of the HP backup system. But when he opened the second tank, it emitted a loud cascading sound. As soon as Hoss heard that noise, he yelled, "Don't plug into that!" The rapid sizzle indicated that the air was being equalized between the first and second tank, suggesting that the first bottle hadn't just been open but was now empty.

This was a devastating discovery. There should have been four full bottles on the roof—Hoss had checked them himself before the day's mission began. The fact that one HP bottle was already spent meant that the divers now could not be sure which system—HP or mixed gas—was responsible for the bad air that had flowed to their face-masks. The sequence of events was unclear.

Given all the earlier problems with the exotic mixed-gas breathing system, they naturally assumed that it had been responsible for knocking down Tim and Billy. After all, that had been the breathing supply they'd all been relying on when DJ and Riggs went down. But since they found Billy and Tim in the process of activating the HP backup supply, and since one of those HP bottles was already empty, what if, in fact, the real disaster had struck only *after* the Humvee crew had switched the main system from mixed gas to HP? The backup air couldn't be trusted, and therefore it couldn't be used. DJ thought to himself, *We're screwed.*

While Hoss went back over to work on Tim, DJ returned to Billy, trying chest compressions again. He thought back to how, when Tap had asked him to change places with Billy earlier that morning, he'd said he was worried about Billy, not just because of his aching back but also because of his fatigue, and he'd asked DJ to look after him. On top of everything else, DJ now felt a surge of guilt. Not only had he failed to look after Billy, but he realized that if they hadn't switched positions at the last minute, it would have been his limp body lying there. Although it was clear that the chest compressions and facemask purging were having no effect, DJ repeated the steps again and again, not wanting to stop and not knowing what else to do. Finally he realized he had to let go of his friend if he and the other guys had any hope of saving their own lives. He looked straight into those lifeless eyes and whispered, "Sorry, Billy."

Distraught, DJ looked up at Riggs, who was behind him, pacing back and forth. Because Riggs was still forced to kink and unkink his hose for every breath, he was severely limited in how much he could help. DJ saw Riggs looking at the bodies and then looking down the long black tunnel to the shaft. He thought to himself: *Riggs wants to leave now. He wants to run!* He could understand that basic human instinct for self-preservation, because it was taking all he had not to give in to it himself. But he also knew that running would get them nowhere.

That's when it dawned on DJ: they were so far from civilization that they might as well be on the moon. Two divers were already down, and they could very well be next. And nobody was coming in to rescue them.

As comfortable as he had become with Hoss and even Riggs, they were still basically strangers to him. The only guy on the team he knew well was now lying lifeless in his arms. DJ thought back to the mine rescue training they'd had in New Hampshire, when his buddy Ron Kozlowski had pulled him aside before walking off the job. DJ put a lot of stock in what Kozlowski had to say, considering that he'd been a diver for so long and had seen combat. Is this what Kozlowski had been talking about, when he warned DJ about how men can turn on each other in life-or-death situations? DJ had no boogie board, no .45. He thought to himself: *Are these guys going to stick with me?* If they didn't, he knew they had no hope of making it out alive.

11

Trapped in Black

Riggs knew what it must have looked like to DJ, but he actually wasn't looking to run. He just didn't want them to delay confronting reality so long that it left them all dead. "We need to go!" he shouted through his facemask.

Billy was lying lifeless, pinned between the Humvee and the tunnel sidewall. Tim was slumped over behind the wheel. It had been more than half an hour since Hoss's last communication with Tim, when he had frantically reported that the oxygen content in their breathing system was a consciousness-claiming 8.9 percent. Riggs assumed Tim and Billy had been felled by the blended gas from the MAP Mix 9000, the same air that had apparently caused DJ to black out and Riggs to go down after him. He and DJ had survived, thanks only to Hoss's quick instincts. But he knew there was little cause for celebration. Their prospects for making it out of the tunnel alive were dwindling fast.

While DJ was working on Billy, and Hoss on Tim, Riggs had gone over to the trailer that held the main gas supply. He was startled to find the cylinder of liquid oxygen now as frozen and glistening as a

Sno-Cone, with a thick layer of ice choking the regulator and coating most of the bottle. To him, it looked less like an oxygen tank than an ice sculpture of one. Though he couldn't fully explain all that ice, it struck him as pretty powerful evidence that the mixed-gas system was the source of their problems. But because they couldn't rule out the possibility that the bad air had come from the empty HP bottle strapped to the Humvee roof, they couldn't trust any of that emergency supply either. The only air they could trust—the HP cylinder they had dragged back with them on the boat—was about to run out, hastened by Riggs's free-flowing mask.

Riggs looked over at DJ, who was doing everything he could to try to revive Billy, short of the suicidal move of attempting mouth to mouth. He looked inside the Humvee to see Hoss trying to revive Tim. As much as he didn't want to believe it, Riggs was sure that Billy and Tim were already dead—probably had been since right after Hoss had lost communication with them. He was just as sure that if he and Hoss and DJ didn't get in the Humvee and get the hell out of the tunnel right now, it wouldn't be long before they were dead, too. If they made it out alive, he figured, a rescue crew could come in and get Billy and Tim's bodies.

Riggs was a veteran diver, a member of the brotherhood. It went against every fiber in him to leave a brother behind. But in this frenzied trauma, his mind flashed back to the mine rescue training that he and the other divers had sat through six weeks earlier. Back then, the instructor had drilled the divers on the protocol for handling a man down: Don't try to resuscitate him, especially if he's been down for a while. Focus on safely evacuating yourself, and leave the rescue effort to others. Otherwise you're bound to add to the body count.

Mine rescue trainers work hard to try to break down the powerful "band of brothers" instincts that miners and people in similar high-risk professions tend to share. As noble as those instincts seem, they simply cannot hold up to the lessons drawn from more than a century of mine accidents. Government incident reports catalogued by the U.S. Mine Rescue Association show with depressing repetition what usually happens in accidents when workers focus on trying to rescue their comrades rather than getting themselves to safety. They die.

Take, for example, the Woodbine coal mine in southeastern Kentucky. In the summer of 1985, a four-man team of young miners spent the afternoon hauling coal, operating three "scoop" vehicles. Two of the miners, Robert and Ricky Bauer, were brothers. As the miner in the lead scoop penetrated an abandoned shaft, he was suddenly overcome. The three miners behind him rushed to his aid, working to revive him, only to lose consciousness themselves. Somehow, though, twenty-one-year-old Robert Bauer managed to regain consciousness. He dragged his twenty-two-year-old brother to an area of the mine where the air was better, then rushed back to try to rescue the others.

After a long period of radio silence, a separate rescue team headed into the mine. There they found an unconscious Robert Bauer cradling one of the other miners in his arms. All four miners were rushed to the surface, but only Ricky Bauer survived. An investigation determined that after the miners had penetrated the abandoned shaft, that shaft had acted like a vacuum, sucking the good air out of that part of the mine, leaving the men with the oxygen-starved air known in the trade as blackdamp.

Zeroing in on the instincts that take over during a crisis, one mine rescue trainer would frame his scared-straight talk this way: "Are you going to stay there or are you going to get yourself out? If you stay there, you are committing suicide. There is nothing noble about suicide. It's immoral."

· · ·

Again, Riggs called over to DJ and Hoss. "We've got to leave them!"

Still, Hoss was the foreman of the crew. Despite his youth and occasional brashness, Riggs respected him, as well as the preternatural air of authority that seemed to match his imposing height. After suspending his work on Tim so he could quickly take everything in, Hoss let out his verdict. "We're *not* leaving these guys."

Hoss's decision was so firm, so calm, so *right,* that it had a clarifying effect on Riggs. Despite what his brain told him, his heart agreed with Hoss. Instantly, he felt good about it.

Hoss turned to Riggs. "Get those rebreathers ready."

"I'm on it," Riggs replied, hustling to the rear of the Humvee.

Although the main and primary backup air supplies were now useless to them, the divers did have several other emergency devices. There were the bailout bottles that they each wore on their hips, which offered ten minutes of air. There were one-hour Ocenco self-rescuers that had been the sandhogs' emergency supply, which Kiewit had loaned the divers for this job. And finally there were the Dräger rebreathers, the ones Hoss was referring to, which were designed for mine rescue operations and now offered the divers their best shot at survival. On paper, these rebreathers promised each diver a total of four hours of air. But Riggs remembered what the instructors had said back during their training, about how people tend to burn through their air supply much faster during crisis situations. So he knew that he and the other guys should probably take each of those totals and divide them in half. That didn't leave them much wiggle room, since just the drive back to the shaft would normally take about two hours.

When Riggs opened up the Hummer's back doors to grab the rebreathers, he heard a constant, high-pitched beeping coming from several different and dissonant devices. These devices included the second Dräger Multiwarn gas analyzer, which Harald had handed the divers in response to Riggs's request for an in-line analyzer. To Riggs, the back of the vehicle sounded like a stage full of second-graders playing their recorders in a tedious school concert.

The rebreathers, which resembled roller suitcases and contained filters to absorb the carbon dioxide from exhaled air, were the same devices the other divers had used during the fact-finding mission. Their name said a lot about how they functioned, since they "scrubbed" and recirculated air, allowing it to be breathed more than once. Because the scrubbed air burned hot, each rebreather needed to have an ice pack installed in it. Riggs pulled out the first rebreather and began assembling it. It turned out to be quite a struggle, partly because it was a complicated process but mostly because Riggs still needed to devote one hand to kinking his hose after each breath. Standing there, shoving the hose between his legs and trying to use his knees to do the kinking, Riggs began quietly praying. *Lord, have mercy on us. Help deliver us.*

As Riggs looked up, he could see Hoss watching him. Riggs knew that Hoss wasn't religious. But he could see a wistful look on his face. "Pray for us, Riggs," Hoss called.

Hoss headed back to the boat and grabbed three of the one-hour Ocenco self-rescuers. Each of these devices had its own face piece, which sent the short supply of breathing air directly into the diver's mouth while pinching off his nose. Switching to this supply required some real choreography, making it essentially a two-person job. With Riggs focused on assembling the rebreathers, Hoss and DJ helped each other get the one-hour devices attached and activated.

Hoss counted off, "one, two, three," then took a deep breath and held it. In one swift motion, he ripped off his facemask and attached the new contraption to his nose and mouth. There was a lag time before the unit kicked in. Only when he knew it was working did Hoss exhale. Then he helped DJ get his going. Like Riggs, Hoss recalled the trainer's warning about the devices providing a shorter-than-advertised air supply during periods of high stress. This situation certainly qualified. So Hoss hit his stopwatch. To be on the safe side, he planned to have everyone out of the Ocencos within half an hour.

After gesturing for DJ to go help Riggs get his on, Hoss turned his attention to the logistics of driving out of the tunnel. By the time DJ approached him, Riggs had managed with one hand to get the first of the four-hour rebreathers set up, though he hadn't yet had a chance to retrieve its ice pack from a cooler in the Humvee.

To lighten their load, Hoss decided to unhitch the trailer from the Humvee. That meant cutting the hoses that tethered the liquid oxygen and nitrogen tanks to the controls inside the vehicle. They certainly didn't need the liquid tanks or the cursed MAP Mix 9000 anymore. He looked over to see that Riggs was now wearing a one-hour device, finally released from his free-flowing hell. Hoss called him over, asking to borrow his knife, while DJ took over the task of assembling the other rebreathers with ice packs.

As Hoss was about to cut the first hose, he looked down and was startled by what he saw. His hand was shaking uncontrollably. It felt as though his brain were no longer in control of his fingers. He turned to Riggs and asked him to do the cutting. The panic may not have overtaken Hoss's mind, but it had found its way to his fingertips.

Before Riggs began slicing through the hose, DJ yelled over to them. "ALL STOP on cutting the trailer loose!"

Hoss shot him a puzzled look.

"We've got to get these guys out of here, and we can't load them into the Hummer," DJ explained. Although they had used the pair of opposite-facing Humvees for the drive into the tunnel, they had decided to leave behind the white Hummer, which was pointed to the east. Still, with all the equipment jammed inside the gray wagon, it would have been tight to try to fit all five guys into one Humvee—in fact, with three of them wearing the bulky rebreathers, nearly impossible. "We can put them on the trailer," DJ said.

After scanning the setup in front of them, Hoss agreed to leave the trailer attached.

By now, DJ had the other rebreathers ready. Hoss grabbed one and DJ helped him maneuver it onto his back and get him switched over, an even more complicated process than donning the Ocencos. Then Hoss helped Riggs do the same. DJ put his backpack on himself, by laying it upside down on the tunnel floor, sticking his arms into it, and then flipping it over his head, like a preschooler learning how to put on his jacket.

They moved Billy's body first. DJ slipped several times on the slick tunnel surface as he tried to pull Billy out from under the Humvee, his own legs sliding under the vehicle. When they finally got Billy to the trailer, DJ tried one last time to revive him by purging his mask.

It had taken DJ longer to admit to himself what Riggs and Hoss had sensed fairly early on. But he realized now that he couldn't hold out hope any longer. Billy and Tim weren't just down. They were gone. Rather than discussing it, DJ simply removed Billy's mask. Looking at Billy's wide-open eyes above his trademark mustache, he whispered to Hoss, "Tap would not want to see this."

Hoss did what he sensed DJ could not bring himself to do and closed Billy's eyes. They then moved to the driver's side to get Tim. In the tight quarters of the tunnel, it was a challenge to extricate Tim's 220-pound body from the Humvee and then carry him along the slippery tunnel floor over to the trailer. After they managed to lay his body down, they once more tried to revive him, even if, by now, they knew they were just going through the motions.

With Tim out of the driver's seat, Hoss motioned for Riggs to move behind the wheel to get the Humvee started. Hoss stood at the

back of the vehicle and turned on the oxygen-injector system. Then
he gave Riggs the sign to turn over the engine.

Nothing. They tried it again. Nothing.

That was when Hoss, whose poise and quick thinking had kept
the three of them alive, hit his lowest point. He knew they had nowhere near the air they would need if they had to walk the entire nine
miles out of the desolate tunnel. And he knew how long it had taken
to get the Humvee going when they'd had trouble starting it previously. He looked back at the trailer, fixing his eyes on Billy and Tim's
inert bodies. And he thought to himself, *Is that our future?*

Then he answered his own question. *We're fucked.*

DJ knew his party-boy reputation pretty much defined him, but he
prided himself on being nothing but serious when it mattered most.
His mother had always said there was nobody better in a crisis than
DJ. And he had reprised his crisis cool so far on this wrenching afternoon. Still, like Hoss, he felt panic set in when the Humvee wouldn't
start. *How many more things can go wrong?*

He knew there was a standby team of divers back on Deer Island,
but how would they help? The central message drilled into them during their mine rescue training had been to avoid endangering additional workers, so it wasn't clear whether that backup team would
even be sent into the tunnel. If a rescue operation was authorized, it
would take those standby divers time to mobilize before they could
begin the nearly two-hour trek. DJ, Hoss, and Riggs weren't about
to sit around in a panic waiting for a possible ride, burning through
their dwindling supply of air. If they didn't get the Humvee started,
DJ figured there was little hope.

He wouldn't reach his thirtieth birthday for another two months,
and he began to wonder if he'd ever see that milestone. All three of
their lives seemed to be slipping away. Suddenly, a grim vision from

his future flashed into DJ's mind. He saw people climbing the stairs to the duplex apartment he shared with his mother. They were somberly offering their condolences to Lorraine, to Dana, to his brother. "I'm so sorry," they all said. In this vision, DJ's four-year-old son, Cody, was also there, though he seemed to have no idea why everyone around him was so sad.

Yet in this moment of darkness, he was able to find light from an unexpected source. Riggs—the guy DJ had thought was too uptight, the guy DJ had thought wanted to bolt on foot out of the tunnel—was now sitting behind the wheel of the Humvee. And in the moment when DJ and Hoss found themselves most despondent, Riggs was projecting nothing but steely calm. He was going to get the damn thing started.

Wearing the bulky rebreather on his back, Riggs could barely fit into the driver's seat, but that didn't seem to faze him. He appeared to have gleaned from the previous day's engine troubles exactly which steps he needed to follow to trick the oxygen injector into working, and how to do it without draining the battery. He waited for the electronic sensors to reset; then, to override the injector, he flooded the intake with oxygen and ground the starter for longer than seemed wise. All of a sudden, in a thunderous groan, the engine turned over, belching a plume of diesel smoke that filled the tight section of the tunnel as though it were a shotgun barrel.

It was 1:55 p.m. "Keep it running!" Hoss yelled from behind. DJ rounded up some of the equipment they might need for the ride, tossed it into the back of the Hummer, and then jumped into the rear seat, struggling to sandwich himself, and his roller-suitcase rebreather, into a space that was already crowded with gear. By the time he was settled back there, Hoss had climbed into the passenger's seat. Riggs stepped on the gas.

As they drove off, a new worry washed over DJ. Because they had kept the trailer attached, that meant they were still carrying the tanks of liquid nitrogen and liquid oxygen. If the Humvee took a hard turn in the unforgiving tunnel with all that standing water, the trailer might jackknife. DJ knew there was a chance that a collision could turn those tanks into bombs. As much as the divers wanted to hit the gas pedal to race out of the tunnel, they all knew they could only go so fast—probably not more than five miles per hour—if they didn't want to court further catastrophe.

"Riggs, stay on your toes," DJ said.

Tap knew something was wrong. Harald wasn't just being sparing with details, as was his custom. Now he wasn't saying anything at all. Tap had stopped by his trailer for one of his periodic check-ins, asking Harald what was going on. Harald didn't mention that he had been concerned when, just after 1:40 p.m., Tim had failed to call in with his regular update. He didn't mention that he had then tried phoning Tim, and, when he got no answer, decided he needed to alert a Kiewit supervisor about the potential problem. And Harald didn't mention that, at 1:45 p.m., DJ had phoned in with the unthinkable update that two men were down.

Tap saw Harald sitting like a statue at his wooden table now, staring at his laptop and radar screen, saying nothing. Finally, at 1:55 p.m., a few minutes after Tap had shown up, the mine phone rang again. Watching Harald pick up the receiver, Tap moved in closer to try to hear the voice on the other end of the line. The diver said something about how they were "coming out," something about how the guys were "still unconscious."

Tap suddenly felt unmoored, a surge of dread and confusion building up inside him. Some kind of disaster had struck in the tunnel, though he couldn't tell how bad it was. Instantly, he decided he wasn't going to wait around the trailer, hoping Harald would toss him a few more crumbs of information. He grabbed one of the VHF handheld

radios, rushed over to the man cage, and signaled for the crane operator to lower him down the shaft.

At the base of the shaft, he tried to get through to the divers on
their radios. Since he'd been the one to purchase those handhelds, he
knew their range was limited to somewhere between three and four
miles. So he'd need to wait until the Humvee got closer before he had
any hope of connecting with the divers.

A few minutes later, Tap heard a noise and turned to see the
man cage being lowered once again down to the base of the shaft.
The metal door swung open and a pair of paramedics hustled out,
pushing a stretcher and a cart with breathing tubes and other life-
support equipment. The paramedics looked almost like twins: a pair
of young, fit guys with short brown hair, though one of them kept his
hair spiky and the other, closely cropped. The guy with the military
cut approached Tap, introduced himself as Keith, and asked for the
essential information.

Keith Wilson and his partner had been driving around Boston in their
ambulance, having just cleared a case at Massachusetts General Hospital, when the call came through on their radio. The fact that the call
had been preceded by an extremely long tone, followed by a lengthy
list of units being told to respond, immediately conveyed the gravity
of the situation. All the dispatcher had said, however, was that there
were people "trapped, below grade," on Deer Island. He and his partner, Steve Fleming, had been out to Deer Island a few times over the
past couple of years, but those calls had been for routine matters. This
one sounded a lot more serious. They made a U-turn and headed for
the island, knowing they'd need to navigate around a bunch of narrow side roads before they could get there. Along the way, Wilson
found himself wondering what they were going to find.

They arrived on Deer Island a few minutes after two o'clock, with the sun strong and high in the late-July sky. Several units from the Boston Fire Department were already there. Wilson spotted a deputy fire chief huddling with some "white hats," construction supervisors wearing white hard hats, surrounded by sandhogs in yellow hard hats. There was an air of barely controlled panic.

The chief brought Wilson up to speed. Down in the tunnel, hundreds of feet below grade and miles out under the ocean, two divers had been found unconscious by three other divers who were now trying to escape with their lives. "We're not sending personnel down there," the fire chief told the paramedic. There was no fire, no smoke, no wall collapse, he explained, so essential medical personnel were the only ones needed. "It's a medical case."

Wilson was in no position to question authority. He was just twenty-eight years old and had been a licensed paramedic for only a couple of years, after having started as an EMT. During his time working for Boston Emergency Medical Service, he'd responded to calls in the sketchiest sections of the city. He once showed up to the scene of a reported homicide even before the cops got there, finding a warm body slumped over in a recliner and the smell of gun smoke still hanging in the air. Another time, when he'd bent down to tend to a stabbing victim, the guy had lunged at him with a knife. As scary as those situations had been, Wilson had always felt equipped to take care of himself. But *this*?

He was being asked to go down a 420-foot shaft leading to a forbidding tunnel to care for two men who were unconscious. At least everyone was saying they were unconscious. Wilson knew that if these divers had been deprived of oxygen for long, they weren't unconscious. At best, they were in cardiac arrest. As seriously as he took his job, he couldn't avoid asking himself: *Is there some mystery gas filling up the tunnel? Is there a chance that I'm going to be the next victim?*

He and his partner knew that once they went down the shaft,

there would be no coming back up for more equipment. So into the man cage they packed whatever they thought they might need— endotracheal tubes, defibrillators, IVs, a cardiac monitor, as well as a stretcher. The crane operators had set up a bigger man cage than the cramped one that had been used most often during the plug-pulling job. This one was roomy enough to fit Wilson and his partner and their equipment as well as another pair of paramedics from Boston MedFlight, whose chopper had just landed on Deer Island. If there turned out to be more than two victims, they would have enough paramedics to handle them. If it was just the two, each of the fallen men would have a pair working to revive him.

The white hats on site assured Wilson and the other paramedics that there was plenty of ambient air at the base of the shaft, so they would be fine to wait down there without relying on any breathing apparatus themselves. Still, as he and his partner were lowered in the cage, Wilson was concerned. *What if some bad air makes its way to the shaft? What if the cable holding up this man cage breaks?* He knew better than to put words to these worries. Instead, he looked over to his partner and said, "You all right? You ready for this?" When his partner nodded, Wilson decided humor might ease the tension. As the metal cage began its slow descent, he cracked, "Going down? What floor?"

When they made it to the base, Wilson spotted a white hat standing on the ramp, cradling a handheld radio. He went over and introduced himself. The manager, who looked distraught, said his name was Tap.

Wilson was struck by how cold and damp it was at the base of the shaft. The temperature was about thirty degrees cooler than topside. As bad as the indications were for the survival of the unconscious divers, given the ticking clock and the miles the Humvee still had to travel to make it back to the shaft, Wilson took the uncomfortable conditions as a good sign. Cold temperatures can help to preserve a body, providing a shred of promise for even those rescue attempts with the longest odds. As the saying in paramedic circles goes, "They're not dead until they're warm and dead."

Paramedics are trained to remain optimistic, even in the face of dire indicators. And for good reason. There were real cases in the medical literature that supported the "warm and dead" aphorism. They fell into a category called cold water immersion, where people submerged in frigid waters managed to survive. In those cases, facial contact with cold water triggered something called the mammalian diving reflex, which shuts the body down, effectively freezing the second hand on the stopwatch of death.

The most unbelievable of these cases had occurred just two months before the divers showed up on Deer Island. It involved a twenty-nine-year-old skier in Norway. As she made her way down the slopes, she crashed through the ice into a flooded gully. With her head wedged under thick sheets of ice, and her fellow skiers unable to free her, she found an air pocket and managed to stay conscious for forty minutes before giving out. After another forty minutes, her friends were finally able to extricate her, though by then she had long been clinically dead. By the time doctors managed to revive her, her heart had been stopped for nearly three hours. But the frigid water had saved her, lowering her internal body temperature by almost 50 percent. The mammalian diving reflex protected her by rapidly redirecting blood flow away from her limbs and skin and toward her vital organs. Sending more of the remaining oxygen to her brain and heart had staved off the neural devastation that accompanies oxygen deprivation. The skier, whose stunning story would be detailed in the British medical journal *The Lancet*, would go on to spend more than a month breathing off a ventilator. But her recovery would be so complete that eventually she even returned to the slopes.

Statistically speaking, however, she was an outlier. And as uncomfortably cold and wet as it was in the tunnel, even the optimistic Wilson knew it would have to be far colder for the reflex that saved the skier to do the same for these divers.

Heading west, back toward the shaft, Riggs, Hoss, and DJ passed the numbers spray-painted onto the sidewall, charting the tunnel's descending ring numbers. On the drive out in the morning, they had stopped at every thousand to two thousand rings, wherever there was a mine phone hanging from the wall, and called in their updates to Harald. On the drive back, however, the divers had no interest in Harald's protocol, no interest in wasting time on updates. Hoss told the others he was worried that if they stopped, the Hummer might conk out.

But as Riggs continued to guide the Humvee along, Hoss suddenly realized there was a problem with his nonstop, do-not-pass-go approach. If they didn't pause to phone topside, he told Riggs and DJ, Harald might dispatch the standby crew, driving the pair of backup Humvees. That would create a dangerous logjam in the narrow tunnel. "It's going to fuck us if we meet them head-on," Hoss told the guys. "We don't have time for that."

So at 2:40 p.m., as the Humvee approached ring 6000—with about 5.75 miles left between them and the shaft—Hoss got out and picked up the receiver.

"What's going on?" Harald asked.

"Tim and Billy are gone," Hoss said, curtly.

"How?"

Hoss was furious at Harald. Even after the divers had complained repeatedly in previous days about problems they were experiencing with the breathing system, Harald kept sending them back into the tunnel on the same system, reassuring them that they'd be fine. Hoss knew no good could come out of staying on the phone with Harald now.

"They're dead and we're at six thousand," Hoss said, "and driving in with these two guys." Then he hung up.

The Humvee resumed its slow westward chug to the shaft. After a few minutes, though, Hoss started yelling. *My lungs are burning up!* He thought he'd become numb to any more crises, but here, apparently, was a new one. And Hoss found himself freaking out.

DJ quickly figured out what was happening. The chemical reaction going on inside the rebreather made wearing one the equivalent of breathing air from a hair dryer. That's why it needed the ice pack to cool it down. But Hoss must have grabbed the first rebreather, the one Riggs had set up without an ice pack before moving on to another task. From his position in the back seat, DJ reached into the rear of the Hummer to find the cooler and grab an ice pack. Then he pushed Hoss forward in the front passenger's seat, pulled the back off his rebreather, and shoved the ice inside to cool him off.

Hoss calmed down. "Give me some cold water, DJ," Hoss said.

After fumbling around in the back, DJ thrust a bottle between the two front seats and Hoss grabbed it. He lifted the facemask of his rebreather and poured the bottle onto his eyes. The stinging pain was instant. "My eyes are burning!" Hoss shouted.

DJ started chuckling, the first sound of laughter any of them had heard since the horror of finding Billy and Tim. "Relax," DJ said soothingly. "Your eyes are fine. You just got some soda in them."

Hoss was confused until DJ explained that he thought Hoss had asked for the water because he was thirsty, not because he needed to clean out his fogged-up facemask. So when DJ couldn't find any water, he'd grabbed a bottle of ginger ale from his own cooler, part of his private stash that had also included the peanut butter crackers he'd been snacking on the day before.

. . .

Sometime after passing ring 4000, as they approached the three-mile mark from the shaft, the VHF handheld radio lying in the back seat began to crackle. DJ immediately recognized Tap's voice and picked up the radio to respond.

"What happened?" Tap asked.

"I don't know what happened," DJ said, "but everything went to hell in a handbasket down here, and we're trying to get out."

Tap explained that he had headed down the man cage as soon as he heard about the trouble, and had been standing at the base of the shaft ever since, waiting for the Humvee to get to within range of the radio. It soon became clear to DJ that Tap hadn't been with Harald when Hoss had phoned in their last—and bluntest—report about Billy and Tim. DJ couldn't muster the strength to tell Tap that the news was much worse than the initial report. He knew how close Tap and Billy had become ever since they'd all met working in Vermont. While DJ considered them both very good friends of his, he knew the bond between Billy and Tap was stronger. He sometimes thought Tap and Billy were like an old married couple, with that uncanny ability to anticipate each other's thoughts and finish each other's sentences. They'd even developed their own language when they were on a job talking to each other on walkie-talkies, converting the standard "Roger" reply to "Rog-oh." No, DJ wasn't about to deliver news this devastating over the radio.

During their slow trip back, DJ and Tap talked a couple more times on the handhelds. Each time, DJ shied away from telling Tap the crushing reality about Billy.

Given all that had befallen them already, DJ, Hoss, and Riggs took the precaution of continuing to wear their rebreathers, even after the ring counts spray-painted on the tunnel wall suggested they had reached areas where the ambient air was fine. But as the end drew nearer, they decided to remove the cumbersome contraptions, so

DJ had taken off his facemask by the time Tap's last radio call came through.

"How's Billy?" Tap asked.

"Well," DJ stammered. "Billy's gone."

"What do you mean Billy's gone?" Tap asked in an agitated voice. "Where did he go?"

DJ was exhausted, physically and emotionally. He couldn't continue with the dance any longer. "Tap, there's no lefts or rights down here," he said. "He's gone. He's expired. Billy's dead. And so is Tim." DJ asked him if he understood.

There was a long pause. Then Tap answered softly, "Rog-oh."

Topside, shortly after Harald received the initial distress call from DJ, he called Roger Rouleau back at the Norwesco office in Spokane. "We've got a problem," Harald told the company owner, breaking the news that two men were down at the end of the tunnel. Just days earlier, when Roger had been profiled in the Spokane business journal, he'd boasted about how challenging and wacky dive jobs like Deer Island were Norwesco's stock-in-trade. Now he could do nothing but sit at his desk and wait for Harald's follow-up call. When it finally came, with the shattering news that Tim and Billy were believed dead, Roger called his wife and asked her to pack a bag for him because he would be taking the first flight out to Boston.

Following the emergency response triggered by DJ's first call, news of the accident began radiating to other key players in the project. MWRA chief Doug MacDonald received a call from an aide who said gravely, "I've got some bad news. We have two divers in the tunnel who are unconscious." MacDonald was incredulous. *No!* he told himself. *Do I get to wake up from this dream now? This is* not *happening.* His vow hadn't simply been to guide the tunnel and harbor cleanup project to completion. It had been to do it without losing any more lives. After calling Judge Mazzone, the cleanup overseer, to notify him of the accident, MacDonald headed straight for Deer Island to get his updates firsthand.

Dave Corkum, the tunnel manager who had jousted for years with contractor Kiewit over the plugs and insisted that they not be removed before the main utilities, had taken a rare day off on this Wednesday. He had just finished coaching his son's hockey game when he got the call on his two-way radio. The message said only, "There's been an accident in the tunnel." He hurriedly dropped his boys off at a friend's house and then headed straight for the island. In neighboring Winthrop, he turned onto a street to find two un-marked police cars, with their lights flashing, traveling ahead of him. Corkum sensed those cops were heading to the same place he was. So he pulled up behind the second car, keeping pace as it barreled along the narrow side streets leading to the island.

When news of the accident came in to the Local 56 union hall, business agent Dan Kuhs was the one who took the call. Instantly, he remembered his visit to Tap's yard. That day was the first and last time Kuhs had ever seen Harald, but he recalled how the engineer had proudly showed off the exotic breathing system, pointing out all the redundant safety mechanisms he had incorporated into it. The bad news coming in now from Deer Island was that two men were down in the tunnel. But as a veteran diver himself, Kuhs was sure of two things. First, that exotic breathing system, despite all its redun-dant safety features, had to be at least partly to blame. Second, given how far out into the tunnel the two unconscious guys had been, they both had to be dead.

Riggs knew they were getting closer, and making better time than expected, even if he still couldn't see any lights in the distance. Just a little while earlier, all three of the guys had felt a fog of what seemed

to be their certain demise settle over them. Riggs sensed his fog lifting, and he started thinking about what would await them once they
made it back to the shaft.

With his hands on the wheel, he turned to Hoss and DJ and said,
"Let's stick together." When they get to the shaft, he said, "don't let
anyone separate us." The others nodded in agreement.

The base of the shaft might have been near, but it still wouldn't be
easy to get there. Riggs spotted a deep dip in the path of the tunnel,
where the standing water was at its highest. Going through it was
like driving along a low-lying roadway after it had been washed out
in a torrential downpour. The driver needed to operate with a deli-
cate balance of gusto and restraint. Too little power, and the Humvee
would stall. Too much, and it would spin out. Riggs carefully made it
through, although water splashed all around the vehicle.

Half a mile later Hoss turned to Riggs and asked him to hit the
brakes. "We need to check on the guys," he said. With all that water
they'd just gone through, Riggs agreed it was a good idea.

Hoss was thinking about more than just the water. His mind had
flashed back to a job a couple of years earlier, when Roger had dispatched him to a hydro plant in Montana. When he arrived, he had
learned the nature of his assignment. He had the dreadful task of
searching for the body of a worker who had tried to jump out of a collapsing crane, only to get hit in the head by the boom and never make
it out of the water. After two days of searching, Hoss and his fellow
divers had managed to find the body. If that chore hadn't been draining enough, Hoss was taken aback by what was waiting for them
when they returned to the plant. A scrum of photographers from the
local media had arrived to snap pictures of the recovered body. The
experience had left Hoss nauseated.

Fearing that there might be camera crews waiting on Deer Island,
he got out of the Humvee determined to preserve Tim and Billy's
dignity. DJ quickly followed, while Riggs remained behind the wheel,

keeping the Humvee running. Hoss found the bodies lying on the trailer more or less where he had left them, though they had been jostled and sloshed around during the hellish ride back. Tim's leg was hanging over the side of the trailer. As Hoss bent down to move it, his heart sank. On the side of Tim's left boot, a long gash of the tan leather was missing. Inside of that, a two-and-a-half-inch-long section of flesh from his big toe was gone. Part of the bone literally had been ground down, like a freshly sanded piece of oak.

Hoss felt like crying. Instead, he straightened Tim's body and closed Billy's mouth. "Let's cover them," he told DJ. "I don't want them all splayed out." So he and DJ unfolded a couple of blankets and gently placed them over the two bodies. Then they climbed onto the trailer and yelled to Riggs that they would ride back there, holding the bodies, for the rest of the drive to the shaft. They weren't going to let anything else happen to Tim and Billy.

Paramedic Keith Wilson stood on the ramp next to Tap. It seemed like hours since Tap's last communication with the divers, even if it hadn't been very long. Finally the headlights of the Hummer came into view. Still, given the necessarily slow pace of the vehicle through all the standing water, it took a while for it to make it to the ramp.

Wilson had been a paramedic long enough to know how infinitesimal the divers' chances of survival had become. He had been waiting with Tap for the Humvee for nearly an hour and a half. Still, he refused to give up hope altogether. As the Humvee drew closer, he expected to see the other divers hunched over the fallen guys, furiously doing chest compressions. Instead, when the trailer came into view, Wilson saw the bodies covered up with blankets. He was upset. *Why had these guys given up?*

At 3:36 p.m., Riggs navigated the Humvee up the ramp, threw it into park, and cut the engine. Wilson and his partner rushed toward the trailer with their backboards and equipment, and Tap followed along. "What the hell happened?" Tap asked the divers.

Even if there was almost no shot of reviving these guys now, Wilson knew it was always crucial for the friends and family of a victim to see evidence that the paramedics had done everything in their power to try to save their loved one. Only then, Wilson had learned, could they hope to begin the healing process. For him, the hardest emergency calls were those when he'd show up at a home to find a baby dead in her crib, evidently from Sudden Infant Death Syndrome. As devastating as those cases were, he'd learned how important it was for grief-stricken parents to see their baby removed from the house, worked on fully and aggressively, and pronounced dead at the hospital, rather than having the paramedics cover the baby in resignation right there in the family's nursery. Otherwise the parents would probably never set foot in that nursery again. Similarly, in this case, he figured these divers might never enter another tunnel if they didn't see the paramedics doing everything they could to revive their buddies and get them to the hospital.

He and the other paramedics rushed to give the two fallen divers the full assault of epinephrine, atropine, sodium bicarbonate, IV fluids, a breathing tube, and chest compressions. But they couldn't hide their disappointment. As Hoss was climbing off the trailer, one of the paramedics complained that the divers should have never covered the bodies. They didn't have the authority to pronounce people dead.

Hoss heard the paramedic's complaint as an accusation, and it enraged him. He had just come within a hairsbreadth of dying, and these were the first words he had to hear? He looked down at his stopwatch, which he had started back when they'd put on their first emergency air devices—the Ocencos—which had itself been at least a half hour after Tim and Billy had been struck down. The stopwatch now read: 1 hour, 40 minutes. If those paramedics only knew the near-lethal lengths that he, Riggs, and DJ had gone to in the hopes of trying to save their friends. By now, Tim and Billy had surely been dead for at least two hours. They were simply beyond being saved. Hoss was utterly drained, but he knew this much: he shouldn't have to defend himself to anyone. *Go fuck yourself,* Hoss muttered, before Riggs grabbed him by the sleeve and pulled him away.

The paramedics moved Tim and Billy from longboards to stretchers and got to work, with Wilson and his partner focused on Tim and the other pair on Billy. Despite their intensity, none of the resuscitation efforts produced any sparks of hope. They now wheeled the two stretchers along the ramp to the bottom of the shaft, then into the bigger man cage. They continued their chest compressions as they began their slow climb up the shaft, the cage swaying slightly as it made its ascent.

When the sunlight met them at the top of the shaft at 3:53 p.m., and the paramedics began wheeling the stretchers off the cage, a strange thing happened. The thick crowd of workers, managers, and emergency personnel who were arrayed around the shaft began to applaud. Maybe they were applauding the effort. Or maybe they thought Billy and Tim were still alive, even if the survivors—and by now the paramedics—had already bowed to reality.

Like Hoss and Riggs, DJ felt shattered. All three of them had just cheated death, but there was nothing to celebrate because two of their brothers hadn't made it. The paramedics could work over the bodies all they wanted, but they couldn't reverse the cruel fate that DJ had been forced to confront when he'd held Billy's lifeless body in his arms nine miles into that tunnel. He watched as Billy and Tim were taken up the shaft in stretchers. They were headed for the hospital, but DJ already knew they were going to end up in the morgue.

DJ stepped into the man cage, but just before the gate shut, he remembered something and hopped out, bounding back toward the

Humvee. It was his hard hat. When he finally made it to topside, he was surrounded by people who all wanted to know the same thing: *What happened?* He was in no mood to talk. Despite the divers' pact to stay together, once they were out of the tunnel, forces quickly conspired to pull them apart. Tim and Billy had been loaded into separate helicopters, to be transported to separate hospitals. The paramedics who'd been in the tunnel had left with the bodies. But the swarm of emergency vehicles on Deer Island included several other ambulances, and different medics swooped in to work over DJ, Hoss, and Riggs separately. Before he'd even noticed it, one of them grabbed DJ's arm and hooked him up to an IV.

Nearby Harald stood looking shell-shocked. Dave Corkum, the tunnel manager who was enrolled in law school, approached him and said, "Harald, you'd better get yourself a lawyer."

Someone standing next to DJ told him he was lucky to have made it out alive. The comment punctured the surreal haze that DJ had felt hovering over him ever since he'd come upon Billy's body. He turned over his hard hat and caressed his grandfather's Virgin Mary medal that was hanging on a piece of twine inside. "This," he said, "is what got us out." Then he cut the twine and put the medal in his pocket.

12

"Why, God?"

"Potential fatality at Deer Island."

Mary McCauley hung up the phone and looked at her watch. It read three-twenty. McCauley was a Massachusetts state trooper assigned to the Suffolk County district attorney's office, focusing on death investigations. The plainclothes detective did not fit the typical trooper profile. An athletic thirty-four-year-old with shoulder-length dirty-blond hair, McCauley had gone to law school and served for several years as an assistant district attorney before enrolling in the police academy. A big reason for her career switch was that as a prosecutor, she had often felt frustration in trying to understand a case simply by reading the police report and reviewing photos. That approach never satisfied her type-A personality. She regularly spent lots of time reinterviewing witnesses and visiting crime scenes, so she could be more persuasive when she had to stand before a jury. After three years of doing this, working sixty-hour weeks as a prosecutor for pay so low that she had to continue waitressing on the side, McCauley decided that if she became a cop, it might be simpler for everyone involved.

This afternoon's call about the Deer Island incident, which had been routed from a different state police office, contained few details. She most likely wouldn't investigate the case unless someone actually died, but even if that happened, her involvement wasn't a sure thing. Massachusetts was so provincial that there were blurry jurisdictional lines and frequent turf battles among its various law enforcement agencies. On more than one bridge-jumper investigation, McCauley had been able to take the lead only after establishing that the suicide victim had jumped from a state-owned bridge and had landed in a spit of water that fell under state jurisdiction. At 3:53 p.m., her supervisor gave her clearance to head to the scene.

McCauley had been born and bred in Boston, as evidenced by her toughness and accent, which rendered a word like *arson* as *AH-sun*. Yet the areas around the north of the city remained somewhat foreign lands for her. Coming from a tight Irish family, she had grown up in Mattapan, a gritty neighborhood on the city's southern edge. Although she had married a guy from the North Shore and was now raising a family there with him, McCauley was still finding her way around this band of boreal boroughs ringing Logan Airport. The only thing she knew about Deer Island was there was some kind of sewer plant there, and she had once gone for a run along the coastal route leading to it.

When she and her partner, John O'Leary, arrived on the scene at quarter past four, they found lots of commotion. She saw fire engines, ladder trucks, and ambulances from multiple departments, marine rescue units, and mobile command units. So many emergency vehicles were clogging tiny Deer Island that she had to park her unmarked cruiser near the entrance and hoof it all the way to where the bustle was, around the entrance to the shaft. As she approached, the last of the ambulances was clearing out. She tracked down a deputy fire chief who seemed to have a handle on the chaos, and he quickly filled her in. He pointed to the ocean and told her that three divers had been working as part of a "dry penetration" job at the end of the nine-and-a-half-mile-long empty sewer tunnel. Around 1:45 p.m., when the divers had returned to their pair of Humvees, they'd found the other two members of their team lying unconscious. Around four o'clock, the unconscious divers had been airlifted, one to Boston Medical Center and the other to Massachusetts General Hospital, where the three surviving divers had just been taken by ambulance.

As McCauley tried to digest the jumble of information, the deputy chief pointed out a few of the major players on the scene—managers from the various entities involved in the job. Then he added another crucial detail. Deep into that endless tunnel, there was no light or oxygen to speak of.

The detective found herself overwhelmed. As an investigator, she prided herself on her ability to remain calm no matter what she found at a crime scene. One Thanksgiving she'd responded to a call at a house where a body had apparently been left rotting inside for weeks. She could smell the stench a block away. She had arrived to find a bunch of firefighters bent over outside, most of them retching, but she had remained standing. Here on Deer Island, the details the fire chief was sharing weren't grisly. But they were so confusing, they made her head hurt. *Divers traveling in a pair of opposite-facing Humvees along a tunnel that stretched under the ocean for more than nine miles? A tunnel that had no air and no light in it? Was he serious?* Even the terminology— *dry penetration*—seemed like a foreign language. McCauley nodded along, hurriedly jotting down notes onto her white steno pad. But to her, it all sounded like the setup for some half-baked science fiction novel.

As she talked to others on the scene, her confusion continued to mount. The MWRA owned the tunnel, but there was a contractor building it for the water and sewer agency, which also had a separate company called a construction manager that was supervising the job. The two unconscious divers worked for two separate diving companies, one from Spokane, Washington, and the other from

Portsmouth, New Hampshire. People from OSHA were present now, and because that federal agency often investigated workplace accidents, there was some thought that the OSHA people would run this investigation. As bewildered as McCauley was, though, she knew that if she was going to be involved in this case, she wanted to make sure it was done the right way. So she wasn't about to let herself get boxed out. Within fifteen minutes of her arrival, at least one aspect of the investigation became clearer. Word came in from both Mass General and Boston Medical: Billy Juse and Tim Nordeen had been pronounced dead. They were now actual, rather than potential, fatalities. McCauley called her boss, and he gave her the confirmation she needed. This was now officially a state-police-led investigation, and she would take the lead.

She knew that in these critical early hours, she had to focus on controlling what she could. That boiled down to preserving all the evidence and documenting everything she and her partner heard and saw. As she did this, more and more questions began throbbing in her head. The first was the most fundamental: How had these two men died? Was the cause a freak accident? Could it have been suicide? Negligence? Homicide? If it was a crime, who was responsible?

McCauley had probed enough stabbings and murders to know that finding answers to the big, befuddling questions required keeping your eye on the smaller, more elementary tasks—like protecting the crime scene from any tampering. But in this situation, even that basic point morphed into a big, baffling question: Where *was* the crime scene? She knew it wasn't where she was standing on Deer Island or even down at the base of that shaft 420 feet into the earth. The crime scene was where these two men had fallen unconscious and presumably died, and that was nine miles into an unlit, unventilated tunnel. That prompted only more questions: What evidence did she need to collect? How could she collect it? And how would she ensure that it was securely stored to preserve what cops call the "chain of custody"? It wasn't as if she could toss a couple of Humvees into the trunk of her cruiser and bring them back to an evidence locker at her office.

From her early days as a prosecutor, McCauley knew that to do an investigation right, you had to be able to see and touch the crime scene. Even before the full outline of the probe came into view for her, she decided that's what she needed to do. She was no fan of heights, either going up the elevator of a skyscraper or descending down a skyscraper-length shaft, but she felt she had no choice. If the crime scene was nine miles into a tunnel buried below the floor of the ocean, she would have to find a way to get there. But was that even going to be an option? Would some state trooper with no background in engineering or diving be allowed to go into that tunnel? For that matter, would *anyone* be allowed in, given what had happened? Something in the tunnel had left two men dead and very nearly killed three others. Could there be some kind of mystery gas down there, hungry for more victims?

After a round of brief interviews and some basic evidence collection, McCauley realized she needed to get to the people who really knew what had happened, and they were now being treated at the hospital. Good police work requires talking to eyewitnesses as soon as possible after an incident, when their memories are fresh and before other forces can cloud their heads or shut them up. She closed up her notebook and headed to Mass General.

It was the between-meals, late-afternoon stretch at the Poor People's Pub, and the often-bustling New Hampshire establishment was empty save for the half-dozen regulars who sat at the bar, drinking their way through another afternoon. As manager, Michelle took advantage of the lull by heading into the back room to do some restocking.

Before she could make a dent, one of the waitresses found her back there. "Michelle," she said, "you have a phone call."

The waitress knew Michelle had been on edge all week, preoccupied by the job Billy was doing in Boston. The pub staff loved Billy as much as they loved Michelle, so when she saw a look of concern play over her boss's face, the waitress quickly sought to reassure her. "It's Tap," she said. "Everything's okay."

"Tap?" Michelle replied, startled.

"Yes."

"No," Michelle said, "it's *not* okay."

"What do you mean?" the waitress asked.

"Tap doesn't call me at work. It's not okay."

With her stomach completely knotted, Michelle raced to the phone. Tap told her there'd been an accident in the tunnel and Billy had been involved. He stressed that they didn't know much yet. "You need to get to Mass General Hospital."

Shaken, Michelle went to hang up the receiver but pulled it back at the last second. "Tap," she said. "Is he okay?"

There was a pause, and then he answered. "No."

Michelle knew she had to rush to Boston, but she also knew she was in no condition to drive. She surveyed the bar. The regulars who cradled their drinks without budging from their stools had maintained enough of their faculties to detect that something was very wrong with the woman staring at them, but still were in no shape to drive her where she needed to go. Michelle called the pub owner and left her as calm a message as she could muster. "Billy's been in an accident. I don't think it's good," she said. "I'll probably be out of work for a while."

Moments later the owner called back: "I'll drive you."

As she waited for her ride, Michelle decided she needed to break the news to Billy's New Hampshire family, Deb and Ken.

· · ·

DJ's mother, Lorraine, was working as a home health aide, keeping an elderly woman company in the lady's house in Brookline, an affluent community on Boston's southwestern border. They had spent much of the afternoon watching the endless television news coverage of the effort to recover John F. Kennedy, Jr.'s body. The rest of the nation always assumed that the Kennedys could do no wrong in the eyes of every Massachusetts voter, but in fact there was a small but vocal portion of the local population who, because of one transgression or another, had absolutely no use for any member of the political royal family. That subset included this elderly woman in Brookline.

"I'm sick of the Kennedys," the woman told Lorraine in the late afternoon, before turning off the TV.

Lorraine, a woman of strong opinions herself, was fond of the man she referred to as "young John Kennedy." Although she wasn't Irish like his family, she was Catholic and working class and felt the Kennedys had looked out for people like her. She even had the kind of brogue and fair skin that made people mistake her for a native of the Emerald Isle. Still, Lorraine knew better than to argue politics with the woman who owned the home. So she picked up a basket brimming with laundry and carried it upstairs. Rounding the top of the staircase, she came across a small television set, which happened to be on and tuned to a station that was also covering the JFK Jr. story. She put down the laundry basket and listened to the anchorman.

Her eye caught a news bulletin scrolling along the bottom of the screen: ACCIDENT AT DEER ISLAND PROJECT. TWO DIVERS AIRLIFTED TO HOSPITAL. Lorraine suddenly felt weak. *Oh my God,* she told herself, *that's where DJ's at!*

She sprinted downstairs and into the kitchen, where she announced: "My son was on a job in Boston, and I need to get out there! He had an accident. I don't know if he's involved, but I know he's on that project." She bolted out the door and climbed into her car. She shifted it into drive, but then quickly threw it back into park. *Where in God's name am I going?* She wasn't sure how to get to Deer Island, and anyway, she figured that by now all the divers had been rushed to the hospital. Determining which one would be no simple task. With a medical reputation that attracted patients from all over the world, Boston heaved with hospitals. In desperation to go somewhere, she raced home to Waltham. As she was putting her key in the front door of her apartment, she could hear the phone ringing.

"Is this Donald Gillis's mother?" the voice on the phone asked.

"Yes," Lorraine said.

The man said he was calling from the union hall.

"Don't tell me anything about DJ!" Lorraine cried. "I know what happened. I don't want to know if he's alive. Just tell me where he is." She knew she would have buckled under, right then and there, if he'd told her that her son was gone.

"Mass General Hospital," the man replied.

Lorraine called a friend, who hurried over to pick her up. Throughout their ride along the Mass Turnpike to the hospital, she prayed and prayed that DJ would somehow be saved.

Judy Milner had just wrapped up a session with a patient in her child psychiatry practice outside Seattle. Even though her days were filled with heavy matters, she used her quirky sense of humor and irresistible laugh to keep the mood in the office light. Now, as the petite psychiatrist walked out of her office, she found her secretary, Janice, weeping along with another assistant. Judy noticed that one of her longtime friends, Robin, who was a social worker with an office nearby, was visibly shaken. "What's wrong with you all?" Judy asked. "What happened?"

Instantly, she felt she knew the answer. Janice had a sickly dachs-

hund at home, and for some time the secretary's husband had been
saying the dog needed to be put down. Janice had steadfastly refused,
a stance that had won Judy's full backing. "Let's save the dog," Judy
had told Janice more than once, offering her assistance. Now Judy
turned to her secretary and tenderly asked, "Did your dog die?"

"Can we talk?" Janice asked, wiping away tears. She and Robin
guided Judy into her office and closed the door.

Judy turned to Robin. They had jointly treated a chronically suicidal patient, and now Judy asked if the patient had gone through
with it.

"No," Robin said apologetically. Standing with her back to the
closed door, she told Judy, "It's Tim."

Judy immediately stiffened and responded brusquely, "Tim's just
fine. Now get out of my office." When they refused to budge, Judy
pushed open the door and tried to shoo them away. "Just get out!" she
repeated. As a psychiatrist, she would have diagnosed her reaction as
a textbook case of denial, but right now she had no interest in diagnosis. She simply wanted to erase the previous and most shattering
minute of her life.

Janice stepped forward to close the door, positioning her body to
block Judy from the exit. With that, Judy started to sob. "Oh, I told
him not to go in that tunnel if he thought it wasn't safe!" Her two
friends moved in to hug her, but Judy's denial had yet to lift fully. "You
all have to leave," she told them. "I have patients to see now."

"I canceled all your patients," Janice said. She explained that she
had taken that step immediately after Roger had called to share the
devastating news, a call that had come while Judy was in her session.

"You can't do that. They're expecting me," Judy protested, before
finally breaking down.

When she began to regain her composure, she realized that she
needed to call Tim's parents. She didn't think the pain could get any

sharper, but somehow having to call Shirley and Bob Nordeen, a couple she adored, and tell them their firstborn was dead sent her into a new spasm of grief.

As she entered the emergency room, Lorraine spotted a huddle of men, several of them crying. She didn't know any of them, but she knew their look. "Are you boys divers?" she asked.

They said they were.

She introduced herself as DJ Gillis's mother. Overhearing her, a state trooper approached. "Don't worry," he told Lorraine. "Your son is alive." But he said she wouldn't be able to see DJ for quite some time, because he was being evaluated and then he would be questioned by another trooper. Lorraine had no quarrel with any of that. She thanked God that the words she had dreaded hearing had not passed the trooper's lips. She felt all that bottled-up fear suddenly flow out of her.

The trooper told her they'd give DJ a ride home later that night, so Lorraine decided there was no sense waiting around. On her way out, feeling relieved yet still somewhat dazed, she saw a young brunette walking toward her, wearing the same panicked expression that had been frozen on Lorraine's face ever since she'd seen that bulletin crawl across the TV. Lorraine introduced herself. The young woman said she was Michelle, Billy's Michelle. Lorraine tried to smile reassuringly, praying that she, too, would find good news here.

Olga Juse was a woman of precise routines. Several times a week, after leaving her job at the law office, she would meet her friend Sandy at the gym and they would work out together. Afterward she'd head home to her condo and make herself dinner. But this Wednesday was impossibly hot, even by the standards of south Florida, so after they'd finished their workout, she invited her friend over for a swim at her condo complex. Olga changed first and headed into the pool. When Sandy appeared in her bathing suit a few minutes later, she told Olga that the phone in her condo had been ringing. Olga said she'd pick up the message after their swim.

As it turned out, it was from Michelle. "There's been an accident," Michelle said, sounding distraught. "You need to call Ken and Deb's house." Olga was so nervous that she replayed the message three times but still couldn't make out all the digits of the number Michelle had left on her machine.

After all the years Billy had spent bouncing around, Olga was happy that he'd found a soul mate in Michelle. And it comforted her to know that he'd found a tight extended family in New Hampshire. While other mothers might have resented hearing their son call another woman "Ma," Olga hadn't minded it, and she had hit it off with Ken and Deb during a visit to Wolfeboro two years earlier. Since she lived too far away to keep tabs on her son, she was relieved there were other good people who could.

Finally Olga thought to look in her address book for Ken and Deb's number. Apprehensive but hopeful, she dialed. Ken answered.

"I know there's been an accident," Olga said as calmly as she could. "Is Billy still alive?"

Ken spoke slowly. "I'm afraid not."

Olga screamed and dropped the receiver.

Not long after that, Billy's father, Bill, showed up at her door. He and Olga had been separated for several years now, though they continued to live in the same town of Bonita Springs. Bill had left the trucking business and opened a small company distributing coffee and bottled water. Olga had never seen her husband look so hollowed out. For him, the boom had been lowered in a call he had received while watching the Kennedy coverage on TV. "Mr. Juse," the voice on the other end of the line had said. "My name is Dr. Brown from

Massachusetts General Hospital. I'm afraid your son has been in an accident." As soon as he'd heard those words, Bill had gone completely numb. He later told Olga that, earlier that afternoon, something strange had occurred. He'd been making deliveries to clients when his truck ran out of gas. In all the years he'd spent on the road driving a tractor-trailer, that had never happened to him. He had run out of gas just after 1:30 p.m. As it turned out, that was around the same time that, in a tunnel some fifteen hundred miles away, his only son had run out of air.

Riggs stood in the exam room, in a gown, waiting for someone to return his clothes. Back in the tunnel, he'd been the one to remind DJ and Hoss about the importance of not letting anyone separate them. But sure enough, in the commotion that met them as soon as they'd made it to topside, he'd been pried away from the others. Now, after the divers had undergone one round of examinations in the hospital, they had managed to reunite. Riggs watched as DJ approached one of the nurses and asked if she could get a pill to help him calm down. When she returned and offered pills to the three of them, Riggs accepted. But when a psychiatric nurse approached them a little while later with a pill that was designed to lessen the effect of a traumatic event, Riggs declined. He was worried about what might come next for all three survivors, and he wanted to make sure his wits weren't dulled.

Riggs found a phone and called his wife. "I wanted to let you know that I'm okay," he told Karen by way of a greeting.

She was caught off guard. "What are you talking about?"

"Well, I wanted to let you know that I'm okay in case you heard it on the AP, about this accident in Boston."

In fact, the Associated Press had moved a story on the wires about the accident. But it didn't appear until 6:16 p.m., once the hospitals

had confirmed the two diver deaths. And with the Kennedy coverage dominating the news that day, it didn't make a big splash even in
Boston. In Nevada, where Karen Riggs was, it wasn't even a ripple.

Riggs had no interest in unloading a lot of details on his wife. His
only goal in the call was to reassure her that he was alive. Before he
could get into it any further, a group of doctors and nurses entered
the exam room. "I have to go," he told Karen hurriedly. "I have to get
some testing done." He hung up.

Over the next couple of hours, Riggs came to feel like a piece of supermarket produce in the hands of a discriminating elderly shopper—
prodded and squeezed and probed and held up under the lights. There
were blood tests and lung tests and brain tests. And an endless supply
of questions.

Ever since they'd made it back to the shaft, the divers had heard
the same question everywhere they turned: *What happened? What
happened? What happened?* But by nightfall in the hospital, that question had become more insistent. And it was now being asked by people with badges.

They started with DJ. Just before seven o'clock, two state police
detectives, accompanied by the union's Dan Kuhs and a few other
people Riggs didn't recognize, brought DJ into a different room for
questioning, while Riggs and Hoss continued to be examined by the
doctors and nurses. By nine o'clock, as the doctors seemed to be ruling out any signs of lasting physical injury, Riggs had to brace himself
for the second wave of scrutiny. Given the late hour and the fact that
the doctors still weren't finished evaluating Riggs and Hoss, the state
police agreed to interview them together.

The two detectives—a man and a woman—introduced themselves, as did their boss, a sergeant, who had joined them at the
hospital and would be sitting in on the discussion. There was also a
representative of OSHA, in addition to Kuhs from the union. It didn't
take long for Riggs to conclude that the female detective, who said

her name was McCauley and who spoke with a Boston accent, would be the one leading the discussion.

Riggs offered up the details that had most stood out in his mind: how Harald had blown off his suggestion for an in-line analyzer, and how immediately after the accident he'd found the liquid oxygen tank as frozen as an ice sculpture. Hoss recounted his last communication with Tim, when he'd sounded panicked and disoriented and shouted that the oxygen percentage had fallen to 8.9. Riggs knew he and Hoss and DJ had nothing to hide. They'd been given a lousy system and seen their complaints ignored. The direct result had been the death of two guys and the near-death of the rest of them. Yet if all of that was obvious to the three survivors, despite the shattered condition they currently found themselves in, it didn't seem so obvious to this lady detective and the other investigators staring at them from across the room. As the interview dragged on, changing in tone from conversation to interrogation, Riggs began to feel as though he and the other divers were being put on the defensive. It wasn't as though they were being treated as hostile witnesses. It was more that in the cops' questioning, there was no acknowledgment of the hell the three survivors had been through, no sense that they were victims in all this. As the detectives drilled down with their questions, pressing the divers to explain why they had used a breathing system that seemed so risky, Riggs marveled that the survivors were essentially being asked to defend the very system that they themselves had complained about before the accident. *What the hell is going on here?*

After about an hour of questioning, the session ended. Riggs and Hoss were given a lift back to their HoJo's motel, while DJ was driven home to Waltham. Back at the motel, Hoss called a friend and veteran diver who served as his mentor. The guy advised him to write down everything that had happened during the day, while it was still fresh in his mind. After hanging up, Hoss suggested Riggs do the same. This mentor had lived through enough accidents to see that sometimes the responsible parties try to blame the victims in order to cover their own asses. Hoss now understood that the best way to get the truth out would be to commit it to paper.

It was late when Lorraine heard her front door opening. She'd been upstairs in her bedroom for hours, but she knew she'd have no hope of falling asleep until she saw DJ cross her threshold.

He came into her bedroom and leaned against the wall. Clearly, he was struggling to keep himself together. Then he simply dissolved, letting his body slide slowly down the wall and to the floor. In giving out, his legs seemed to be speaking for his heart. He lowered his head and started to sob. "Why did it happen?" he cried, smacking his fist against the floor. "Why did this awful thing happen, Mom?"

Lorraine rushed to comfort him, to reassure him that he'd be okay. In her mind, though, she was struggling to reconcile how the son who had been such a source of strength for her all these years could suddenly seem so broken.

DJ regained his composure enough to excuse himself. "You may as well go to sleep," he told his mother, "because I don't think I'm going to be sleeping for days."

As soon as he closed the door to his bedroom across the hall, Lorraine heard him begin bawling again. His cries punctured the air, closing the space between them, making it seem as though he were still hunched over on the floor next to her bed. "Why, God, did this have to happen?" she heard him saying. "Why did you have to take these guys?"

Stepping into the funeral parlor, DJ was relieved to have his girlfriend at his side. The combination of sleep deprivation and indescribable sadness had left him feeling as though he'd been sleepwalking for the four days since the accident. As much as Dana and his mother tried to comfort him, though, DJ was racked by survivor's guilt. Why had that tunnel taken the lives of Billy and Tim but spared his? He was haunted by the memory of Billy asking him to change places at the last moment.

His mother, a faithful Catholic who'd attended her share of wakes over the years, had handed DJ a prayer book to give Billy's family as a gift. As close as DJ had been with Billy, this awful occasion marked his first time meeting Billy's parents, Olga and Bill, and his younger sister, Jolene. DJ instantly took note of Bill's mustache. It was more tapered—and grayer—than his son's, but it must have been the inspiration for Billy's. Yet DJ was struck by how much more Billy favored his mother physically. He wasn't as tall as his father and had his mother's dark complexion and oval face. Remarkably, they even shared an unusual birthmark on their brown eyes—Billy's was beside the pupil of his left eye, the mirror image of the one Olga had on her right.

As he scanned the packed room, DJ noticed the unusual alignment. Billy's parents and sister were arrayed on one side, while Michelle stood with Ken, Deb, and Tap on the other. As he knelt before the open casket, DJ was surprised to find Billy wearing a nice shirt and a tie that featured purple, yellow, and mauve pansies. He was used to seeing Billy only in coveralls. When he walked over to pay his respects to Michelle, he struggled with what to say. So he tried to lighten the mood, telling her, "You should have buried him in his Carhartts."

Bill Juse felt dazed as he stood in the receiving line next to his daughter and estranged wife, listening to the kind words being said about his son. He had to wonder what it meant that all these people who had been so close with Billy were nothing but strangers to him. He felt like an outsider, standing alone in Times Square on New Year's Eve, except in mourning rather than in celebration.

Bill had always thought he had a good relationship with his son, unlike the stormy one he'd had with his own father. He knew he'd missed a huge chunk of Billy's childhood because he was on the road so much. In his mind, he'd done that to provide for Billy and his sister so they wouldn't want for anything. He had to admit that the breakup of his marriage to Olga had deprived the kids of a single home base they could return to. So he understood the appeal for Billy in forging a second family here in New Hampshire. Still, Billy's own family had managed to remain close in their own ways—Olga as a nurturer, Bill as a provider, and Billy as a dutiful big brother to Jolene, never failing to send his sister pink and purple flowers on her birthday. Bill had intended to keep providing, hoping to turn over his small Florida beverage business to Billy someday, when his son finally tired of the brutal New England winters.

Bill liked to think he had taught Billy everything he needed to know in the world, and that they understood each other intuitively. So he was crushed when he discovered that Billy had somehow felt the need to conceal from him the news that he had sold his share of the Black Dog business. Had Billy really thought his father would have been any less proud of him because he'd left the ranks of small business owner? Billy should have known that his trust in his father mattered to Bill a whole lot more than whether he was signing paychecks or punching a clock.

Standing in this receiving line, listening to the comfort of strangers, Bill was growing more agitated. When he spotted DJ chatting with a couple of other divers as they headed out the door, Bill bolted out of line to follow them. In the parking lot, he approached the divers with the question that had burned in his brain for days: "Guys, what really happened?"

There was an awkward silence. Bill could tell how drained and wounded the divers were. He also sensed that they had been told not to talk, or maybe they were just too exhausted and despondent to relive their hell. All they really said was that there were "things that should not have happened."

The next morning Bill returned to the funeral home to prepare for the procession. Ken, Deb, Michelle, and Tap had made all the decisions about Billy's arrangements—the church, the cemetery, the fund they'd set up in his memory. Bill appreciated all their effort and obvious affection for Billy. Still, this was *his* son, his *only* son, the last male to carry the Juse name. At the last minute, Bill walked over to the cluster of men who had been selected to be Billy's pallbearers, and he delivered the news: *I'm going to be a pallbearer for my son.* He knew it was uncommon for a father to serve in that capacity. But to him, the more important point was how uncommon and unnatural it was for any father to bury his own son. Unbearable, really.

He had one reason for wanting to be a pallbearer, for wanting to carry the casket from the church and then to the gravesite. He wanted to be the last one to let Billy go.

Amid all the heaviness, Ken tried to inject lightness where he could. A few days after eulogizing Billy as the son that God gave him later in life, he accompanied Michelle and Tap to Boston to retrieve Billy's pickup truck from the racetrack parking lot and drive it back to New Hampshire. Once they arrived, Ken found a big Tupperware container with the remnants of one of the meals that only Billy could have enjoyed. It wasn't as off-putting as a deer heart, but there were some bits of Vidalia onions and cucumbers and some spice-heavy sauce that had to have been disgusting even when it was fresh. When Ken flipped off the lid, he was attacked by an odor fouler than a carnival Porta-Potti. "Oh my ever-loving God," Ken cracked. "He should have died of *this!*"

. . .

They sat in silence, on folding chairs arrayed in a circle in the driveway. Every few minutes, someone would stand up and offer a re-
membrance about Tim and then sit back down. More silence. Then
someone else would stand.

Hoss didn't know what to make of this Quaker-style memorial
service, which was being held a month after the accident, at Tim and
Judy's house in Washington. He knew Tim was neither a Quaker nor
a churchgoer of any stripe. But Judy explained that Tim had liked the
serenity of a Quaker meeting they had once attended. And Hoss could
see how this low-key approach would have suited his friend's style.
The setting sure seemed right: the secluded five-acre swath of God's
country where Tim had spent countless hours toiling and taking care
of the many creatures that also called it home.

Still, Hoss could find no peace himself. He was distracted by two
of the other people sitting in folding chairs. Ever since the accident,
as he had struggled to assign blame, his mind had returned again and
again to Harald and Roger. He was furious with Harald for saddling
the divers with such a half-assed system and for ignoring all the concerns they raised about it. And he was angry at Roger for having let
Harald put their lives in jeopardy, despite the warnings the divers had
given him about how poorly things were going on the job. Hoss was
also pissed at the Norwesco owner for the smaller but inexplicable
slights in the days immediately after the accident. He still couldn't get
over Roger's response when Hoss had told him back in Boston that
the Norwesco divers planned to stay there so they could attend Billy's
wake: *Who's going to pay for the motel rooms?* Hoss knew Roger wasn't
an unfeeling man. When the Norwesco crew had gone to the Boston
funeral home to pay their respects before Tim's body was cremated,
Hoss had seen both Roger and Harald break down in sobs. Still, Hoss
had trouble reconciling that with the other instances of tone-deafness.

On the flight back to Washington, as Hoss had crammed his six-foot-five frame into coach, Roger had reclined in first class.

Roger would later say he had no memory of the motel bill remark, insisting it would have been out of character. But he was the first to admit that he hadn't been feeling like himself in the days after the accident. On the same day Hoss remembered the Norwesco owner balking at the motel costs, Roger had been summoned to meet with several high-ranking Kiewit executives who'd flown in on the company's private jet. Someone had handed Roger a list of three lawyers and told him to choose one. The whole exchange had felt so coldly matter-of-fact that as he left, his head was spinning.

Hoss could tell that Roger was now trying to make amends. He was covering some of the cost of this memorial service for Tim, including the spread of salmon that had been laid out and the hotel rooms for him, Riggs, and DJ. Even Harald appeared to be trying, giving Judy and Tim's parents copies of a photo he had taken of Tim back in Tap's yard, looking rugged and relaxed as he leaned against the white Humvee.

Hoss couldn't get over how forgiving Tim's parents, Bob and Shirley, and his younger brother, Todd, managed to be. Like Tim, they were clearly just kind, easygoing people. Todd delivered an uplifting eulogy, saying his first thought after hearing the horrible news of his brother's death was "What a great life Tim had." Part of Hoss wished he could be so positive in his outlook. Then again, maybe he knew too much about what had gone down on the tunnel job to offer that kind of blanket pardon.

Hoss watched as Riggs stood to deliver a special reading. The reflection had actually been published as a letter to the editor in *The Boston Globe* two days after the accident. Written by a stranger named Parker Pettus, it contrasted the "lavish" wall-to-wall coverage of the death of John F. Kennedy, Jr., who had been granted a mariner's funeral aboard a warship, with the "unadorned" news report about the deaths of Billy and Tim.

> *These men were not rich or famous or privileged. Certainly they would have preferred not to have been in a dangerous tunnel hundreds of feet below the surface and miles from any help.*
>
> *They died while doing a hazardous, unheralded job, and their contribution to a clean, revived Boston Harbor will last for generations. They will not be immortalized in the media, they will not be buried at sea from the decks of a warship.*
>
> *These workers are the kind of heroes who are so often taken for granted. We would do well to think of Boston's clear, blue, living harbor as a monument to the courage and sacrifice of the ordinary heroes who made it a reality.*

After Riggs finished, there was more silence. Then another diver friend stood and declared, "I think Tim would just want us to go drink a beer."

Even if Hoss's mind had fixated on the failings of Roger and Harald, he knew there had to be others who bore some responsibility for what had happened. Toward that end, he, Riggs, and DJ met the day after Tim's service with two lawyers who had flown in from Boston.

Judy had already signed up with one of the lawyers, a forty-year-old by the name of Bob Norton, who had been a competitive hockey player in college and still had the build to show for it. DJ had signed up with a friend of Norton's from another firm, a tanned and slim forty-four-year-old named John Prescott, and had recommended that Billy's parents do the same.

Hoss attended the meeting with his wife, Heather, bringing along his own healthy suspicion of lawyers. His main motivation in suing wouldn't be to get money but rather to have the satisfaction of seeing the responsible parties stand up and acknowledge their failings. After a three-hour meeting with the lawyers, Hoss had to admit he was impressed.

While DJ and Riggs flew home, Hoss and Heather stayed with Judy for several days, helping her go through Tim's things. Hoss knew that as close as Tim and Judy had been, there was one secret that Tim had kept from his wife.

"Judy," Hoss said finally, "I have something to tell you."

"What?"

"I don't know if you know it, but Tim has some guns."

"Oh, I know," Judy replied nonchalantly. When it came to their dueling views on guns, she and Tim had tried to keep things light. Tim would threaten to sign her up for the National Rifle Association as a Christmas present. Judy would counter with a threat to send a check in his name to gun-control activists Jim and Sarah Brady.

"No," Hoss pressed on. "Tim has a *lot* of guns."

Judy's ears perked up. "What's a lot?"

Hoss retrieved a cache from the plywood compartment Tim had built under the mattress in the guest bedroom. He helped Judy find other stashes in the deep recesses of closets. At one point, after he had built a pile in the living room containing more than fifty weapons, a couple of Judy's neighbors came by to drop off a casserole. Seeing them nervously eye the giant cache, Judy had to laugh. It looked as if her house in rural Washington State was the headquarters for some end-of-days homegrown terrorist organization.

The man cage jerked and swayed as it went down the shaft. Since Mary McCauley hated heights—and heights in reverse—the state police detective tried to distract herself, downloading observations into her small notebook. As unsettled as she was, she realized this trip on the day after the accident would be a lot more frightening if she had been granted her request to journey all the way out to the actual crime scene, to the spot where Billy and Tim had lost consciousness. That request turned out to be a nonstarter. Even though she was taking the lead on the death investigation, OSHA had jurisdiction over the tunnel work site. The federal safety agency had ruled that no one would be allowed to venture past the base of the shaft until mechanical ventilation—the kind of bag line that the sandhogs had ripped out before the divers went in—had been reestablished.

McCauley heard people at the MWRA grousing about the decision. How were they ever supposed to finish the tunnel—and by extension, the multibillion-dollar Boston Harbor cleanup—if they couldn't get back in? But McCauley didn't care about this question. She had plenty of others weighing on her mind. Since all five divers had been breathing off the same system, she still couldn't understand how the pair doing the easier tasks had somehow had less time to switch to their emergency air than the three doing the backbreaking work at the end of the tunnel. Even more pressing for the detective was the overarching question of accountability: Who should be held responsible for Billy and Tim's deaths? After all her hours with the surviving divers, she had concluded that they'd been pawns in a much bigger game of risk.

The man cage touched down, and McCauley, wearing a yellow hard hat, headed over to the deck where the gray Humvee and trailer were parked. After a crime scene photographer documented everything inside the Humvee, McCauley climbed in. She couldn't get over how jam-packed it was. Every inch of the interior seemed to have some kind of hose or valve or tool taking up space. There were also everyday artifacts, like the crumpled-up green Nature's Valley granola bar wrapper on the driver's-side floor, which now served as an unnerving reminder of Tim and Billy's presence. Examining the mixer, the manifolds, and the big tanks of liquid oxygen and nitrogen on the trailer, McCauley had two intense and yet contradictory thoughts. First, the breathing system was incredibly complicated, with hoses going in every direction and attached to all kinds of contraptions. Second, this incredibly complicated system seemed so jury-rigged. The gas cylinders looked old and rusty. The manifolds where the various hoses met were made of plywood—*plywood!* And the quarter-turn valves were labeled with pieces of tape. McCauley couldn't believe it. *All the millions that they poured into this project, and this is what these divers were given to keep them alive? Where did all these supplies come from—Home Depot?* It was like an eighth-grade science fair project gone horribly wrong.

13

This Crazy Idea

Standing on a hill in a pristine harborside park, Doug MacDonald let his eyes scan the horizon. The park was on Nut Island, a peninsula in Quincy, the same south-of-Boston city where nearly two decades earlier a lawyer had stumbled across sewage during his early-morning beachside jog and ended up triggering the court-ordered cleanup of Boston Harbor. From the spot where smelly holding tanks had once stood, the MWRA chief now could gaze out at a sprawling expanse offering a mile of paved trails and unobstructed views of the Boston skyline. In place of the foul tanks, there was a single, essentially odorless pump-house building. It received the sewage from toilets flushed across communities south of Boston and then sent it on to Deer Island through the new five-mile-long Inter-Island Tunnel, which had been up and running for more than a year. If only MacDonald could say the same for the other massive tunnel built on his watch.

On this Friday in mid-September 1999, nearly two months after the deaths of Billy and Tim, MacDonald sat under a tent with a hundred people who had gathered to celebrate the rebirth of Nut Island.

One Quincy official told the crowd, "This place is like an instant hospital that offers eco-environmental care." Whether you're walking the trails or just sitting and watching the boats bobbing in the harbor, the man said, "All your cares just drift away."

The magic was somehow not working on MacDonald.

When he had taken the job as head of the water and sewer agency seven years earlier, he had arrived as a neophyte manager unschooled in megaproject public construction. He had been able to surmount one hurdle after another as he steered the unwieldy project to its conclusion. The stress had left its mark on the workaholic whose brown hair was turning gray. But he felt it had all been worth it. Sure, the Deer Island tunnel was on track to cost $100 million more than its original $202 million estimate. But the price tag for the overall harbor cleanup—thanks to some significant cutbacks in scope and better-than-expected contractor pricing—had been shaved from its early $6 billion estimate down to $4 billion. Shellfish beds, after years of having raw sewage dumped on them, were open again. Swimmers, after years of avoiding the city's beaches for fear of contracting hepatitis, were returning in large numbers.

All this was thanks to the dramatically improved sewage treatment process at the new Deer Island plant, and to the decision, years earlier, to end all the stomach-churning sludge dumping into the harbor. Yet the MWRA was still forced to dump the treated sewer water directly into the harbor every day, leaving it cloudy and producing algae blooms. MacDonald knew that this would all change once the Deer Island tunnel was up and running. But that couldn't happen until somebody figured out how to get those cursed plugs out.

Now, as he looked out at the crowd in Quincy, MacDonald couldn't help but think of the ceremony he should have been presiding over. Despite all the headaches, all the interim milestones met and missed over the years, September 1999 had always been the ultimate deadline for the Boston Harbor cleanup. Although the schedule had been set long before MacDonald assumed his post, he was bound by it. Like a prisoner making scratches on his cell wall, he had come to view it as his release date, the time when he and his agency might finally be freed from their status as defendants. Ever since the diver deaths, though, the tunnel had been in limbo. As he looked out from under the tent, the light rain that had been falling turned into a downpour.

MacDonald had been trying to walk a careful line between respecting the loss suffered by the divers' families and relentlessly working to get the project back on track. A few days before this Nut Island dedication, he had told a reporter for *The Christian Science Monitor*, "We may be many weeks away or months away from filling the tunnel." Truthfully, he had no idea how far off the end might be. What's more, the tunnel critics whom he had managed to vanquish years earlier were using the diver deaths as an opportunity to reload. The same *Monitor* article quoted the cofounder of an antitunnel citizens' group reviving concerns about how all that treated sewer water released into Massachusetts Bay might consign the relatively pristine waterway to the same degradation that had ruined Boston Harbor. Several scientific studies had concluded that the whales and the fish and the fauna living in the wide-open waters of the Atlantic should be fine. What's more, the MWRA had agreed to spend millions of dollars every year testing the tunnel discharges to ensure that they weren't harming marine life. The monitoring program, which environmental officials called the most extensive in the nation, would routinely test discharges for 158 pollutants and collect some two million pieces of data annually. Still, the critics remained unconvinced.

With some work, MacDonald felt confident that he could beat back those critics once again. But he had to admit he had no idea how the project would be completed. The answers to even the most fundamental questions—such as who could authorize a new plug-removal plan—eluded him. State police and OSHA were conducting separate probes into the deaths of Billy and Tim, and OSHA wasn't letting anyone back into the tunnel until workers could survive inside it without needing to bring along breathing equipment.

After months of fruitless meetings, MacDonald came to an important conclusion. As much energy as he'd spent aggressively fencing with Kiewit CEO Ken Stinson over the contractor's claims for more money, MacDonald figured that this was not the time to haggle. The price of inaction would be far higher than the price of any plug-removal project. Once the plugs were out, they could worry about who would pay what. Until then, he decided, the MWRA would cover the costs to keep the project moving.

In light of OSHA's stance, MacDonald's team seemed to have no choice but to install a new ventilation system similar to the massive bag line that had supplied air to the sandhogs during tunnel construction. But it had taken more than a year for crews to dismantle and remove that bulky system. It would probably take crews six months to install a new bag line and then remove the plugs. If it took them an additional six months to dismantle the new line, that would mean adding a full year to the schedule of a project that was already woefully late. What's more, for the six months that it would take to remove the new bag line after the plugs were gone, crews would be working in the tunnel with no secondary protection from the ocean above. The MWRA had found that option unacceptable years earlier, when Kiewit had proposed yanking out the plugs while it was pulling down the old bag line. So how could it be acceptable now?

That fall, during a blue-sky brainstorming session with his lieutenants and engineering consultants, MacDonald tossed out a question. Unlike most of the others in the room, he had no framed engineering degree, so he asked it with some tentativeness: *Is there any way we can use one of those diffuser heads to get air into the tunnel from the other side?* He worried that the question might be laughed off for its naïveté. Instead, one of his top engineers said the idea of off-shore ventilation was promising.

It owed its viability, in large part, to the work the divers had done before Billy and Tim were killed. Because the divers had managed to remove three of the fifty-five safety plugs, the first three risers now offered a direct connection between the tunnel and the seabed. Before long, MacDonald's offhand question triggered a rush of sketches and calculations by a team that included two of the tunnel managers who'd battled Kiewit during construction, Kaiser's Dave Corkum and his counterpart with designer PB, Eldon Abbott. No one was sure if this unproven approach could work. How could they even open up one of the diffuser heads to pump in good air without letting ocean water flood into the tunnel? Still, after months of despair, at least there was finally some hope.

In addition to the engineering challenges, they faced a huge potential roadblock: the MWRA would have to get Kiewit officials to embrace the idea. None of them had been present at the blue-sky meeting, and MacDonald was fairly sure that they wouldn't greet the suggestion as enthusiastically as everyone else had. In fact, Kiewit might simply walk away from the tunnel job. If the MWRA couldn't get the contractor on board with some sort of solution, MacDonald knew his agency would be stuck in neutral. Given the enormous stakes of the harbor cleanup, the relentless demands of the calendar, and the revived campaign by the tunnel's critics, he also understood that neutral really meant reverse.

Back in Omaha, Ken Stinson had a lot more on his plate than just one endless tunnel out in Boston that somehow refused to get finished. A year earlier he'd been promoted from CEO of Kiewit's construction group, which had seen the most successful year in its history, to chair-man and CEO of the whole corporation. At fifty-seven, Stinson was running a company that had posted $3 billion in revenue, had no long-term debt, and had just made the *Fortune* 500 list for the first time in its history.

But it weighed heavily on Stinson's bald head that he hadn't been able to help his team get the Deer Island project right. In the decade since the tunnel job began, one of Kiewit's junior partners on the Deer Island project, a San Francisco company called Atkinson, had filed for bankruptcy. On the bright side, partly as a result of all the trips Stinson had made to Boston to negotiate with MacDonald, the MWRA had agreed to pay Kiewit an additional $34 million. But Kiewit's separate $62 million "rock too hard" and $60 million "water too wet" claims were still pending before an independent review board. After the MWRA's auditors reviewed Kiewit's books, they confirmed that the contractor was perhaps $80 million in the hole on the job. Moreover, because of the disastrous end to the diver mission, it wasn't clear how, when, or even *if* this project would ever get finished.

Now word was filtering back to Omaha that the MWRA gang had been spitballing about an experimental approach to ventilate the tunnel from offshore. MacDonald had apparently tossed around the idea with some of the same managers Kiewit had been battling for years. Stinson was a civil engineer to his core, so the challenge of finding solutions to towering problems spoke deeply to him. But so did his common sense. He felt Kiewit had been burned more than once by the MWRA and its consultants, and he wasn't about to let his team get too close to the flame again.

Stinson even felt scorch marks on his own skin. Back during the height of the "memo wars" in 1997, when his managers on the Deer Island job had been warning of catastrophe and planning a public relations assault by PowerPoint against the MWRA, Stinson had been the one to veto that offensive. While he'd agreed with his managers that the only logical sequence would be to remove the safety plugs while the tunnel still had ventilation, it had become clear back then that the MWRA and its advisers were not going to allow it. So Kiewit had the choice of either walking off the job or engaging in a high-wire campaign to shame its client. Neither approach, Stinson had decided, would have been up to Kiewit standards. Instead, the Vietnam veteran had chosen a path of deescalation. As crazy as Kiewit managers had felt the idea was to send a dive team into the tunnel, the Omaha powerhouse's expertise was in construction, not exotic breathing systems. Stinson had decided his lieutenants should seek input from some experienced diving companies. That had led them to Harald Grob and Norwesco. That, in turn, had led to the experimental mixed-gas diving operation, which had left Billy Juse and Tim Nordeen dead.

Now, although Kiewit couldn't even access the tunnel without permission from OSHA, it still bore responsibility for finding a resolution. Anyone who suspected that Kiewit might walk away from the job clearly did not know Stinson. However, with the deep financial hole that Kiewit officials found themselves in, and the enormous legal exposure they faced as a result of having consented to someone else's sequence, Stinson was going to make sure his company was more cautious than ever. He wasn't going to ignore his gut again.

Not long after MacDonald's blue-sky meeting, a thickset middle-aged engineer named Dave Beck began hearing talk about the stymied Deer Island tunnel job. Beck was once again working in Boston after having logged several years on a tunnel in Hong Kong. But he knew all about Deer Island. In the early 1990s, he had served as a project manager for the installation of those fifty-five risers and diffuser heads, logging lots of hours on the giant jack-up barge that had been parked in Massachusetts Bay. Now, searching for his next assignment, he picked up the phone and dialed Eldon Abbott, the veteran project manager for PB who had become a fixture on the Boston job.

Abbott told him there were three ideas being tossed around for how to remove the remaining fifty-two plugs. The first option was to send crews in once again on supplied air, but OSHA was standing firm against that. The second was to install a new bag line along the entire tunnel, but Kiewit wanted $30 million to do that and it would present all kinds of problems relating to its removal. The third was what he called "this crazy idea," which had come out of a brainstorming session. It called for the tunnel to be vented not from the shaft side but out at sea, by tapping into one of those fifty-five diffuser heads. Sitting in his office, staring at the framed photo he kept of a diffuser head, Beck smiled and said, "Eldon, that's not a crazy idea."

Beck wasted no time in reaching out to the joint venture partners who'd been behind the riser-installation job, to gauge their interest in being reenlisted. He asked the owner of the IB-909 if that three-thousand-ton monster barge would be available. The company told him the barge was off the coast of Venezuela at the time, but for the right price, it could be brought back to Boston.

A few months and lots of engineering sketches later, Beck and the reconstituted joint venture sent the MWRA a formal proposal for venting the tunnel from the IB-909 barge parked in the waters over riser number 3. The MWRA liked the plan but stressed that nothing could happen until they figured out a way to get Kiewit on board.

From the bench, the bow-tied judge looked over at the attorney's table. "I need a plumber," he quipped, "I don't need a lawyer." In fact, Armando David Mazzone had himself been playing the role of master plumber as much as credentialed jurist during his fifteen years overseeing the Boston Harbor cleanup case. Whenever relations among the various agencies and actors involved in the cleanup became hopelessly clogged, the judge would step in. Artfully using his federal power to nudge and cajole, he somehow had always found a way to clear the obstruction.

With the tunnel project now at a complete impasse, and the only communication between the MWRA and Kiewit resembling the huffy indignation of divorcing spouses, Doug MacDonald had maneuvered to get both sides before the judge. In his time as a defendant in Mazzone's courtroom, MacDonald had become a big admirer of the man, a smart but self-effacing Harvard-educated son of Italian immigrants. The lanky guy known as Stretch had been a star tight end in high school, but his role quarterbacking the harbor cleanup case had given him a new nickname: the Sludge Judge. By the time of this hearing in March 2000, David Mazzone was seventy-one years old, refusing to retire until he could close the case.

The hearing began with a briefing by the MWRA's team on the two options for ventilating the tunnel: either the reinstallation of a giant ventilation bag line or the offshore venting of the tunnel from the IB-909 jack-up barge. An engineer for the MWRA explained that, since the accident, the tunnel had no lights, no ventilation, and about four million gallons of water seeping into it every day—the reason for Kiewit's "water too wet" claim against the MWRA. Whenever those safety plugs could be removed and hundreds of millions of gallons of treated sewer water could start flowing through the tunnel every day, the hydraulics of that eastward flow would render the current seepage insignificant. But until then, all that water seeping in would naturally continue to flow westward back to the shaft, where it had to be pumped up to Deer Island and then back into the sea.

Installing a new bag line, the MWRA engineer told the judge, would require some kind of "scissors car," a truck with staging that could be elevated enough to allow workers to hang the line from the ceiling of the twenty-four-foot-diameter tunnel. As they had with the installation of the first bag line, crews would ventilate the tunnel in sections, bringing air with them as they moved east. Doing so would also require a new power line with transformers or powerful generators. They estimated it would take small crews five or six months to install these services and establish free air. Workers could then yank out the remaining fifty-two safety plugs, before spending months more to remove all the utilities once and for all.

The offshore option would involve fabricating what amounted to an enormous steel "straw," which would be lowered into the sea to make a vertical airtight and watertight connection to diffuser head number 3 on the seafloor. Once the connection was made, crews would use giant fans on the jack-up barge either to blow enough good air into the tunnel or to suck enough bad air out, drawing fresh air from the Deer Island shaft.

"How do you get air out nine miles?" the judge asked. "Is there a pump strong enough to suck air in from the tunnel shaft?"

The engineer explained that, in theory, the answer was yes, but they still had a lot of number crunching to do.

As for how long this offshore option would take, the MWRA estimated about three months to mobilize and get the tunnel ventilated and then three weeks for crews to get the fifty-two plugs out. The biggest difference was that once the plugs had been pulled, the offshore option would involve almost no cleanup. They'd just pull up the giant steel straw and leave.

The MWRA's lawyer said time dictated that they settle on one option, and the agency elected to pursue the offshore plan. He told the judge that the MWRA saw enough promise in it that the agency had already cut a check for three-quarters of a million dollars to reserve the IB-909. Despite their confidence in this route, the MWRA was seeking the judge's intercession because relations with Kiewit were in such a bad state. "These big contractual issues do loom over us," the lawyer told the judge, "and particularly the question of who will own—who will be responsible for the work."

With that, Bob Popeo, Kiewit's high-powered lawyer, stood up to address the judge. Popeo was a partner with the firm Mintz Levin, and his colleague had been the same lawyer-lobbyist who, years earlier, had put together the devastating-but-shelved PowerPoint presentation about the tunnel plugs.

Even before Popeo began, though, the judge shifted into action. From the gallery, MacDonald watched with delight as the distinguished federal jurist adopted the humble approach of a simple country judge. "I'm not quite sure what to expect from Mr. Popeo," the judge began, "except I'm always flattered to see him."

The judge said that, through the years, "the less I have to do with the project, the better it works." As if it needed repeating, he stressed that the job was supposed to be done by now. "We should have turned the key on this project, and you'd have seen the last of me, and I can't tell you how much I was waiting for that day." Then he hammered home his point. "Mr. Popeo," the judge said, "I really earnestly need your help. I'm asking for it."

The approach was classic Mazzone, solicitous rather than scolding. How could anyone deny such a reasonable plea for help?

"You will have it, Your Honor," Popeo replied.

Still, Popeo hadn't risen to his level of success by allowing himself to get outmaneuvered in the courtroom. While MacDonald had orchestrated this hearing, Popeo was going to use it to get on the record all the failings by the MWRA and its consultants that Kiewit felt had forced the doomed diver plan and thrown the project into crisis. "This wasn't its choice," Popeo said. Yet by reluctantly going along with the plug-pulling sequence insisted on by the MWRA, the lawyer said, Kiewit somehow found itself holding the bag.

"I can assure you, Your Honor," Popeo said, "that we're *not* going to be subjected to that again. The plan's going to be safe. We're going to believe it's safe. Everybody is going to sign on to it or we're not going to do it."

Stressing that the offshore option had never been tried before, Popeo cited a host of unknowns. After they sank the steel straw down to the seabed, he asked, "Will the tunnel handle it?" Resurrecting the contentious matter at the heart of the memo wars, he asked, "Who owns the risk if there's damage to the tunnel when you sink that down? Who owns the risk if this doesn't work?"

Despite all the criticisms that Popeo managed to level at the MWRA, for MacDonald, the takeaway line was the lawyer's comment to the judge that "the contractor intends to do the work." That message was decidedly different from the one the agency had been hearing from Omaha prior to the hearing.

Still, the question about risk hung in the air. After all, a distorted concern over risk had laid much of the groundwork for the diver-mission disaster. The desire of all of the project's players to insulate themselves from liability and to off-load risk onto other parties had made their relationships thoroughly dysfunctional. Unwittingly, that dysfunction had greatly elevated the risk of a bad outcome for everyone involved.

Heading toward the rising sun, the white helicopter glided through the sky, leaving behind Deer Island with its hulking egg-shaped digesters. The chopper flew above the swarms of seagulls that hovered over an unusually large population of sailboats and yachts clogging the harbor. Fifteen miles east of the Boston coastline, and nearly ten miles east of Deer Island, the chopper landed in a familiar spot, on the helipad of the IB-909. The green and white shopping-plaza-size barge, jacked up above the Atlantic waters of Massachusetts Bay, was back where it had been eight years earlier. Deep below the rig, beneath about a hundred feet of water and sitting on the ocean floor, was the domed mushroom-shaped diffuser head for riser number 3.

Out of the chopper stepped Ron Kozlowski, DJ's old buddy. The short, ponytailed veteran diver and ex–Navy SEAL had returned to help finish the job he had started nearly a decade earlier. Like project manager Dave Beck, Kozlowski had been here for the riser-installation operation back in 1991 and 1992. This trip effectively marked Kozlowski's third tour with the project. The cocksure diver was grateful that he had trusted his instincts and quit Tap's dive team just weeks before the plug-pulling operation was set to begin in the summer of 1999. Although the lack of penetration pay had been his loudest complaint back then, Kozlowski had also been sufficiently concerned about the soundness of the plan to give DJ his unforgettable "boogie board and .45" warning. Considering the wreckage of that doomed mission, though, Kozlowski took no pleasure in knowing his intuition had been right.

After the chopper took off, Kozlowski lingered on the helipad to take in the view. Then he retrieved his video camera from his backpack and hit the record button to preserve the memory. "Watching the Tall Ships come in," he said, beginning his narration. "Awesome!" The flotilla of regal ships from ports around the world hadn't sailed into Boston since the last time Kozlowski had been standing on the deck of the IB-909. "There's the Russian vessel," he said, training his lens on a ship with the portholes painted to look like gun ports. "Most people have to pay to see this."

On the deck of a transport barge idling next to the IB-909 lay an enormous green pipe that was seven feet in diameter and about a hundred feet long. Attached to the bottom of the green pipe was a seventeen-foot-long white cone that flared out to a width of just over sixteen feet at its very end. Even if it was to function like a steel straw, the entire package had the look of a giant plunger.

With his video camera, Kozlowski slowly panned the steel pipe and cone. "This is the reason we're here today," he said, continuing his narration. "To install a one-hundred-ton vent tube that will inflate underneath the ocean, on top of a diffuser head to ventilate the tunnel." The engineers had chosen the route of sucking the bad air out, rather than trying to pump good air in.

Despite a length that stretched along the entire deck of the barge, the green plunger was only one piece of the 172-foot-long vent pipe that would have to be lifted upright, as straight as a sequoia, and then be lowered through 120 feet of ocean water, so it could fit over the diffuser head that capped riser number 3. The job of preparing and lowering that pipe would largely be left to the crane operators and riggers. The divers had their own challenges in front of them.

Before long Kozlowski was sitting on a smaller vessel, in front of a bank of monitors and controls at the dive station, right beside a white decompression chamber. As dive superintendent for this job, Kozlowski spent most of his time manning the station, wearing a headset that kept him in contact with whichever worker was in the water.

The divers accomplished several complicated tasks before the giant steel straw was put into place. Making their way down to the ocean floor, they first cleaned the area around the domed diffuser head. Using an underwater vacuum cleaner called an airlift, they sucked up all the gravel backfill and debris around the diffuser head, so the straw's white cone bottom would fit tightly over it. Next up, they had to take off the diffuser's protective dome. The heavy acrylic topper wasn't designed to be removed, so the divers had to cut it off, with all the precision they could manage, using an underwater chain saw. The last seal between the ocean and the tunnel was a manhole cover inside the diffuser head. They prepared to remove it by loosening its bolts and attaching a steel hoop to it.

Crews then lowered the 172-foot-long plunger—the green steel straw—into place, inflating two rubber inner tubes inside the rigid white cone at the base of the plunger. That created a watertight seal with the diffuser head. They pumped out all the water inside that seven-foot-wide pipe and bell, which stretched from the seafloor all the way up to the air above the ocean. Then they rigged a second pipe—this one was white and just four feet in diameter—and lowered it down into the center of the wider green pipe. This smaller white pipe would be the actual conduit for the tunnel ventilation.

With all that in place, a crane operator lowered a cable with a hook at the end, like a long fishing line, to latch on to the steel hoop the divers had installed on the manhole cover. A year earlier Riggs had removed the safety plug from inside riser number 3's elbow. So this manhole cover—sitting just above the diffuser head's discharge nozzles—truly was the final barrier between the tunnel and Massachusetts Bay. Because all the water had been pumped out of the vent pipe, the manhole cover could now be removed as well.

Once it was gone, crews on the IB-909 barge attached a powerful jet-engine fan to a T at the top of the white ventilation pipe. They flipped on the switch, and the waiting game began. The initial readings on the gauges at the base of the vent tube showed how hopelessly starved of oxygen the end of the tunnel and risers had been. Now crews would monitor those gauges, waiting for the oxygen levels to rise.

It was July 14 when they turned on the massive fan from the barge. The first plume of air to exit the giant straw had the disgusting color of rust, the exhaust of all the ferric oxide that had built up at the end of the tunnel. Like the smokestack in some unapologetic third-world factory, the rust-colored plume kept going and going, for more than a day. By July 17, though, the gauges told an incredible story. Shockingly, just three days after they had turned on the jet-engine fan, the tunnel was entirely refreshed. The powerful turbine had managed to remove all of the tunnel's bad air by drawing fresh air from topside on Deer Island, sucking it down the 420-foot shaft, along the nine-and-a-half-mile-long tunnel, all the way up riser number 3, and finally out the vent pipe that was attached to the side of the jack-up barge. There was now plenty of oxygen in the ambient air of the tunnel, allowing workers to travel along its entire length without having to worry about any kind of breathing apparatus.

A team of sandhogs piled into a couple of the same Humvees that DJ and the other divers had used, and they began driving east toward the end of the tunnel. The jet-engine fan on the barge continued to spin, ensuring a consistent flow of good air from Deer Island all the way through the tunnel. The sandhogs' first task, on July 18, was to reestablish the mine-phone communications line. Their second task, the following day, was to remove the white Humvee that DJ, Riggs, and Hoss had left more than nine miles into the tunnel, when they'd begun their race to get out alive. Then, at long last, the plug-pulling could resume. Fortunately for the sandhogs, Riggs and Billy had already done the hardest work, having crawled through the connector pipes to remove the plugs for risers 1, 2, and 3—the ones that were located the farthest from the shaft, in the tunnel's most distant and cramped quarters.

Once all the plugs were out, the MWRA began the slow process of flooding the tunnel. Workers used treated sewer water to do it, opening up the gates to the enormous chlorination basins that had been built near the shaft. When the tunnel was 90 percent full, the manhole cover was put back in place over riser number 3, the steel straw and vent pipes were removed, and the dome was returned to the top of the diffuser head. When the tunnel was completely full, Kozlowski and the rest of his team made a series of dives to remove the sturdy nylon caps from those sprinkler nozzles on the diffuser heads. The divers managed to yank them off with nothing more than a sledgehammer.

On September 6, 2000, Doug MacDonald finally got his opening ceremony for the tunnel, standing alongside Judge Mazzone and other dignitaries. Just before noon, the treated sewer water was officially rerouted from Boston Harbor to begin its 9.5-mile journey deep into the ocean. With that, the cloudy harbor waters began to clear.

The innovative "crazy idea" vent operation had gone better than anyone could have expected. The MWRA's engineers had estimated that once free air was restored in the tunnel, it would take the sandhogs three weeks to remove the remaining fifty-two plugs, an estimate that Kiewit's lawyer had suggested was wildly optimistic. In the end, it took only about three days. Still, the operation hadn't come cheap. While Norwesco's original bid to send in the divers had been under $600,000, the price tag for this vent solution was twenty-five times that figure, clocking in at $15 million.

Before MacDonald and the other dignitaries celebrated during the opening ceremony, they observed a moment of silence for Billy and Tim, as well as for the three workers killed earlier in the decade-long harbor cleanup project. During periods of deep reflection, MacDonald found himself wondering what would have happened if the tunnel had taken the lives of DJ, Riggs, and Hoss in addition to Billy and Tim. If none of those divers had emerged on that awful day, MacDonald knew there was a chance that no one would have been let back into the tunnel, and the mystery of what had happened would never have been solved. He could imagine the questions people would be asking: Did the tunnel collapse? Did the divers drown? Was there a fire? Was there some of kind of sea monster? "I mean," he said, "how would you have ever known?"

MacDonald was relieved to be able to put the throbbing questions behind him. This engineering marvel of a tunnel was finally done, and one of the most ambitious environmental cleanup efforts in the nation's history had finally reached the finish line.

But others in Boston were still on the hunt for answers. Less than a week after the ceremony, the woman who had taken charge of the investigation into the diver deaths submitted a seven-page, single-spaced memo to her superiors. Although the investigation was far from complete, she detailed stunning allegations of malfeasance, of mistakes both cosmically complicated and unforgivably simple. Someone had to be held to account.

14

Justice

For Mary McCauley, the most memorable interview of the investigation took place just two days after Billy and Tim had died. That's when the state police detective found herself in a construction trailer on Deer Island with Harald Grob. By then, she had already heard plenty of complaints from the divers about the controlling Canadian engineer and project manager. But now that she was sitting across from him, she was surprised to find a relatively soft-spoken man.

Working alongside McCauley and her state police partner on the probe were several investigators from OSHA. The arrangement was something of a shotgun wedding between agencies since McCauley had never met the lead investigator, Elena Finizio, or her colleagues from the federal workplace safety agency. Even the job title on Finizio's business card—"industrial hygienist"—sounded strange to McCauley, like a description of someone who cleaned giant teeth. Regardless, the detective was relieved to be able to turn to people who knew a lot more about tunnels than she did.

Given her own lack of technical knowledge, McCauley felt it would be better to let the OSHA rep take the lead in questioning Harald, who had his lawyer beside him. McCauley mainly kept quiet while taking detailed notes and making sure her cassette recorder continued to roll, so that Harald's every word would be captured on tape.

Yet McCauley couldn't resist drawing a few conclusions about him. She sensed that for all his intelligence, Harald wasn't nearly as bright as he seemed to believe he was. The clues came early on in the interview. Knowing that on his résumé and in his conversations with project officials, Harald had claimed previous experience using mixed-gas systems, OSHA's Finizio asked him to elaborate.

Q: What type of mixed air systems have you used?
A: There is a group of us who took mixed gas diving, which is helium and oxygen. And we've been using, over the course of five years now, a nitrox system, which is an oxygen and nitrogen mixture.

Although Harald presented this mixed-gas experience, which presumably began with some kind of training course, as evidence of his familiarity with the system he was using on Deer Island, the OSHA investigator's follow-up question indicated she was dubious:

Q: So it's in a gas form as opposed to liquid?
A: Yes.
Q: Have you ever used a liquid oxygen and nitrogen mixed air system?
A: No.

McCauley had just begun her study into the field. Yet she'd already learned that there was a world of difference between breathing the vapors of liquid gases you'd mix yourself, in as inhospitable an environment as the mind could conjure, and breathing gas that had been premixed and pretested. She felt that for Harald to have implied otherwise was like a guy confidently getting behind the wheel of a school bus full of children even though he'd never driven anything bigger than a station wagon.

The OSHA investigator asked Harald to explain the procedures for using handheld analyzers to test the oxygen in the breathing supply, pressing him for how those measurements could possibly have been accurate. After all, he'd told the divers to hold the analyzer in front of an open valve of breathing air that couldn't help but blend with the ambient air. McCauley noticed how abruptly Harald's manner seemed to change when he was challenged. "It's *certainly* accurate," he huffed.

Several times, in answer to questions about apparent deficiencies in the system, he responded that the protocols he'd given the divers were "industry practice." But after the investigators probed further, he was forced to acknowledge that the whole mission had been too experimental for industry practice to have played a role.

In at least one regard, McCauley was impressed by Harald. Rather than run away from responsibility for the design of the system, he owned up to a big chunk of it. "I was the person who wrote the schematics, the requirements for the system," he said. The OSHA investigator drilled a bit deeper, asking, "Who was the one that was in charge of designing the other aspects, the vehicles to go out there, communications, monitoring?" Harald didn't equivocate. "Me," he replied.

He stressed that he'd had help from his Spokane supplier, A-L Compressed Gases, in securing equipment and that he'd received considerable feedback from the other players in the project. "Between MWRA and Kiewit and ourselves and our people," he said, "we went through it, proposals back and forth, and revisions back and forth." But even with his lawyer sitting next to him and steely-eyed cops sitting across the table from him, Harald did not try to absolve himself of responsibility for the system's design.

Still, McCauley could detect no contrition in him. Instead, she sensed a detachment bordering on arrogance. The interview lasted nearly four hours. As she left the trailer, she told herself, *I definitely need to talk to this guy again.* Yet there were so many other people she felt she needed to speak with first. In the weeks that followed, she began working her way down the list.

Around the same time as that first interview with Harald, McCauley's partner called her from the medical examiner's office. It would take months for the full toxicology reports to come back, but after performing the autopsies on Billy and Tim, the ME's office could now report a preliminary cause of death: asphyxia by exclusion of oxygen. The autopsies found gas bubbles in the vessel areas of both brains, which could suggest excess nitrogen bubbling out of the tissue. McCauley's partner mentioned one strange discovery. The ME had found what she called an ulcer on Tim's left foot, a two-and-a-half-inch-long-by-one-and-a-half-inch-wide section around his big toe where the skin and a good part of the bone had worn away. "What is *that* all about?" McCauley asked. None of the surviving divers had mentioned it. She added it to her growing list of questions that needed answers.

In the weeks after the accident, McCauley became a regular visitor to Deer Island, where she had secured a locked storage garage to hold evidence. She met there with a man named Tim Reading, who was the safety manager for Airgas Northeast, the company that had supplied the tanks of liquid gas and HP air. After inspecting the setup and the Humvees, Reading rattled off a host of examples of its poor design. The valves bumped up against one another. The disabled interior dome lights in the Humvee, while helping to conserve battery power, had made it much harder for the divers to monitor the mixer and accurately note the levels. The process for measuring the oxygen content of the breathing air was unreliable, and the access to the backup air supplies was poor.

McCauley was disturbed by Reading's findings, though she reminded herself that he represented a company that had served as a supplier and therefore held some potential liability. McCauley realized she needed a diving expert who was an entirely neutral party. She and her OSHA counterparts found the person they were looking for at the U.S. Navy. Dr. Marie Knafelc was a leading authority on diving and the senior medical officer of the navy's Experimental Diving Unit, based in Panama City, Florida. It didn't take long for the navy physician, whose Slovenian name was tricky to pronounce (*Kah-NAY-fulls*), to show up on Deer Island. Although Hoss had already returned home to Idaho and Riggs to Nevada, McCauley was grateful that DJ was still close by. He agreed to come back to help Knafelc understand the original operation.

McCauley was impressed that for all her credentials, the navy doctor was a straight shooter like herself. When Knafelc inspected the sophisticated oxygen-injection system that each Humvee had been equipped with, then compared it to the improvised mixed-gas breathing system the divers had been given, she couldn't hide her disgust. "Oh my God," she said at one point. "They cared more about that Humvee than they did about the divers." She told McCauley that she would reserve final judgment until returning to Florida to review her notes and write a full report. But McCauley had little doubt which way it would fall.

Based on her growing knowledge of the events leading up to the incident, the detective knew she needed to reinterview Harald. He was clearly the man behind the curtain. And yet she was becoming just as convinced that ultimate responsibility would not end with him.

. . .

McCauley felt as though she'd been sucker-punched.

At 9:20 on a mid-August morning in 1999, after she'd been work-ing the investigation for nearly a month, she found herself in the con-ference room of OSHA's regional office. Her state police partner was there with her, as was the woman who had become another partner for the purposes of the tunnel probe, OSHA's Elena Finizio. While the state police were in charge of the criminal probe, the federal work-place safety agency was conducting its own investigation mainly to assess employer responsibility for the deaths. Because the witnesses were the same, and because OSHA had expertise that the state cops didn't, McCauley had been collaborating extensively with Finizio, and their styles seemed compatible.

They were in the conference room to interview key managers for contractor Kiewit, construction manager Kaiser, and project designer PB. If it struck someone like McCauley, who'd never gone to engi-neering school or even been scuba diving, that Harald's Norwesco plan should have flunked the common-sense test, how could it not have been more obvious to people with extensive industry experi-ence? So she asked the managers how they could have gone along with Harald's plan, which relied on divers using portable "supplied air," when OSHA regulations made it clear that a "mechanical venti-lation" system like the bag line was mandatory whenever there were workers in the tunnel.

"You're aware," one of the managers replied, "that OSHA was con-sulted long before the divers went in, aren't you?" He explained that officials with Kiewit and Norwesco had met with several OSHA of-ficials, in this same agency conference room, back in July 1998, a full year before the divers drove the Humvees into the tunnel. During that meeting, they had handed OSHA managers copies of Harald's plan. Although the plan listed liquid gas as only one of several sources of breathing air, it was abundantly clear that the divers would be rely-

ing exclusively on supplied air—and not on mechanical ventilation—
for their mission.

McCauley thought, *This is a joke, right?* She turned to Finizio and
asked, "Is this true?"

Finizio hesitated but did acknowledge that OSHA people were involved in some capacity, although she was not aware of the details.

As proud as McCauley was of her detective skills, she knew she
lacked a poker face. Now she was helpless to mask the apoplexy surg-
ing inside her. Granted, no one at OSHA had hatched the crazy plan
that led to the diver deaths. And it seemed clear to her that Finizio
herself hadn't been complicit—she was a good investigator whose
involvement had begun only after the fatal accident. Nonetheless,
unlike McCauley, Finizio wasn't hearing about this early OSHA involvement for the first time.

McCauley had been deferring to OSHA's technical knowledge and
sharing all her crime scene materials. Now she wondered: Who could
she trust? She knew that the longer this investigation went on, the
more lawyered-up the key people were getting. She also knew everyone's deniability would become more plausible if they were able to
show that they had given a preview of their plans to the government's
gatekeepers.

Later that day McCauley fired off an angry letter to Brenda Gordon,
OSHA's area director, demanding an explanation. Gordon's defense—
that OSHA did not have statutory power to approve plans—struck
McCauley as utterly unpersuasive. The detective decided she needed
to keep some daylight between her and the federal safety agency.

In December, the medical examiner's toxicology reports and final autopsy results for Billy and Tim crossed McCauley's desk. No drugs
had been found in either diver's body, and the ME could confirm the

preliminary report from back in the summer. Billy and Tim had died as a result of asphyxia when air rich in nitrogen and starved of oxygen had flowed into their facemasks. They had been killed quickly, while trying to complete a task. Evidently that was why Tim had been found with his shoulders turned to the back seat and why Billy had been found lying on his back on the tunnel floor.

McCauley was determined to conduct an exhaustive investigation. As a former prosecutor, she knew the bar would be high for the Suffolk County district attorney's office to prove criminal behavior beyond a reasonable doubt. She also knew she had more police work to do. But based only on what she and her partner had already uncovered, she had no doubt about their ability to demonstrate probable cause, which the DA's office would require in order to file criminal charges. As far as which players should be held to account, Harald Grob remained first on her list. She continued to believe, however, that the list should probably be longer than just one name.

McCauley hadn't spoken to the Norwesco project manager since his interview two days after the accident, and he had returned to Canada shortly after that. In the five months since, however, McCauley had uncovered mounds of fresh material as well as several shocking findings that made her question key aspects of Harald's account.

Still, McCauley worried how a potential prosecution against Harald might play out in court. After all, during his interview he had stressed that all the major parties—the MWRA, Kiewit, Kaiser, and PB—had not just approved his plan but had weighed in with suggestions for refining it. Given the involvement of all those big players, it would be hard to paint Harald as some kind of rogue actor. With all that shared responsibility and liability, then, the whole diving operation resembled the way patients are often cared for at the highly respected but labyrinthine teaching hospitals that made Boston a medical mecca. The patients see so many interns and residents and fellows and attending doctors and department chiefs and specialists, all swooping in and out of their room, that it's impossible to figure out who's in charge of their care. Everyone's in charge, so no one's in charge.

On top of all that, McCauley worried about the prospect of Harald taking the stand and pointing his finger at OSHA for having failed to raise any objections about his plan, when the agency had received it well ahead of time. She asked herself, *How will that play to a jury?* She knew the answer. *Not well at all.*

Two weeks into the new millennium, McCauley finished her twenty-four-page report on the sudden deaths of Billy and Tim. The nearly six-month investigation she had conducted with the help of her state police colleagues had sometimes felt to her like a series of intense graduate school seminars. The surviving divers had been among her most important tutors, impressing her with their recall and their candor, even when it came to agonizing details, such as what had happened to Tim's foot. The detective's other tutors included equipment suppliers, project managers, and especially the expert analysis of the navy's Dr. Knafelc. In writing her report, McCauley leaned heavily on the navy analysis as she worked to answer all the big questions that had confounded her when she'd started.

How responsible for the disaster was the seemingly slapped-together liquid gas breathing system? Knafelc had answered that question head-on in her report. "The design of the breathing apparatus was inadequate in its ability to support working divers," her analysis read. "Lack of appropriate monitoring equipment contributed to the death of two individuals."

The navy report faulted Harald for giving the divers the Multi-warn handheld gas analyzers rather than installing a far more reliable in-line analyzer, as Riggs had suggested. For short money, that plumbed-in analyzer would have provided rock-solid, continuous monitoring of the breathing supply and would have given Billy and Tim fair warning of impending doom. Knafelc noted Harald's failure to account for the predictable loss in air pressure that took place at each point where the five different three-hundred-foot umbilicals were connected. And the navy doctor complained that Harald had failed to test the breathing system adequately, under full-load conditions—again, something Riggs had suggested—before sending the divers into the tunnel. Those light walk-through simulations that Harald had instructed the divers to do in Tap's yard had been a poor substitute, especially given the absence of the key component of the system, the MAP Mix 9000.

What explained the Sno-Cone effect, the frost that enveloped the liquid oxygen tank and regulator? As the divers worked harder at the end of the tunnel, that put a greater demand on the mixed-gas system back at the Humvee. That increased demand, in turn, required the liquid tanks to produce more gas—to vaporize more product. Liquid gas is extremely cold—cold enough, of course, to burn warts off skin—because the cold allows the molecules to slow down so they can be packed more tightly into the tank. But in the process of being vaporized into gas, those molecules expand rapidly. To do that, they pull heat in from the surrounding environment as their source of energy. As cold as the tunnel was, it was a whole lot warmer than inside those cylinders of liquid gas.

As the liquid oxygen cylinder struggled to keep up with the divers' heightened demand, those molecules raced to expand more quickly. Because they needed more energy to do that, they continued to draw in whatever heat they could get from the outside environment, beginning with the regulator sitting atop the cylinder. That made the regulator colder and colder. And because the tunnel environment was so damp, as the regulator got colder, the moisture on it began to turn to frost. To keep expanding, those molecules of gas inside the cylinder intensified their demand for more warmth from the outside. Eventually that turned the frost covering the regulator into ice. That process kept going, layering more and more ice onto the regulator, until it took on the appearance of an "ice sculpture," as Riggs had so memorably put it. By then, it was about as cold on the surface of the cylinder as it was inside it. The regulator was as frozen as an outdoor faucet in a blizzard, unable to allow anything to get by. It simply gave out. With no more oxygen coming through, the divers' mixed-gas breathing air basically turned into a pure shot of nitrogen.

Judging from Harald's lengthy résumé, the navy doctor was stunned that he wouldn't have known the added risks involved with this kind of system. "Using liquid gases requires closer attention to detail because of the potential for the regulator to freeze," the navy analysis read. "The designer's experience in diving, including nitrox and mixed-gas, and training as a professional engineer should have provided him the necessary tools to design and build a safe, effective breathing gas system."

The navy doctor also faulted Harald for failing to include two fairly basic but essential components, which could have gone a long way toward guarding against the Sno-Cone effect. The first missing device was a volume tank, essentially a way station where the breathing gas could have been stored after the vapors of the liquid nitrogen and liquid oxygen had been blended. That volume tank would have allowed the mixed gas to be tested before it ever reached the divers' facemasks. It would also have buffeted the cylinders of liquid gas from the varying demands of the divers, reducing the chance of the regulator freezing over. Breathing directly off a mixer "with only episodic gas sampling is not acceptable," the navy report noted.

The second missing device was a heat exchanger. This component would have helped the system safely deliver the breathing air at a higher flow rate as the divers' demand increased. It was mystifying why the divers' system had no heat exchanger even though the Humvees did. That tall metal device, the one that resembled the radiator in a trendy condo, had ensured that the liquid oxygen tank on the back of the Humvee was converted into gas at the temperature needed to keep the vehicle humming. The divers deserved at least as much.

Why hadn't Billy and Tim been able to sense the danger they were in? McCauley knew the absence of the in-line analyzer was only part of the answer. Riggs told her about how he had asked Harald what to do with the MAP Mix 9000's AC electrical cord, and how Harald had instructed him to keep it coiled up with its factory-issued twist tie. By now, McCauley understood that although the mixer didn't need any electricity or battery power to function, its built-in pressure alarm did need an external power source. That's what the AC power cord was for. McCauley learned that this built-in alarm would not have measured the oxygen content of the breathing air, but it would have provided a critical warning to the divers that the pressure coming from either the liquid oxygen or liquid nitrogen had fallen below safe levels. Despite the tunnel's lack of electricity, the alarm could have been powered by a battery. McCauley couldn't get over the fact that the damn alarm was simply left unplugged.

Without either an in-line analyzer or a functioning pressure alarm, the divers were forced to rely on the patently ridiculous protocol of using the handheld monitor to test the blended breathing air and hoping that the ambient air mixing with the sample didn't throw off the measurements too much. Harald's response to Riggs's request for an in-line analyzer—namely, to give the divers one additional but equally unreliable handheld device—had been no solution at all. This shoddy measurement system had masked the deepening trouble to the point where the divers' first warning sign hadn't come until DJ and Riggs blacked out at the end of the tunnel. McCauley now understood why those two divers had been affected before the others. They were engaged in the most physical exertion, huffing and puffing as they lugged equipment and crawled into narrow pipes. McCauley

referred to the chart on the effects of reduced oxygen levels that the
navy doctor had included in her report.

12–15 percent: Breathing increases, especially in exertion, pulse
increases, impaired coordination, perception and judgment

Given their strenuous work, DJ and Riggs had been like tennis
players at the end of a long, grueling match, breathing more strenuously to suck in the extra oxygen their bodies were craving. Billy and
Tim, meanwhile, were just sitting inside a vehicle, waiting. Although
they were beginning to suffer the effects of oxygen deficiency, in light
of their sedentary situation in the dark, it would have likely presented
itself, at least initially, as barely detectible fogginess.

10–12 percent: Breathing further increases in rate and depth, poor
coordination and judgment.

Hoss's exertion level fell somewhere between the other two pairs
of divers. And when he called Tim with an urgent request to check
the oxygen reading, that plea appeared to have briefly snapped Tim
and Billy out of their fog: "Shit, it's 8.9!"
    Again, the chart:

8–10 percent: Mental failure, fainting, unconsciousness, nausea

The navy report noted that the 8.9 percent oxygen level Tim
relayed to Hoss must have continued to drop until Tim and Billy's
deaths. It also contained the answer to another mystery, the same one
that had puzzled McCauley from the start: *If all five divers were breathing off the same supply of mixed liquid gas, why did only Billy and Tim die?*
Clearly, Hoss had saved the lives of the excursion crew by switching

their air supply to the HP tank on the boat. But as it turned out, they had more of a cushion.

It all came down to geography. When the weakening oxygen regulator finally froze over, turning the breathing air into something that was mostly if not entirely nitrogen, proximity became Billy and Tim's enemy. That toxic hot shot of nitrogen had to travel up to fifteen hundred feet to get to the manifold in the excursion crew's boat and then out to the facemasks of Hoss, DJ, and Riggs. To get to Billy and Tim, however, it had to travel just ten feet.

If each of those findings had answered an important question that had been weighing on the detective's mind, a pair of related revelations blew her away. The first came from Tim Reading, the safety manager for Airgas Northeast, the company whose New Hampshire facility had supplied Harald with the gas used on the job. He insisted to McCauley that neither Harald nor anyone at his equipment supplier A-L back in Spokane had ever informed Airgas that the tanks of liquid oxygen and liquid nitrogen would be used as *breathing air for human beings*. Reading said company policy and industry practice prohibited both the blending of gases for human consumption and the sale of liquid gases for that purpose. The risks were simply too high.

The more shocking discovery had to do with the MAP Mix 9000 itself. In her attempt to understand why the mixer had failed to do its job, McCauley wasn't content to talk simply with Tony Drybanski, the salesman-president of the mixer's local distributor. She also persuaded the mixer's manufacturer, PBI Dansensor in Denmark, to send one of its top engineers to meet with her on Deer Island. Immediately upon inspecting the MAP Mix 9000 on site, that engineer, Steen Hansen, made it clear how distressed he was.

Through his thick Danish accent, he explained that the mixer's cover had been removed and its factory-set pressure settings had been changed. Even though Harald had done this in consultation with Drybanski, the Danish engineer now said it never should have been tampered with. That's why there was a warning in bold type right there on the first page of the mixer owner's manual: "Important: MAP Mix 9000 should be opened by authorised personnel *only.*"

That led Hansen to deliver a bombshell about the device. The MAP Mix 9000 should never have been used for the divers—or for any human beings. The owner's manual specified that it was intended for "industrial uses," and the Danish engineer gave McCauley examples of what that meant: packaging meat and cheese and burritos, to give them a sort of vacuum seal that would lead to longer shelf life. The mixer was *not* to be used to blend breathing air.

Hansen said he and his colleagues never thought anyone would be stupid enough to use the mixer for such a high-wire act. (Only after the accident would his company add a new warning label to the mixers that read: "!!Danger!! Do NOT use this equipment for air supply for human beings.") The unit had a margin of error of plus or minus 2 percentage points, which might be fine when packaging burritos in plastic containers but was clearly inadequate for trying to keep a team of five divers alive in a moonlike environment under the ocean. When the MAP Mix 9000 was used in food packaging, the Danish engineer said, its built-in pressure alarm was usually hooked up directly to the packaging machine. That way if the pressure dropped on one of the gases as it was being blended, the packaging machine would automatically shut down. Hansen said that even though the mixer should have never been used to supply the divers with air, Billy and Tim would likely still be alive if Harald had taken the simple step of energizing the pressure alarm. Instead, he had instructed Riggs to tie up the device's electrical cord and forget about it. Although there were no working electrical outlets in the tunnel, Steen said, it would have been easy to hook up a twenty-four-volt battery to power the alarm and then arrange for the HP air backup supply to kick in automatically if the pressure dropped.

All this seemed to explain why Harald and Drybanski had gone overseas to find the mixers, resulting in the Danish-made units getting tied up in customs and arriving late to Deer Island. Harald had been unable to locate an approved mixer in the United States because no domestic manufacturer produced mixers for this kind of use. Circumventing that problem by going overseas suggested to McCauley that Harald hadn't just been in over his head. In her view, he had shown willful disregard for the lives of the divers. How else to explain the decision to send them into the darkness without telling them their lives would depend on a device designed to package supermarket burritos and cheese?

Confident that she had demonstrated probable cause, McCauley submitted her report to the district attorney's office.

Around the same time, the Norwesco fax machine in Roger Rouleau's Spokane office started to whir. After wrapping up its own investigation, OSHA was now issuing a raft of citations and recommending a total of $410,900 in fines. Norwesco received the biggest ticket, at just over $200,000. Kiewit and Kaiser each received $91,000 in fines, and Black Dog Divers $25,400. OSHA alleged a series of violations by all four companies, ranking them on an agency scale of seriousness, with "willful" being the most severe. OSHA said the breathing system the divers were given was "woefully inadequate," and the agency slapped Norwesco with eleven "serious" and two "willful" violations, the latter for designing and building a deficient respirator system and failing to remove workers in the face of repeated problems with it. The deficiencies OSHA cited included inadequate hoses and mixers, inappropriately duct-taped fittings, disabled alarms, and the absence of an in-line analyzer. Meanwhile, Kiewit and Kaiser each received one "willful" violation, for failing to provide mechanical ventilation in all work areas in the tunnel, and three "serious" violations, for failing to provide adequate lighting and communication. Black Dog received thirteen "serious" violations, for various failures involving oversight of the divers.

As he looked over the citations, Roger was incensed. Of course, the Norwesco owner had expected some kind of slap from OSHA. But it was the agency's tone of astonishment that rang hollow with him. The citations in no way acknowledged the fact that Roger and Kiewit officials had traveled to OSHA's office a year before the mission began to brief officials on their plan. And yet no one from the agency had raised any red flags. As he scanned the citations now, Roger's eyes stopped on the signature at the bottom. It belonged to Brenda Gordon, OSHA's area director.

The name sounded familiar. Then it dawned on him: *I met that gal back in 1998.* At one point during Roger's meeting with OSHA prior to the mission, a woman had walked in to turn down the office air-conditioning, which was controlled by a thermostat in the conference room. The OSHA assistant director who'd been running the meeting then introduced her to Roger and the Kiewit officials, explaining to his boss that the team was there to share their plug-pulling plan for the Deer Island tunnel. She chatted with them briefly before leaving the room. Roger thought: *And now she's showing shocked indignation?*

The OSHA fines didn't represent a conclusion so much as an opening salvo. All four of the cited companies were allowed to contest the fines, and all four of them quickly signaled publicly that they would. To Dan Kuhs, the union business agent, that response was just further evidence of how shameful things had become. "You have two fatalities," he told a local newspaper, "and everyone involved seems to think they have no responsibility." He pinned more of his hope on the state police's criminal investigation.

As far as Tap Taylor was concerned, the fines against Black Dog were entirely unfair. During the project, Tap had felt belittled by Harald and ignored by Roger and had seen the job claim the life of his best friend and threaten the existence of the business they had built together. Although the comparatively lower fines that OSHA had levied against Black Dog acknowledged his company's lesser role, he was adamant that he had done nothing but provide labor for the job. He felt he was a victim, not a perpetrator.

Meanwhile, Roger, who had taken Kiewit's advice to hire one of the top white-collar defense lawyers in Boston, now engaged a respected diving specialist as a consultant. He began an e-mail round robin with that consultant as well as Harald in his effort to mount an aggressive defense.

In March, eight months after the deaths of Billy and Tim, the Suffolk County district attorney's office announced its decision: it would not be bringing criminal charges against Harald Grob or anyone else in connection with the case. McCauley couldn't hide her frustration and disappointment. The deeper she had dug, the more convinced she had become that it had not been an accident that killed Billy and Tim. "The results were completely foreseeable," she argued. "To suggest it was some accident is *absurd*." Given the black and white prism through which she viewed the world, that conclusion naturally led her to another one: "If it wasn't an accident, and people died, somebody should be held accountable."

Prosecutors in the DA's office offered several reasons for declining to charge Harald. They felt that because he hadn't been up to the challenge of putting the system together, he didn't realize the extent to which his actions were endangering the divers' lives. Also, they felt that because the divers had survived despite all the problems on Monday and Tuesday, it was reasonable for Harald to expect the same result when he sent them back in on Wednesday. Finally, because so many other players owned a share of the responsibility for the deaths, going after only Harald was a risky proposition.

McCauley had enormous respect for the prosecutors on the case, and she knew how difficult their task would have been in getting a jury to convict beyond a reasonable doubt. Still, she thought they might have been overthinking things. "It's not that complicated," she would later explain. "The guy knew that if something bad happened in the tunnel, those guys would die. And yet he went ahead. He didn't send himself into the tunnel, did he?"

Of all the reasons the prosecutors cited for not filing charges, she agreed most with their reluctance to pin everything on Harald. She was personally appalled that so many seemingly bright people, working for so many large corporations and agencies, had let the situation deteriorate to the point where they needed to call in Harald, then had signed off on what was clearly a half-baked plan. She came to believe the managers should have had the sandhogs remove the plugs while the ventilation bag line was still in place, then instituted a policy of evacuating workers whenever a bad storm was forecast, when the risk of an errant anchor dragging along the seafloor would have been higher. If that sensible compromise had become obvious to a cop like her, why had it eluded the administrators and engineers who were in charge? Still, McCauley felt the fact that others owned some responsibility shouldn't diminish Harald's far larger share. While she could see how this shared culpability might lead to the prosecution of more parties—both individuals and corporate entities—she did not want it to be used as a justification for not charging anyone.

Despite the blow of the district attorney's decision, there was one promising development for McCauley. Thanks largely to union pressure from Local 56, the case was being turned over to the state attorney general. Maybe that office would find a more successful path to prosecution.

· · ·

McCauley walked in carrying a bulging box under each arm. The cartons contained memos and reports, photos and notebooks, VHS tapes and audiocassettes. The detective had arrived at this Boston office in the spring of 2000 to introduce herself to the woman who would be taking over the case. Joan Parker was the director of safety in the Massachusetts attorney general's division of fair labor and business practices. McCauley's supervisor had asked her only to brief Parker and share copies of her investigative materials. Right from the start, though, McCauley made it clear that she was too invested to simply dump and run.

Given the uphill battle the case faced, McCauley was hoping to be able to hand it over to a fearless field general who wouldn't back down. As she gave Parker the once-over, she was worried. The woman struck her as a sort of hippy-dippy beatnik—the kind of brainy lefty who carried around organic produce in a reusable burlap sack, or drove a well-loved Volvo with a bumper sticker that read JUNG AT HEART; in other words, the kind of person who wouldn't last long in a tough neighborhood like the one McCauley grew up in.

The two women seemed like complete opposites. With her unmistakable accent, rapid-fire speaking style, and don't-even-try-to-snow-me posture, McCauley was Boston through and through. She kept a gun on her hip and her dirty-blond hair pulled back in a tight ponytail. Parker, meanwhile, was a citizen of the world. Born in Peru, she'd been raised in a succession of exotic locales—Morocco, Panama, the Virgin Islands, and Puerto Rico. Tall, with long wavy reddish-brown hair, she had a light voice and an artsy flair. At fifty-two, she was nearly two decades older than McCauley.

No matter, McCauley thought. You go to battle with the army you have. "This is outrageous!" she told Parker, launching into an overview of how powerful parties had off-loaded the enormous risk of the plug-pulling operation onto Harald and Norwesco, and then Harald had off-loaded it onto the actual divers.

After McCauley's pitch, Parker began peppering her with probing questions that made it clear how much she'd already read up on the case. Despite McCauley's initial fears about the woman and their obvious stylistic differences, she suddenly saw in Parker a kindred spirit. First impressions can be deceiving. Though it was true that Parker was a bit bohemian, she had a real toughness to her. She drove around in a stick-shift Ford pickup truck with no air-conditioning and a big, shaggy mutt riding with its head out the passenger window. The only sticker on the truck's bumper read: MY DOG CAN LICK ANYONE.

When her boss had first asked her to look into the diver deaths, Joan Parker had been cautious. She knew how difficult it was to prove cases like this. She also knew that no special statutory enforcement powers would be available to the state attorney general. The only relevant statutes were those of common law, the same ones that the district attorney had determined were inadequate to charge Harald or anyone else. Prosecutors would have to prove criminal recklessness to support a charge of either homicide or manslaughter, since Massachusetts, unlike many other states, had no lesser statute of criminal negligence on the books. Still, she understood why the case had landed in her lap. The AG's office, with a safety director and team of inspectors, had far greater expertise in workplace safety cases than did the DA. Although Parker was primarily focused on inspection and enforcement rather than prosecution, she could leverage her deep experience to make sure no avenue was overlooked during this second-round vetting. And as she read up on the case, she could feel it awaken a passion in her.

Nearly twenty years earlier, after graduating from Cornell with a degree in horticulture, Parker had been working as a plant researcher in a lab. Inexplicably, she and her coworkers began getting sick. Eventually she learned that a faulty ventilation system was exposing workers to the fumes from pesticides being tested in the lab. With her eyes opened about worker safety, she decided to change careers, ultimately moving to Boston to earn a graduate degree in industrial health and engineering at the Harvard School of Public Health.

In her many years in the field, investigating claims of child labor and unsafe working conditions, Parker had come to believe that too often employers were willing to sacrifice their employees' well-being in pursuit of a dollar. She often felt she had very little power to stop them, besides writing up reports that employers could simply flick away like a fallen eyelash.

But when she found this animated state police detective sitting across the table from her, Parker felt invigorated. *I love this woman!* she told herself. Nonetheless she knew that a successful prosecution could not be built on passion alone.

As she began to marinate in the volumes of tunnel material, she had one important advantage. She'd actually been in the tunnel twice before—not just to the base of the shaft, like McCauley, but miles and miles in. Several years earlier she'd ventured in to evaluate complaints from sandhogs that the tunnel conditions were giving them respiratory ailments. It wasn't lost on her that as utterly remote as she had found the tunnel to be, her visits had taken place when it still had good light, ample air, and a working transportation system. She could only imagine how the divers must have felt down there in dark desolation at the moment when their air supply went bad.

When she came across the audio recordings of Harald's four-hour interview with McCauley and OSHA, she could scarcely believe what she heard. Despite taking heavy responsibility for the design of the system, Harald seemed unwilling to acknowledge the connection between the failings of his system and the fact that Billy and Tim had not come out of the tunnel alive.

An eye-opening example was his response to the OSHA investigator's question about how quickly things could have gone wrong without the five divers realizing it.

> HARALD: But the people, once the oxygen level drops, they would
>   become aware of it, and they have a backup air supply to
>   switch to.
> INVESTIGATOR: So why didn't Bill and Tim switch over to their backup?
> HARALD: I don't know, but the excursion crew did.
> INVESTIGATOR: Basically, these people were overcome quickly. They
>   obviously didn't realize what was happening, because they didn't
>   switch or they weren't successful in switching to another system.
> HARALD: You're speculating that, because we don't know that.

Stunned, Parker rewound the tape and played it again. After dispatching the divers into a hostile underworld with a system that independent experts had since determined was hopelessly flawed, after downplaying or ignoring the divers' concerns and simply sending them back in while he remained safe in his topside trailer, was Harald *really* trying to suggest that it could have been Billy and Tim's fault that they had died? Just the suggestion made Parker irate.

The more she read and listened, the bigger the gap she found between what Harald said and what the investigation had found. He insisted that he had been working to devise and refine the system for over a year. But Parker had seen the letter sent by his contact at A-L Compressed Gases just a month before the job was to begin, asking another supplier to quote a "turn-key system for new equipment." Parker wondered: If the design of the system was still so wide open at such a late date, what had Harald been doing for a year? Before the mission, he had assured tunnel managers that he had ample experience using systems blending liquid oxygen and liquid nitrogen, notably in "small submarines." But when pressed after the fatalities, he had to admit that the tunnel job had been his first time using such a system. He acknowledged that he hadn't told his New Hampshire gas supplier that the tanks of liquid oxygen and liquid nitrogen would be used for human consumption. But he didn't seem to grasp the recklessness of that decision. He insisted the system was sound because of all the redundancies and backups he had marbled into it. But he seemed unwilling or unable to confront the fact that whether the system had three or thirty or three hundred backups, it wouldn't have mattered if there was no reliable mechanism to alert the divers when the main system failed—then automatically switch to a backup supply. Even machines packaging burritos had that mechanism.

Finally, he insisted that he and A-L had settled on a liquid oxygen system because it would give the divers plenty of breathing air. But a follow-up navy report by a different analyst, which the Experimental Diving Unit had undertaken after McCauley finished her investigation, told Parker a different story. The report showed that, because of crucial miscalculations, Harald's system had, in fact, been undersized, leading to a dangerous undersupply. "It appears," the report concluded, "that a poorly designed breathing system in general caused the failure."

The navy analyst noted that human beings breathe not at a steady flow rate but in a pattern of inhalation and exhalation. When people inhale, their demand climbs to three times its average. When they exhale, there is no demand at all. But the calculations for Harald's system were based on *average* flow rates. Because there would be times when all five divers were inhaling simultaneously, the calculations should have been based on combined *peak* flow rates. "It might appear to be unrealistic to assume that every person will be requiring their maximum flow at the same time, especially since they are breathing at different rates," the navy analyst, Kirk VanZandt, wrote. In fact, it can happen with surprising frequency. Through an evaluation that he conducted, he found that three divers doing strenuous work reached their combined maximum flow rate requirements several times an hour, and at some point in every minute they required more than 95 percent of their maximum flow. Plugging in the figures for maximum—rather than average—flow rate requirements showed that "much higher pressures and flows are required by the system than the gas mixer is capable of providing." Harald, the navy analyst concluded, was either "misinformed or inexperienced."

As for just how close Hoss, Riggs, and DJ had come to dying, the navy analyst now put a figure on it. He based his calculations on how long it would have taken the hot shot of nitrogen to travel from the mixer and through the umbilicals to the end of the tunnel: thirty seconds.

Had Hoss delayed switching to the emergency supply just half a minute longer, there would have been five dead divers.

Parker desperately wanted to talk to Harald herself, but she knew his return to Canada made it difficult to demand his cooperation without getting involved in the thicket of extradition. She wondered if the cold detachment she detected in him from his Deer Island interview could have been a function of shock on his part. But then she uncovered other documents that made her think differently.

These were copies of the e-mail correspondence that Roger had initiated with Harald and his diving consultant in Norwesco's effort to challenge the OSHA citations. The volley of e-mails, coming just seven months after the divers' deaths, displayed an odd, often nonsequitur mix of bubbly pleasantries, nuts-and-bolts questions, and steel-belted defense strategy. In one note to Harald in early February, Roger wrote, "Skiing has been good? Pretty warm here but still great to go up and crank up some speed. On the connection from the hose to the mask, was there a QD or 1/4 turn or was it held together with duct tape?"

A week later Roger e-mailed Harald about the MAP Mix 9000 units, asking, "I'm trying to get an idea of how much they are worth as salvage."

Harald replied, "How was skiing on Friday? There has not been any new snow here for more than a week, but it's supposed to start tonight. I don't know if the mixer guy knew what they were going to be used for. When they were delivered I asked him if he was interested in buying them back or reselling them after we were done. He mentioned that he might be able to find buyers for them and we should be able to get a good resale price."

By late March, Harald was busy sightseeing. "Just finished a trek in the jungles of Borneo last week and am now in Kuala Lumpur," he e-mailed Roger on March 31, 2000. "Will be here for a few days then it's off to Thailand. Have not had access to Internet for a while so this is my opportunity to catch up with everyone. How are things in Spokane? Is there still snow? Any new projects? Any news on Boston?"

Nothing in Harald's tone suggested to Parker that he had been grappling with his role in a double fatality. Nonetheless Roger's strategizing with him appeared to have been worth the effort. Consistent with its pattern in other cases, OSHA ultimately reduced the fines against Norwesco by more than half.

If, at the outset, there had been some doubt in Parker's mind about the wisdom of pushing this case forward toward prosecution, that doubt was now gone. She had become as much of a true believer as McCauley. Although Parker wasn't a lawyer, she was well versed in the relevant case law, and it would be her job to recommend to the attorney general and his litigation team whether to press ahead. Like McCauley, she felt strongly that people, beginning with Harald, had to be held to account. As they grew closer, McCauley joked to Parker that she must have been a cop in a previous life, given her dogged detective approach. If that was the case, Parker was still struggling to answer the pivotal question that drove the narrative in all those

TV detective shows: What was the motive? Why would Harald have taken these steps that had put the divers at such extreme risk, far above even the high threshold that they accepted as an occupational hazard?

Parker began with the most obvious assumption: money. But as she combed over the contracts and financial records, she was stunned to learn that Harald was simply a worker for Norwesco, being paid just thirty-five dollars an hour. Unlike Kiewit's contract to build the tunnel, Norwesco's subcontract actually had no late penalties built into it. As far as Parker could tell, Harald stood to gain no bonus, no piece of the pie. Any financial gains for Norwesco would have accrued most directly to owner Roger Rouleau—one more reason Parker kept him in her sights for possible prosecution as well. (Complicating any case against Roger was the apparent shallowness of his knowledge of the plan's details. When asked during the OSHA investigation why he had leaned so heavily on Harald to design Norwesco projects, Roger admitted, "Because I don't know how to do that stuff.")

In time, a more likely motivation for Harald dawned on Parker. From his exhaustive résumé to his often pedantic responses, he seemed in her estimation to be someone who was deeply egotistical. Kiewit had viewed him as a boy wonder for his innovative work on Lake Mead. And when the powerhouse contractor had found itself in even more of a bind on Deer Island, Harald had jumped at the chance to be its savior once again. Parker envisioned Harald delivering the keynote address at a professional conference, waxing about how he had led a team to accomplish a seemingly impossible task, using a dazzling, experimental breathing system, and making history in the process.

As it happened, when she was knee-deep in her research, Parker was herself asked to speak to a professional gathering. The aim of this occupational health event was to remember workers who had lost their lives on the job over the past year. Parker decided to focus her talk on Billy and Tim, stripping from it any references to regulations or statistics. Her only goal was for the sixty or so people in the audience to feel as though they were in the tunnel with the divers.

"To bury a body," Parker began, "you need to dig a grave six feet down into the earth." She was referring to the depth that is widely believed to have been set in London during the Great Plague of 1665, in a futile attempt to limit the spread of the outbreak. "It's surprising how deep that is," she continued. "Then, imagine digging another grave, below the first one, and another one, and another one, until you've dug seventy graves, one under the other. Then, after you've dug down these 420 feet, imagine digging straight out from there, parallel to the ground, for nine miles. Down there it's deep. It's dark. It's wet. It's cold. And there's no easy way out. For some, there was no way out at all."

Looking out at the crowd, Parker felt a searing sadness wash over her. She had been so focused on the legal challenge of assembling the pieces to the tunnel puzzle that she realized she'd been suppressing the emotional side. Without knowing it, she had come to identify with Billy and Tim, and now she was determined to find justice for them. But what would justice even look like?

The lawyers stood nervously in front of the Hyatt Hotel's bank of elevators, looking expectant each time the silver doors opened—and disappointed each time they closed without delivering DJ.

"Where is he?" John Prescott asked his law partner, Nina Pelletier, as he paced the lobby. Pelletier could see that Prescott's usually broad smile was now as absent as their client.

DJ was late for the breakfast meeting they had scheduled to go over last-minute strategy, and he wasn't answering his phone. The stakes were high for the divers and their families, to say nothing of their lawyers. At the other end of the Hyatt, about seventy attorneys and insurance adjusters were filing into a chandeliered banquet room.

Prescott and Pelletier were involved in the civil lawsuits that had sprung from the case. They were representing DJ in his personal injury case as well as Billy's parents, Olga and Bill, in their wrongful death claim. Other lawyers were representing Tim's widow, Judy, in her death claim and the injury claims by Hoss and Riggs. While the lawyers believed Harald's conduct had been inexcusable, they felt even greater fault lay with all the powerful parties that had let things deteriorate to the point where they needed to call in someone to engineer a dangerous escape maneuver. Accordingly, they had sued the MWRA, the Kiewit joint venture, Kaiser, PB, and Norwesco as well as A-L Compressed Gases, based on the critical role that the Spokane supplier had played in helping Harald design the system. There were slight differences in the lawsuits because of quirks in the law. Hoss, Riggs, and Tim's widow weren't able to sue Norwesco because of restrictions in state law against workers filing a claim against their own employers. And Michelle wasn't able to sue anyone on Billy's behalf. Even though she and Billy had been buying a house together and had spent four years building their lives around each other, they weren't married, so Michelle didn't really exist in the eyes of the law. Filing Billy's wrongful death claim fell to his parents, who despite being devoted to their son hadn't seen him in over a year.

Prescott and Pelletier, working closely with Judy's lawyer, Bob Norton, had spent more than a year building their cases—and, in many ways, running their own investigation. To bring a successful punitive damages claim, the lawyers would need to prove gross negligence, demonstrating that the defendants had shown willful disregard for the divers' safety. It was a high standard. But the many hours they had logged sifting through boxes upon boxes of project files had produced enough damning memos to give them confidence in their chances.

Over the course of her year working with DJ, Nina Pelletier had come to love his innocent honesty mixed with roguish charm. By now, she considered him her favorite client of all time, even if time was something he often struggled to keep. Pelletier was not much older than DJ and looked considerably younger than her thirty-eight years. But her relationship with him was maternal, drawing on the same nurturing instincts that found her packing peanut butter sandwiches for herself and Prescott to snack on during breaks when they were on a long trial.

Although they were prepared to go to trial in the tunnel case, the lawyers had agreed to meet with the defense attorneys and insurers at the Hyatt on this autumn morning in 2000. The two-day mediation session would determine whether the alphabet soup of defendants would agree to a settlement that the divers and Tim and Billy's survivors felt was fair, or whether Pelletier would have to start making a lot of peanut butter sandwiches.

While Prescott and Pelletier paced near the elevators, Olga and Bill Juse waited in the hotel restaurant, sitting across from Judy and Hoss. The lawyers had warned their clients that there would be lots of downtime during the mediation session. So the divers and the families had commandeered a table in the restaurant to be their waiting area, with its windows offering views of Boston Harbor and the airplanes taking off from nearby Logan Airport.

The lawyers needed everything to go right. But DJ was AWOL, as was Riggs, and having two of the surviving divers miss the mediation session would be very bad. Eventually Prescott hustled over to the front desk, persuaded the hotel clerk to give him DJ's room number, and then hopped into the elevator. Standing outside the room, he rapped on the door. Finally, it swung open, and DJ appeared in his underwear, moving in slow motion. He squinted at Prescott, then moaned, "Fucking Riggs." Prescott peered into the room to see Riggs asleep on the floor.

When the pair made it down to the restaurant for a rushed breakfast, they confessed they'd been out late the night before. Riggs had
been crashing with DJ and his mother in Waltham for several weeks.
The profanity-averse teetotaler and devoted father, whose Bible Belt
sensibilities had driven DJ to distraction before the accident, had been
struggling with marital problems. Like Hoss and DJ, he had found
himself unable to return to diving since the accident. After many
years of sobriety, Riggs was now drinking and cursing heavily, though
DJ still managed to outpace him. One late-night escapade had ended
with Riggs watching DJ get a large tattoo on his right arm. Because
tattoo parlors were outlawed in Massachusetts, the inking had taken
place in some guy's basement.

Now DJ sat at the table in the restaurant, guzzling coffee, still a
bit dazed. When a young blond waitress with beautiful eastern European features approached, he jolted to attention, turning on the
charm. In her accented English, she seemed to respond in kind. After
she stepped away, DJ turned to Pelletier, asking his lawyer for one
of her business cards. He scribbled his phone number on it and then
handed it back to Pelletier, asking her to arrange a date for him with
the waitress.

Pelletier knew DJ had been on something of a wild streak since he
and his girlfriend Dana broke up, although she hadn't pressed for details. Rolling her eyes, she refused his request. "DJ, I am your lawyer,"
she told him, "but I don't represent you in *all* aspects of your life."

They headed into the banquet room, where the army of lawyers
and insurance adjusters sat behind tables hugging the cream-colored
walls. The size of the crowd spoke to both the tunnel project's complexity and the depth of concern among the defendants about their
legal and financial exposure. As the surviving divers took their seats,
Pelletier worried how they would hold up.

Judy's lawyer, Bob Norton, addressed the crowd first, laying out
the key clauses in the contracts and subcontracts that tied all the

defendants to the project. Then Prescott took everyone through the correspondence that told the tortured history of the plugs and, in some cases, foretold the divers' horror. Together the two lawyers hammered home all the systematic failures that, they argued, left each of the parties in the room liable for the diver deaths.

One pair of letters was particularly damning. All during the "memo wars," when Kiewit had lobbied for permission to pull the plugs while the bag line was still in place, and Kaiser-PB-MWRA had refused, both sides had claimed worker safety was its paramount concern. Those arguments persisted even after Kiewit floated its plan to remove the plugs while the tunnel had ventilation and then install a temporary blockade at the venturi, which would have given workers several extra hours to evacuate in the event of a flood. However, PB had rejected that proposal. A memo from the tunnel designer explained that if a diffuser head sitting atop a riser became compromised, air in the sealed-off diffuser tunnel would try to escape up the failed riser at the same time the water was flowing down it. "The result would be a choking of the flow similar to what happens when a 2-liter bottle of soda is inverted," the memo read. That would jack up the temperature in the failed riser, and "these elevated temperatures of greater than 300 degrees Fahrenheit would endanger the integrity of all 54 undamaged riser pipes and place the outfall tunnel's future ability to operate in jeopardy."

Kiewit had then altered its proposal to reduce possible temperature spikes. Instead of installing a blockade at the venturi, the contractor suggested erecting a fifteen-foot-high temporary dam in the main tunnel, about six miles from the shaft. But the designer had rejected that proposal as well, based on concerns that the tunnel could still sustain serious damage. A follow-up memo from PB said that even though the temporary dam would give workers ample time to make a safe escape, "this only serves to help evacuate the tunnel and does not address the recovery of the tunnel so that it can be placed into service after such an event." So the rejection of the plan stood.

While it was understandable that the designer and owner would have little interest in being left with a damaged or even unusable tunnel in the event of a freak flood, this memo suggested that their "paramount" concern had not, in fact, been worker safety. It had been the integrity of the tunnel.

When Norton and Prescott finished speaking, Pelletier exhaled, feeling the first part of the presentation had gone well. For the second part, the lawyers had commissioned a sophisticated animation sequence to convey how impossibly remote and confined it was at the end of the tunnel. They asked Hoss to narrate as it flashed on the screen.

Hoss spoke in an even-keeled, matter-of-fact manner, like some tough but taciturn John Wayne character. Pelletier felt his narration was much more poignant because he had experienced the horror firsthand and barely survived it, yet he was recounting it in the most understated way. There was silence in the room as everyone strained to hear his low voice. Hoss recalled his final conversation on the comm wire, when a disoriented Tim confirmed how dangerously low the oxygen percentage had dropped. Hoss described finding Billy under the Hummer and Tim slumped over in the driver's seat. As Hoss neared the end of his narration, he mentioned what had happened during their desperate drive back to the shaft, when they had stopped to check on Billy and Tim. He had been devastated to discover that in their race to make it out alive, driving through up to three feet of standing water, Billy and Tim's bodies had been jostled around so much that Tim's leg had been thrust over the side of the trailer, causing his foot to drag along the tunnel wall. This had resulted in his foot being worn down, through his boot, right to the bone.

There was something aching, almost confessional, about the way

Hoss relayed this information. Despite the heroism he showed in saving the lives of DJ and Riggs and then assuming more risk by refusing to leave Billy's and Tim's bodies in the tunnel, Hoss seemed haunted by the notion that he had somehow failed his close friend by allowing his foot to be damaged. There was nothing rational about this feeling, but it didn't make it any less palpable or honorable. Instead of trying to improve the size of the settlement by attacking the defendants for their wrongdoing, Hoss seemed to be using his audience with these seventy attorneys and insurance adjusters to admit to what he felt was his own failing. As he described discovering the deep wound on Tim's foot, his voice finally cracked.

Instantly, Pelletier could feel her synapses overloading and her eyes watering over. Fearful that she would begin to sob, she put her head down on the table, using her arm to protect her eyes and ears.

The divers and families retreated with their lawyers to the hotel restaurant to begin their long wait. At one point, the union's Dan Kuhs arrived to show his support for the divers and to introduce Joan Parker to them. The official from the AG's office had talked to a couple of the divers over the phone, but this was her first chance to attach faces to the names that had preoccupied her.

In their opening salvo, Prescott, Pelletier, and Norton had pegged the value of the wrongful death claims for Billy and Tim, factoring in the impact of punitive damages, to be $7.5 million each, and the value of DJ's injury case to be $1.75 million. As was typical for negotiations like this, they had deliberately started high. In particular, they knew that the claim for DJ, like those for Hoss and Riggs, was inflated, considering there was no evidence of lasting physical injuries and considering that state law allowed punitive damages only for death cases.

Now the defense side made its first offers, and they were offensively low. The plaintiff's lawyers knew not to take it personally. This

is how the process would unfold. For the divers, however, it was hard
not to take umbrage.

When the second offer came in, only slightly higher, Hoss decided
he had seen enough. "I'm not interested in playing these games," he
said to the others. He'd been in counseling for many months now,
struggling to get his head straight. The costs for those sessions were
covered by his worker's compensation policy, but he'd been told that
if he received any kind of settlement, his counseling tab would automatically be deducted from his settlement check. The lowball figures
the defense was throwing out were scarcely enough to cover those
costs. Hoss's biggest goal with the lawsuit, aside from ensuring that
Tim and Billy's families were taken care of, was to see the responsible
parties stand up and admit that they'd done wrong. That comported
with the code he lived by: you work hard, strive to do your best, and if
you fall short, you man up. The last thing he had wanted was to have
to recount his hellish experience in front of a banquet room full of
suits. Rather than try to amp up his story, he had told the unvarnished
truth. He expected the other side to respond in kind, not act like they
were haggling over the price of a key chain with some street vendor
in Cancún.

Hoss expressed his frustration to Judy, who had signed on to the
lawsuit for similar reasons. From her medical training, the psychiatrist had been taught that the best way to respond to a preventable
death was what doctors called the "morbidity and mortality conference." That was a regular session conducted at teaching hospitals
where doctors gathered with their colleagues to review the care of
a patient who had died or been harmed, trying to identify errors so
they could be avoided in the future. The origins of this practice could
be traced nearly a century earlier to Mass General, the same hospital
where the divers had been treated after the accident. That postmor-
tem was exactly what Judy wanted to see now, in response to her
husband's death.

While the lawyers were doing a lot to uncover what had gone wrong, Judy had come to realize that the responsible parties were never going to admit their mistakes unless they went to trial. Even then she understood that, given the number of well-heeled parties involved and the coiled nature of the case, a trial would take years, and still there would be no guarantee of either contrition or reform. During her fourteen years with Tim, Judy had relied heavily on denial, kissing him goodbye when he left for a dive job and going about her own life rather than being paralyzed by worry. Now, as much as she wanted to force all the players to do both soul-searching and root-cause analysis, she didn't think she could bear to spend years having to hear, again and again, the awful events that had led to Tim's death.

So when Norton approached her on the second morning of mediation and reported that the offer had climbed into the reasonable zone, to around $3 million, Judy knew what she needed to do. "Bob, I don't care," she told her lawyer. "Settle for anything."

Norton could tell Judy was in pain, and ultimately it was nobody's decision but hers. Yet he also knew it was his job to help guide her through the process. "I think there's probably more to be gotten by negotiation," he told her. He asked for her patience just until the end of the day, to see what he could do. She agreed.

Meanwhile, Olga Juse sat next to Nina Pelletier, holding her hand. Billy and Tim's lawyers had put the same dollar figure on both death-benefit claims. After all, Billy and Tim had met the same horrible end in the tunnel. But the lawyers also understood that the other side wouldn't see things the same way. Although justice may be blind, in the emotion-free eyes of the law, Billy's life was worth less than Tim's, primarily because Tim was married and Billy wasn't. Tim's wife would be entitled to his future earnings, based on the presumption that he would have helped support her (even if, as a physician, she earned considerably more). No one was entitled to Billy's future earnings, however, because the law didn't recognize his relationship with Michelle. It was only because he had no wife and no children that Olga and Bill had been entitled to file the lawsuit.

Olga and Bill would have paid any price to get Billy back, but of course, that wasn't possible. While Judy had responded to her pain with passivity, telling her lawyer to take whatever the defense offered just so she could be done with the ordeal, Olga had responded in the opposite way. She wanted to honor her son's life by making sure the defendants felt a real sting for their sins. Her complaint about unfairness had nothing to do with Tim, a man she had never met but whose soul she now prayed for every night. Her anger was with the corporations and agencies and insurers that she saw as hiding behind the law. They hadn't put sufficient value on Billy's life before he went into that tunnel. Now she would insist that they acknowledge his worth in the only language they all understood. Once she got a sense of the likely figure Judy would be offered, Olga drew her line in the sand. She informed her lawyers—as well as Billy's father—that the defendants would have to meet her number of $2 million, and not a penny less, or they would go to trial.

Then Olga and Bill, who had checked out of their hotel room, retrieved their suitcases. Olga wheeled hers to a bench in the lobby, where she sat with Pelletier. "Do you think I'm being unreasonable?" Olga asked.

"You lost your son," Pelletier responded reassuringly. "You've got to do what you feel is right."

Pat Riley, the defense lawyer for the MWRA, stopped by to talk to Olga's lawyers. Because Kiewit's contract to build the tunnel had included an indemnification clause protecting the MWRA from lawsuits, Riley was representing the sewer agency, but he was actually being paid by Kiewit's insurance company. He had been pushing for a settlement ever since he'd reviewed the damning memos and decided a strong argument could be made that his client and their consultants had placed a higher value on the integrity of the tunnel than on workers' lives. Chatting in the lobby, he told Prescott he didn't think the other defense lawyers would agree to meet Olga's number, but he might be able to get close.

A little while later he returned with an offer that had been goosed to $1.9 million. Prescott was inclined to take it. Olga, however, stayed true to her word, saying she wouldn't accept it and would hop in a cab so they could make their plane. Riley asked Prescott for one more chance. Olga remained on the bench and didn't let go of Pelletier's hand.

Eventually Riley returned, relaying to Prescott what he had told the other defense lawyers: "Let's put these poor parents out of their misery and let them go." After some resistance, he said, they had agreed.

Olga and Bill thanked him as well as their lawyers, then rushed to make their flight.

Once Norton succeeded in getting the settlement figure for Judy bumped up to $3.25 million, and the Juses settled for $2 million, the defendants quickly followed up with an offer of $725,000 for DJ's claim, a number that fell on the high side of the range that Prescott and Pelletier had put forward in their last conversation with the mediator. DJ said yes. With those three claims settled, Hoss and Riggs agreed to the same figure that DJ had accepted. As part of the settlement, those amounts would be sealed in court records for ten years.

In the end, the dollar figures for the tax-free settlements were not huge. After deducting legal fees, expenses, and worker's comp payments, Judy received about $2 million, the Juses split about $1.3 million, and the surviving divers got somewhere between $400,000 and $500,000 each.

Although the divers and families were spared a draining trial, as it turned out a long process would go on in civil court without them—

independent of any criminal charges the attorney general might decide to bring. That's because although the many defendants had agreed to the settlement figures with the plaintiffs, they had failed to agree on how much each of them would have to pay toward that settlement. So the total settlement payments to the families were fronted primarily by the insurers for Kiewit and Norwesco, with a small remaining slice paid by the insurer for A-L Compressed Gases. Then those insurers went to court to demand reimbursement from the other parties, slugging it out for years.

One week after the civil case had settled, Joan Parker took her seat in a cavernous conference room on the twentieth floor of a building across the street from the gold-domed State House. Seven months after she'd been handed the Deer Island case, she was finally getting her audience with Attorney General Tom Reilly to advise him on whether to go forward with criminal prosecution.

In the last couple of months, Parker had received some conflicting messages from her superiors, especially after she sent them a detailed memo in September offering her strong argument for prosecuting Harald. Parker's immediate supervisor was now sitting at the table with her, as was his own supervisor. She was grateful to have Mary McCauley, the detective who had become her friend, seated at her side. But the only person who mattered now was Reilly, the fifty-eight-year-old who had been elected attorney general two years earlier. As confident as Parker was about the merits of the case against Harald, she knew there would be plenty of other people appealing to the AG's inner caution, arguing all the reasons he should avoid a high-stakes prosecution in this case.

She also couldn't help but wonder if Boston's power brokers, who were desperate to end the tortured saga of the harbor cleanup, were applying pressure to avoid a potentially messy criminal prosecution into the diver deaths. Tellingly, although no one from MWRA had ever been seriously considered as a potential target of criminal prosecution, deep in the file there was a court transcript of the MWRA's lawyer cheering the district attorney's decision not to charge *anyone* in the case. He told Judge Mazzone that the decision "was fortunately a relief."

Parker knew the odds were against her. During her career, she had investigated probably fifty workplace fatalities. Of those fifty, she felt that only about five had sufficient evidence to demonstrate criminal recklessness. And of that handful, the office had chosen to prosecute just one. In that case, the previous attorney general had charged the owner of a scrap metal processing company with manslaughter in connection with the deaths of two workers. That man had eventually pleaded guilty to two lesser counts of assault and battery and been sentenced to probation and community service. Parker felt that, in this case, a manslaughter charge against Harald would be even stronger because there was a more direct line between his actions and the resulting worker deaths.

As the attorney general opened the meeting now, he gave a clue to his thinking. "I understand that this is a very complicated case," he said. "I also understand that it is very unusual to pursue this kind of case under common law. It's risky."

Parker tried to avoid letting her facial expression convey her sinking heart. To his credit, though, the AG said he would hear Parker's presentation with an open mind.

Parker spoke for more than half an hour. As the AG's director of safety, she was one of the few people in the room who did not have a law degree. She realized how that might put her at a disadvantage, but she leveraged her fluency with the statutes to make a forceful argument. As she methodically took the AG through the contours of the investigation, she hewed closely to the detailed memo she had written to her supervisors, arguing that Harald "took full responsibility to design the system, intentionally and blatantly keeping his plans from others, precluding any meaningful participation or assistance." She conceded that other parties, from the contractor to the construction manager to the MWRA, had failed miserably in their oversight of the mission. But because Harald had kept key details of his plan from everyone, she argued, he alone was responsible for the circumstances in the tunnel during the three days in which the divers were relying on the mixed-gas system, and he alone understood and controlled this intricate system that the divers were counting on to stay alive. She rattled off all the unnecessary risks she believed Harald had taken, the grave danger he had subjected the divers to, and his pattern of dismissing the valid concerns that the divers had raised with him. All this, Parker argued, made Harald's conduct willful and put him on a different plane of blame.

When Parker was finished, the attorney general paused for a minute. Then the soft-spoken, cautious man turned to her and said, "I'm willing to take the risk."

The attorney general, however, wanted something from Parker. In order to build the strongest possible criminal case, he asked her to deepen her investigation, leaving no facts unexamined. For five months, she did just that, although her effort was complicated by a family crisis. Her mother was battling end-stage renal failure, requiring Parker to travel frequently to Florida to care for her.

On the last Friday in March 2001, Parker returned to the office from Florida to find a Post-it note attached to her desk chair. It had been written by her supervisor. His message was short but devastating: *We've decided not to prosecute the Deer Island case.*

Parker was enraged. And shocked. Having found Harald's behavior to be so egregious, she had told herself more than once, *If we can't prosecute a case like this, there's no case we can prosecute.* After spending a year of her life putting all the pieces together, she was now being told, essentially, *Thanks, but we're dropping it and moving on.* Via a Post-it note.

She stormed into her supervisor's office. Resisting the urge to swear or scream, she made it clear just how strongly she disagreed with him. He told her the decision had been an extremely difficult one. Although he agreed with Parker that Harald had ignored basic safety precautions and showed an obvious disregard for the well-being of the divers, he didn't feel the Canadian engineer's actions rose to the level of a manslaughter charge.

To prove involuntary manslaughter, the prosecution would have to establish beyond a reasonable doubt that the killings of the divers had been unlawful and unexcused, and that they had been caused by wanton or reckless conduct on the part of Harald. In other words, it would have to prove that Harald had intended to commit the act that caused both divers' deaths, although he had not intended to cause the deaths themselves. Her supervisor explained that the office had to be mindful of the "blame the victims" defense strategy hinted at by Harald and Norwesco's lawyer—the suggestion that because Hoss had acted quickly to switch to the backup system and save the lives of the three divers on the excursion crew, somehow Billy and Tim should have been able to do the same. While Harald's breathing system had clearly been a disaster, Parker's supervisor said he didn't think the divers' deaths could be solely and definitively attributed to the system's failure. Ultimately, he said, he did not feel it would be fair to prosecute only Harald when the long list of parties—Kaiser, Kiewit, MWRA, OSHA, and others—shared responsibility. It might also create perception problems for the attorney general, if people believed he was going after one employee of a small company while turning a blind eye to the conduct of several large corporations.

Parker's supervisor had already written a thirteen-page memo to his bosses explaining his recommendation. She wasn't going to be able to change his view.

After stewing about it all weekend, she fired off an e-mail to her supervisor's boss, the chief of the bureau that housed her division. She acknowledged the concerns about the perception of the office "picking on the little guy." But, she asked, what about the perception of letting everyone go scot-free? The answer, she argued, shouldn't be to drop charges altogether. The answer should be to charge Harald individually and then draw up corporate charges under the criminal statutes against Kaiser, Kiewit, and Norwesco. Because those corporate charges involved a lower standard, she felt there was sufficient evidence to prove them.

By Monday evening, she received her reply. The bureau chief reaffirmed the decision not to prosecute Harald or anyone else.

Parker would not give up that easily, and she was successful in getting one more audience with the attorney general. This delayed the final decision for a few more months. Eventually, though, Parker had to confront the reality that the office was simply not going to prosecute. It all had the finality of an extinguished cigarette, ground into the dirt. She knew she had no choice. She tendered her resignation.

15

The Long Tail of Trauma

It wasn't a big dive, but the sounds and especially the darkness left Hoss rattled. He had taken this Norwesco assignment at a hydro plant in Washington State right after leaving Deer Island, to try to get past the haunting reminders of the tunnel. It hadn't taken long, however, for him to realize he'd made a big mistake. The trauma he'd experienced was too severe for him to simply dust himself off and splash back into the water. Later that night he was alarmed to discover that he couldn't remember how to tie basic knots, the type of routine task he'd learned as an eighteen-year-old during his first week of dive school. He told himself: *Something's wrong with me.*

Capable, gung-ho, cocky Hoss was demoralized by the thought that he might be forced to find another line of work. Diving was what he excelled at. Was he just supposed to accept the fact that at twenty-five he had to relinquish the career he was meant to have? When he returned home to Idaho, he found himself feeling irritable and depressed. He kept having the same harrowing nightmare where he was trapped in a maze, opening every door he saw, frantically looking for the way out. Instead, behind each door, he found a flood of water threatening to wash over him.

He realized he needed help, so he went to see the counselor his health insurer recommended. The guy looked even younger than Hoss. "I need a counselor," Hoss said.

"For what?"

So Hoss told him his story.

"Whoa," the young counselor said, "that's definitely out of my league."

Next came a more experienced therapist, who told Hoss, "You have post-traumatic stress disorder."

"I think you're wrong," Hoss replied. He'd heard about PTSD, of course, but he associated the diagnosis with veterans returning from the battlefield. He had always had a lot of respect for combat vets. As wrenching as the tunnel accident had been, he somehow didn't think it put him in the same category. It took him some time to accept the fact that even though the disorder was yoked to the battlefield on account of its history, war wasn't necessary to trigger it.

The researcher who helped lay the groundwork for PTSD in the 1970s was a former Marine by the name of Charles Figley. He'd volunteered for combat duty in Vietnam and came back disillusioned and angry. At protests organized by Vietnam Veterans Against the War, he got to know lots of vets struggling with nightmares, rage, terror, and addiction. The wreckage he saw was unmistakable, and the similarities striking, yet the Veterans Administration was both unwilling and ill equipped to understand what was happening. Figley returned to school to study psychology and human development. As an assistant professor at Purdue University, he interviewed combat vets and established the Consortium on Veterans Studies to pull together others doing similar work. In his 1978 book *Stress Disorders Among Vietnam Veterans,* he described the markers for "catastrophe-related stress," in which vets felt helpless to stop reliving a traumatic event.

As soon as word about the book began circulating, a young psychologist named Terry Keane, who was working for the VA in Jackson, Mississippi, flew to Indiana to meet with Figley. Keane had begun as a substance abuse researcher but shifted his focus after noticing that his patients who were Vietnam vets were all showing the same kinds of symptoms. When he would ask them why they got drunk or high, they'd invariably reply, "Because I can't go to sleep."

"Why not?"

"I'm afraid of the nightmares."

"What are they about?"

"What I saw in the war."

Figley opened all his files to the young psychologist from Mississippi, who was stunned at how closely they matched those of his own patients. Keane in turn described for Figley the novel approach that he and his colleagues were using to treat the psychological wounds of veterans. The technique, an early form of what would later be called exposure therapy, focused on helping desensitize patients to their most paralyzing memories by having them recount them, in full detail, again and again. These painful discussions, balanced with relaxation techniques, were designed to help patients gain control over what haunted them.

Two years later the American Psychiatric Association gave the disorder its new clunky four-letter acronym and included PTSD in the third edition of the *Diagnostic and Statistical Manual of Mental Disorders,* the bible of the mental health field.

Meanwhile, Keane continued his intense therapy. During one excruciating session, a Vietnam vet lurched over to Keane's trash can to throw up. Keane learned that his patient was reliving the horror of seeing his buddy blown up by a rocket-propelled grenade just a few feet away. That explosion had sent his friend's flesh flying into the patient's mouth. Vomiting had become the patient's way of coping with that hideous memory—and ridding his mouth of the lingering taste of burnt flesh.

In the late 1980s, Keane became director of the National Center for PTSD's behavioral science division in Boston. When he collaborated with psychologists treating victims of rape or industrial accidents, he was once again struck by how much the symptoms resembled those of combat vets. All of these patients relived the trauma in flashbacks and nightmares. They became hyperaroused: simply watching the nightly news or seeing a place that looked like the scene of the trauma could trigger it all over again. But mostly they were numb, shutting down as caregivers or spouses.

Many victims turned to alcohol and drugs to relieve their symptoms. Even for people who previously had been heavy users, the character of the substance abuse fundamentally changed after the trauma. Whereas they had once partied to feel good, they now used drugs and alcohol as joyless self-medication.

One of the mysteries of PTSD is that people exposed to the same trauma are often affected in profoundly different ways. Keane interviewed a host of high-achieving business and civic leaders who had served in Vietnam. He wondered what explained their resilience when members of their same platoons had turned into the scary-looking guys lying on park benches, swigging from bottles and swearing at passersby.

Keane found that it wasn't a matter of intelligence or capability. (Temperament and predisposition appear to play important roles.) Most of the vets he worked with were savvy guys who had no trouble finding jobs. What was hard was holding on to them. Plagued by feelings of abandonment and vulnerability from the original trauma, PTSD sufferers often resort to substance abuse and anger. While booze and most drugs numb the sense of powerlessness, the anger displaces it, allowing a reassertion of control. But both introduce serious new stressors. It's hard to hold down a job and keep a marriage together if you're constantly drunk or angry.

Keane eventually became one of the nation's leading voices on PTSD. He developed some of the most widely used assessment measures for the disorder, which is now believed to affect 24 million Americans, or about 8 percent of the population, at some point in their lives. And it afflicts not just combat troops but all sorts of civilians, from cops and paramedics to victims of rape and childhood trauma. While a wide range of treatments have emerged over the years, all of them come with drawbacks. Keane admits there is no magic bullet, but he continues to argue for cognitive behavioral therapy—which includes exposure therapy—since it has the most evidence to support its effectiveness, as well as the backing of the Institute of Medicine. Figley, meanwhile, came to believe that the mental health field's emphasis on evidence-based treatments is too restrictive, since a significant number of people hate exposure therapy and some even appear to be retraumatized by it. Still, both men agree that those suffering from PTSD need to find a way to deal with the trauma head-on, if they are to have any hope of returning to a normal life.

Although PTSD was birthed by the Vietnam War, it describes psychological wounds that are as old as military history itself. One of Keane's colleagues at the Boston VA won a MacArthur Foundation "genius grant" for his work identifying PTSD parallels in the ancient Greek epic poems the *Iliad* and the *Odyssey* and using them to inform contemporary treatment. That psychiatrist found in his combat-vet patients the same violent tempers, danger seeking, and aimlessness that Homer ascribed to Achilles and Odysseus. The road home after trauma can be as treacherous as any battlefield.

. . .

Once Hoss accepted that he had PTSD, he asked for some pills to make it go away. He soon learned that while drugs could treat conditions associated with the disorder, like depression and anxiety, there was no medication for the PTSD itself. Therapy wouldn't just be important—it would be indispensable. Hoss found the exposure therapy to be especially brutal. Having to retell his therapist the tunnel horror story again and again was bad enough. But in addition, anytime some shard of memory about the tunnel flashed into his mind, no matter where he happened to be, Hoss had to force himself to think about the accident, in its entirety, from start to finish. The goal was to provoke the anxiety often enough that the nightmares plaguing him would be defanged. Like other victims of PTSD, he was taught to think through five questions: What happened? Why did it happen? Why did I act the way I did at the time? Why am I acting the way I'm acting now? (In other words, Am I crazy?) And will I be able to survive if this happens again? Because having a sense of hope is critical to recovery, Hoss's therapist helped him confront the absolute worst memory from his trauma. For him, that moment had come after they'd loaded Billy and Tim's bodies onto the trailer and went to start the Humvee, but the engine wouldn't turn over. That was when Hoss had felt the weight of uncertainty collide with the fear that his close friend's blank stare and lifeless body were a preview of his own likely fate.

With that nadir established, the therapist next employed other forms of cognitive behavioral therapy. He instructed Hoss to suit up in his dive gear and go to the bottom of a swimming pool, adding a few minutes to his "bottom time" each time he did the exercise. Hoss learned that when people suffer a traumatic event within a known territory, that territory can become toxic to them. It isn't uncommon for pilots who survive a plane crash or firefighters who survive a roof collapse in a tall building to become phobic about some aspect of their occupation, developing, for instance, a crippling fear of heights or enclosed spaces. That phobia is usually just a manifestation of the overall fear that their job—which they may have done heroically for years—has become inherently unsafe. The time Hoss spent suited up at the bottom of the pool helped him understand that it wasn't the diving, or the air tank, or the act of being underwater or below grade that was dangerous. He came to appreciate that what had been unsafe was the particular job he'd been asked to do in the Deer Island tunnel, and the appallingly inadequate system he'd been given to do it with.

As depressed as Hoss was, he committed himself to seeing his therapy through. He and his wife, Heather, had plans to build a beautiful house on the property in Coeur d'Alene they had purchased before the tunnel job, yet Hoss drew little satisfaction from that. While she continued to work full time, Hoss spent much of his day in bed, sometimes getting up only to attend his therapy sessions. He could tell that his wife was trying to be supportive, but one of the sad byproducts of PTSD is that people suffering from it, in their hopeless and self-absorbed state, can often come across as ungrateful to the people trying to help them. Hoss found that he and Heather were speaking to each other less and less.

Eventually, though, he began to see slow, modest improvements in his condition. He tried to recapture more of the person he'd been before the accident, the guy who loved doing thoughtful things for others, the good sport who picked up the tab when no one was looking. He took two of the antique guns he'd found in Tim's stash and had them mounted in handsome display cases. He sent them as gifts to Tim's parents and brother. Then he had Tim's dive helmet thoroughly restored, arranging for Tim's name to be painted in black cursive onto the side of the yellow hat. He sent that to Judy.

On the first anniversary of Tim's death, Hoss and Heather once again stayed with Judy in rural Washington, where they resumed the tall task of helping to declutter her house. In one closet, where Judy kept Tim's clothes, Hoss found a box. When he opened it, he felt his stomach drop. "Judy, I want you to get rid of that box," he said somberly.

She didn't have to ask why. "You know I don't want to."

The box contained the outfit Tim had been wearing on the day of the accident—his gray sweatshirt, his red sweatpants, even his red underwear. The clothes, which had been cut open by the paramedics as they tried to revive him at the base of the shaft, had been returned to union managers in Boston, who then mailed them to Judy. As was the case with so many of Tim's things, she hadn't been able to part with them.

"I think you should," Hoss told her. He understood her sentimentality, but he thought this was taking it too far. "Maybe it's harder for me," he acknowledged, "because when I saw that box it just sent me all the way back a year." It was, he said, "as though I had never been through any therapy or treatment or anything. It was just like I was *in the tunnel again.*"

There was something else Judy kept in that closet, a box she herself had brought back with her from Boston. It contained Tim's ashes. Tim had once offhandedly told her that when he died, he wanted to be cremated and have his ashes scattered at sea. She'd honored the first part of his request but had found it impossible to deal with the second part. Once when Tim's parents had come for a visit, Judy had removed the box with his ashes from a closet shelf. But scattering them seemed too final, so the box had gone back on the shelf. Their cat Matta, whom Tim had rescued and who loved nothing better than sitting on his lap, had taken to hanging out in front of that closet door.

For nearly two years after the accident, Hoss stayed out of work and in therapy. Eventually he felt strong enough to give diving another shot. He reconnected with one of his earlier mentors from Norwesco, the same diver who had cautioned him against taking the Deer Island job. The guy had signed on with a Seattle company called

Global Diving & Salvage. As he explained it to Hoss, the new place was very different from Norwesco. The whole business model was built around safety. The owners might charge their clients a bit more, but because of their zero tolerance for unnecessary risk, the job was far less likely to end in costly injuries or death. Therapy had helped Hoss understand that while his employer had shortchanged his safety on the Deer Island job, his profession wasn't inherently unsafe. Those assurances were enough to persuade him to make the leap again.

Hoss was relieved to find that his confidence in the water quickly returned. But he also recognized that he was a different type of diver now, more willing to speak his mind when he saw something that didn't seem right, more demanding of genuine solutions, and less tolerant of pat answers. The best part of his settlement money was the security it gave him to walk off the job whenever he felt things were unsafe, without giving even a moment's thought to how it might affect his career or his wallet. That confidence also made him more of a leader, since he was always willing to speak the truth.

Hoss's return to diving went so well that, in time, he ascended to the highest ranks of his profession. He became a saturation diver, which required him to remain underwater in a pressurized bell for up to a month at a time. The work took him from Latin America to the Middle East. He was enormously grateful to Heather for staying with him during his lowest periods and felt he couldn't have made it without her. Still, the strain of his recovery had taken its toll on their relationship. During one lengthy stint at sea on a Norwegian vessel, he fell for the vessel's secretary, Karolina, a bright and determined woman from Poland. After returning to Idaho, to the dream house that he and Heather had finally managed to build, he confessed to his affair. He and Heather soon divorced.

. . .

Three years after the accident, Riggs traveled from his home in Nevada to Deer Island for a ceremony to dedicate a pair of memorial plaques and benches. He loathed the idea of returning to this dungeon of dark memories, but he had agreed to attend because he felt Tim and Billy should be remembered properly. Now that he was here, though, something felt wrong. In truth, ever since the accident, not much had felt right.

The weather was beautiful—warm September sun—and the bronze plaques looked classy. Billy's mother, Olga, had worked closely with the agency that owned the tunnel to plan a tasteful ceremony. Still, Riggs had to wonder what they were really all doing here. Hoss couldn't come because of work, but DJ was there with Riggs. Standing alongside them were many of the key players behind the tunnel project, all solemnly commending the ultimate sacrifice of two divers as though they were war heroes, all ignoring the many preventable screwups that had combined to rob Tim and Billy of their lives. Riggs thought to himself, *What the fuck?*

He found himself thinking, and saying, *fuck*, a lot lately. He recognized the irony of his situation. The good Christian who, before the accident, had admonished DJ and Hoss to watch their tongues, now swore like the sailor he had once been. It wasn't the only echo from his navy days. He'd responded to the tunnel accident the same way he had responded to the deadly crash aboard the USS *Nimitz* two decades earlier, dulling his pain with beer. As he saw it, he had been living the right way, doing the right things, busting his tail to provide for his family, being a good diver and a good father. Yet where had it gotten him?

In the immediate aftermath of the tunnel accident, Riggs had sought guidance from the same structures that had served him well for so long. He had showed up at Lake Mead in Nevada, where Norwesco was finishing up its long project, and gotten right back in the water. He knew that people managed to recover from all sorts of tragedies. Like Hoss, he refused to see himself as a victim. He would tough it out. Once he realized the folly of that approach, he left Lake Mead and Norwesco for good.

He tried therapy, something Hoss encouraged him to stick with whenever they spoke by phone. But Riggs wasn't as fortunate as Hoss in finding a therapist he could trust. Besides, like some PTSD patients, Riggs found little good in talking to some stranger over and over again about the horror he had witnessed. Words did little to fix the hollowness he felt from knowing that all the good in the world had disappeared when he and four other guys were trapped down in the tunnel. And words seemed powerless to curb the nightmares that returned him to that hell at least once a week. He knew he was the kind of guy who worked best when he was able to figure out his problems on his own. So he took the meds the therapist offered and tuned out the rest, chasing down the pills with beer. He knew that intake would probably do little more than sand down the edges of his problems, but that was better than nothing. More and more he and his wife, Karen, alternated between long silences and occasional arguments.

Their worst fight had taken place shortly before his last trip to Boston, for the mediation session in the civil lawsuit. One morning, furious that Karen was riding him for what felt like no good reason, he got in her face and yelled, "Why are you trying to fucking provoke me?" He was so infuriated, and got in so close, that the bill of his baseball cap hit her face. From her perspective, that was no accident—it was an assault. And it was something she could have never imagined happening before the tunnel disaster, when she'd always seen her husband as a kind man and a doting dad. Later that day she told him he needed to leave the house before the cops got involved. He decided to head to Boston earlier than planned, crashing with DJ and Lorraine for several weeks before flying home and forging a precarious peace.

Once Riggs accepted that he wouldn't be returning to diving, he set about getting qualified as a certified welding inspector. He was good at this job, inspecting bridges and other structures, and it kept him close to the water without requiring that he get in it. But prior to getting his certification, he'd been out of work for so long that he had begun to worry his family might lose their house. He was grateful for the lawsuit settlement money because it had spared him the shame of foreclosure. In many other ways, though, Riggs regretted having agreed to it. Like Hoss, he'd wanted the responsible parties to admit to their guilt. But the settlement felt like just another transaction. And the depositions that he, Hoss, and DJ had been forced to undergo just to get to the mediation phase had ripped the scabs off their mental wounds. He had tried not to let the pressure of those proceedings get to him. When one defense lawyer had asked him what he did to treat the searing headaches he'd reported suffering since the accident, Riggs had said, "Take some ibuprofen and drink a couple of beers." When the lawyer asked if that worked, Riggs replied, "If it doesn't, I drink a couple more." As the deposition wore on, though, he'd been unable to hide his distaste for the well-dressed defense lawyers peppering him with accusatory questions. If their goal had been to get him angry, they had succeeded. Ever since the accident, Riggs had anger to spare.

The more he'd learned about advanced breathing systems, the more he'd understood how hopelessly and inexcusably flawed Harald's setup had been. One saturation diver friend pointed out that Harald's setup violated a cardinal rule in having the excursion team breathing off the same system as the Humvee crew. Riggs burned with rage toward Harald for having failed them, and with Roger for having failed to police Harald.

Still, in his most introspective moments, he was tougher on himself, and that felt a whole lot worse. He knew some people might blame the divers for not having walked off the job in the face of all the warning signs. But just like in the military, if you couldn't trust the people you were working for, you were probably a casualty before the battle began. He took responsibility for entering that tunnel, but he felt guilty about something he had done once inside it, something he believed might have contributed to the deaths of the other guys in his foxhole. At one point during his deposition, when a defense lawyer was mortaring him with questions that implied errors on the part of the divers, Riggs had expected him to bring up this point. To his surprise, neither that lawyer nor any of the others had ever mentioned it. But that didn't mean it weighed any less heavily on Riggs's mind.

When he and DJ had been removing the plugs at the very end of the tunnel, they'd felt the need to work as fast as possible to compensate for the aborted missions of the two previous days. Still, they knew that they could make up significant ground once they cleared out of the final five-foot-diameter section of the tunnel, and that would happen after they'd finished with plug number 3. Once they could stand up straight, they'd be able to proceed much more quickly. And by the time they had pulled the plugs all the way up to number 12, where the Humvees were parked, they'd be able to move with real dispatch. They just needed to get over this one last, hellish hump. So he and DJ had worked hard, *really* hard, dragging plugs, carrying sandbags, and duck-walking as fast as possible, all in pursuit of that goal. They had been so damn close. Riggs couldn't help but think that, in doing all that intense work, he and DJ had strained the mixed-gas system to the point where they had overbreathed it, causing it to fail. In other words, Riggs couldn't help but wonder if his actions, however well intentioned, were what had caused Billy and Tim to die.

Riggs had discussed the guilt he felt with just two other people, DJ and Hoss, and both of them had assured him that it wasn't their fault. The system had been crap from the start, they said, and should have never been approved. On an intellectual level, Riggs understood that. But on an emotional level, he couldn't get past feeling that he owned some of the responsibility. If only someone had thought to share the navy analyses of the accident with him, he might have better appreciated how the divers never really had a chance.

Hoping a change of scenery would help him get unstuck, Riggs and his wife moved the family from the desert of Nevada to the verdant mountains of Plumas County, California. There was no work for him there, though, so he would spend the week on job sites in the San Francisco Bay area. Each weekend he'd make the nine-hour round-trip drive home to be with his son and daughter, as well as with his wife, although their estrangement continued to grow.

It was around one in the afternoon when DJ threw off the five-hundred-thread-count sheets and climbed out of bed. In the bathroom, he splashed water on his face before drying with one of his Ralph Lauren Egyptian combed-cotton washcloths. He padded through his leased condo in an upscale Waltham complex, past the boy's bedroom with the PlayStation that he'd set up for his son's visits, past the crystal stemware in the dining room, past the high-definition TV in the living room. He opened the blinds to take in the view. In the parking lot, DJ spied two of his newest and most prized possessions. There was the silver 2001 F-350 power-stroke diesel Ford pickup truck that he'd special-ordered, with every possible add-on, to the tune of $43,000. And sitting on a trailer hitched to the truck was his black $36,000 custom Bourget Low Blow motorcycle, which turned heads on the highway every time he opened its throttle.

It was a beautiful day, and for the first time in his life, DJ was surrounded by beautiful things, the kinds of material status markers he had dreamed about when he'd been the high school transplant from Oklahoma who showed up in Massachusetts embarrassed by how little he had. As he stared out at his acquisitions now, though, DJ could feel no happiness. Instead, he felt an overpowering depression.

With tears forming in his eyes, he began talking to himself. *Why don't I just get on my motorcycle and go for a ride?* He felt immobilized, and he couldn't figure out the reason. *Why am I so depressed?*

In moments of deep reflection, DJ began to understand what made him incapable of truly enjoying his new things. He'd ask himself, *Where did that motorcycle come from? That truck? Do I have them because I worked hard and put my money in the bank?* Then he'd answer his own rhetorical question with cold candor: *I have them because those two guys are dead.*

It hadn't taken long after the accident for DJ to get back on the partying circuit with Dana. For her, it was a way of celebrating the fact that DJ had cheated death. It took her a while to notice that, for him, there was a different motivation. He no longer seemed to take pleasure in his partying, even as he pursued it more vigorously than ever.

She thought that getting him away from Boston might help. Two months after the accident, they went on vacation together in Jamaica. They threw themselves into the high life, but the week ended on a low note, as they both sensed the drift between them.

Back home, they became more detached. To Dana, DJ had grown hopelessly sad. All the things that had made him so attractive to her—the lightness he brought to everyone around him, the pride he took in his appearance, the irresistible sparkle in his eyes—all seemed to have disappeared. It was as if those qualities had somehow remained trapped in that hellish tunnel even as his body had managed to escape. His reliance on pain relievers had begun to worry her. One night he held out his palm to reveal a handful of pills of various shapes and in a rainbow of colors. "Take one," he told her.

Dana hoped that, if nothing else could help him, maybe their Catholic faith might. For his thirtieth birthday, she gave him a Jesus medal on a chain. On the back, it read, "Dear Lord, have mercy on us." Under the prayer, she had it engraved: "Love Dana, 9-27-99."

But his descent continued. One night three months after the accident, DJ came over to her house, carrying a pair of steak-tip dinners in Styrofoam takeout containers. He was unkempt and unshaven, and as he sat across from her, attempting to cut his steak, the knife was unsteady in his hand. She had sensed for some time that the guilt he felt from having survived the accident was leading him to believe that he didn't deserve anything good, and causing him to let himself go. Now she concluded, *He's just gone.*

She told him she couldn't be with him anymore. She feared that if she stayed with him, she'd slip down the same self-destructive path. She didn't feel strong enough to save them both, so she decided she needed to save herself.

Five years earlier, when Dana had broken up with him the first time, DJ had settled into a long, deep funk. In perhaps the clearest indication of the accident's lingering cloud over him, he met this second breakup with resignation. He handed Dana back the Jesus medal and headed home.

He was consumed with panic. The air around him was a dense mix of smoke and fog. It was so dark and he was so disoriented that he couldn't tell which way was left and which way was right. But he knew that if he made the wrong choice, he would die. They would all die.

DJ turned to the two guys who were with him. It was hard to make out their faces in the dark, but he knew it was Hoss and Riggs. They began arguing over which way to turn to get out. DJ could feel the sweat soaking through his clothes.

He awoke with a start. Just like in the nightmare, his body was drenched in sweat, but now he noticed that so were his bed sheets.

He was breathing heavily, and there were tears in his eyes. Ever since the accident, he'd had the same recurring nightmare. There were occasional variations—Billy was sometimes there, joining the debate over which way to go—but DJ always came away feeling completely drained.

Before the accident, whenever he'd heard people talk about traumatic flashbacks and nightmares, he could never understand what the big deal was. Even if it was a bad nightmare, it wasn't real, so couldn't you simply splash some water on your face and get on with your day? He'd never grasped how paralyzing a nightmare could be. Every time he had one, he awoke consumed with guilt. Guilt that he had switched positions with Billy at the last minute, even if Billy had been the one who requested the change. Guilt that maybe, if he had just continued chest compressions a little longer, he might have been able to resuscitate his buddy. Guilt that, no matter how negligent Harald or anybody else was, he had personally failed to save Billy and Tim.

Although Hoss, Riggs, and DJ had all experienced the same trauma and were all racked by nightmares, each one of them was responding in a fundamentally different way. While Hoss bowed to the power of his PTSD in the hope of conquering it, Riggs was trying to outsmart it and, when that seemed impossible, to displace it with anger. As for DJ, he just kept trying to outrun it.

Even though he'd found himself unable to return to diving, the settlement money had bought him plenty of time. After Dana, he'd moved on to a series of head-turning girlfriends. He knew he probably should return to the therapy sessions he'd undergone in the year after the accident, when he'd been diagnosed with PTSD and depression and had been prescribed Klonopin and other antianxiety medications. The pills had side effects, and the draining sessions did little to keep the soul-sapping nightmares and waves of guilt at bay. So he came up with his own treatment regimen—one built largely on continual motion. If he resumed the role of being the life of a never-ending party, he wouldn't have to confront the ways he was dying inside. Not long after the accident, DJ took a few Percocets to try to ease the pain. He found they helped take the edge off, so he added them to his diet, one that he later fortified with Vicodin and other opiate painkillers. Still, he hated to go to sleep at night because he knew that's where the nightmares would find him. So he began deliberately trying to stay up all night. This was done most easily in loud bars. He found that when he mixed his prescription painkillers with lots of booze and a line or two of coke, he'd eventually pass out, giving him a few hours of peaceful sleep. He liked the combination of cocaine, which gave him a feeling of euphoria and increased sex drive, with opiates, which relaxed his muscles and eased the paranoia that the coke often introduced. But to maintain its effectiveness, the regimen demanded that DJ be forever on the go.

He felt better when there were lots of people around. When his closest friends couldn't keep up with the pace, DJ began filling their slots with hangers-on who viewed him like an open bar at a wedding, a bottomless source of free top-shelf drinks that demanded nothing in return. He started hanging around with the Hells Angels. He bought a nice mobile home parked near Weirs Beach in New Hampshire. It quickly became a favored partying spot for him, his club friends, and the many women who spend their lives chasing Angels.

He used a chunk of his remaining settlement cash to open a welding repair and fabrication shop in Waltham, renting a three-bay garage and equipping it with all the best tools. His newest girlfriend, who was named Michelle, encouraged him to do more to reach his potential. She used her skills as a lawyer in the financial industry to help him set up business accounts and establish credit with suppliers. DJ had always been a capable welder. If he couldn't return to diving, he figured running a welding shop would be the next-best thing. Leveraging contacts he had around town, he got his business off to a good start.

For years, DJ had counted on the built-in structure of the diving industry to keep his partying impulses in check. He knew that the next job, along with its mandatory drug test, was always just around the corner. Now, with only himself to answer to, he began to lose all self-discipline. He kept two stocked refrigerators in his welding shop, and the place became a magnet for friends and acquaintances.

By 2004, DJ started coming to terms with the fact that a life of non-stop partying was not sustainable. In his shop, while jobs backed up, he'd fall asleep at his desk with the air conditioner on full blast. After three years of throwing around wads of cash like a sailor on leave, he'd blown through most of his settlement money. He'd already lost the high-end condo and moved back in with his mother. Now he fell behind on his rent at the garage as well as his child support payments. Troubled by his behavior, Cody's mother restricted DJ's access to their son. Searching for a new source of discipline, DJ turned to the military. He decided he would become an Army Ranger. With all the initiative he had shown when he'd started out in commercial diving, he threw himself into the application process. He rounded up a stack of exemplary recommendation letters and included a note from a Waltham cop commending DJ's spur-of-the-moment assistance years earlier in chasing down a fleeing suspect. In the end, though, the army denied his application.

Dispirited, he returned to his partying ways with abandon, causing so much drama that he found himself banned from a couple of the Waltham bars where he'd been a popular fixture for years. He moved from Percocet and Vicodin to the more powerful opiate OxyContin. Before long he found himself unable to function without those powerful little green pills called OCs. Although OxyContin had arrived on the scene only in 1996 as a treatment for cancer patients and others in serious pain, it quickly became the hottest prescription pill in the underground drug trade. Once the time-release coating was removed from the little pill, it took just a few seconds to deliver a head-spinning high that was packed with a twelve-hour punch of morphinelike pain relief. Oxy was so damn powerful that tolerance quickly built up, demanding more and more milligrams to deliver that same high. And any addict who tried to stop feeding the addiction was sentenced to debilitating withdrawal sickness. Oxy was also incredibly expensive. A single 80-milligram pill could cost eighty dollars or more on the black market, opening up the drain on DJ's rapidly vanishing bank account.

Eventually, his erratic behavior cost him his relationship with his supportive girlfriend Michelle. Now not only did he have to keep running to stay one step ahead of the nightmares, but he had a worsening addiction that was beginning to dictate his life. One day in 2005, after his stash had run dry and he was suffering miserable withdrawal sickness, he called a friend who shared his opiate addiction and asked, "You got any OCs?"

"No," his friend said. "Nobody's got anything."

DJ wasn't buying that story. "You don't sound like you're sick right now."

"No, I'm not," the guy admitted. "But I don't want to be the one to turn you on to something you don't need to get into."

DJ knew exactly what he was talking about. "All right."

Then he hopped on his motorcycle and, five minutes later, was standing at his friend's door, telling him, "You gotta give me some of what you've got."

So began DJ's introduction to the cheaper, deadlier opiate of heroin.

Among the many cruelties of heroin is that it functions like a doorbuster deal at a big-box retailer's Black Friday sale, perverting the concept of value to incentivize the reckless purchase of things you don't need. DJ found that for what he'd been paying for a single OxyContin pill, whose high wore off after several hours, he could now get a gram of heroin. That generous supply could keep him feeling good for up to three days. Before long the addiction was consuming his life. Gone was his sense of initiative. Gone was the muscular physique he had long prided himself on but which had become impossible to maintain. Gone, most surprisingly, was the irrepressible charm that had made DJ so attractive to the parade of women he'd met and seduced over the years. The hunt no longer held any appeal for him. All he wanted to do now was get his drugs so he could get high.

DJ worked hard to keep his habit from his mother, but one day he'd been unable to hide it. As disconnected from reality as he'd become, seeing that lethal look of heartbreak and pity on her face punctured his haze. He'd always cherished the idea that he was her rock. Now what was he to her but another burden in a life that had already been saddled with too many?

Not long after that, he got a call from an old diving buddy. It was early 2006, and there was a desperate need for divers in the Gulf to tackle the rebuilding work following Hurricane Katrina. DJ's friend told him he could make up to a thousand dollars a day. All he had to do was show up—and bring along his own diving helmet because hats were at a premium in the Gulf.

DJ stopped to consider the offer. On one hand, he hadn't worked as a diver since the tunnel accident. And on his worst days his heroin addiction made him barely able to handle a walk to the convenience store, never mind a dive from an offshore rig into three-hundred-foot waters. On the other hand, he was broke and desperate, and maybe this was the chance he needed to jump-start his life. He sure didn't want to see that look on his mother's face again.

He drove to a local hospital and checked himself into a quickie detox program. After that, he had to find a dive hat. He had sold his helmet years earlier, and he certainly didn't have six grand lying around to buy a new one. He hopped into his truck and drove north to New Hampshire.

"Tap," DJ said, "do you have a hat I could buy or borrow?"

Nearly a decade earlier, DJ had loaned Tap his helmet, and now, standing in the same spot where they'd trained for the tunnel job, he was asking Tap to return the favor.

"Here, DJ," Tap said, handing him one. "Take this, and just give it back when you're done."

He was 311 feet under water, using a large impact wrench to help install a caisson, and he was working strenuously. He began to hyperventilate, taking rapid, short breaths rather than the long breaths necessary to make his demand regulator work properly. Divers have a term for what DJ found himself doing. It's called overbreathing your hat, and it could be fatal. It meant he was failing to fully breathe his oxygen and fully exhale his CO_2. He grew more anxious. His mind flashed back to the tunnel, to Billy and Tim, as anxiety turned into panic. He worried he wasn't going to make it. "All stop!" he yelled into his comm wire.

He realized that, given his depth, if he let himself get to full-scale panic, he'd be a dead man. He leaned forward, put his head down, and cracked the free-flow valve on his hat wide open. A rush of air filled his helmet and began circulating. He took deep, steady breaths. He got himself together. He would be all right.

Once he finished that project, DJ was relieved to discover that even though he hadn't worked as a diver in seven years, his skills came back to him easily. Before long he was finding peace in the water. Outside of it, he was feeling pretty good, too. Before leaving Boston, he'd bought some Suboxone pills on the street, semisynthetic opiates used in detox to wean addicts. When he'd arrived in Louisiana and prepared to head to the offshore rig, he'd brought the pills with him. He had some heroin left, which he knew he should get rid of, but chose instead to stash it in the glove compartment of his truck. He'd chopped up the Suboxone pills, taking one piece every day to help him stave off the sickness.

After working offshore for a week and a half, he called his brother. Just before he'd left the bitter cold of Boston, he'd been broke and dope-sick. Now he was in sunny Louisiana, back diving and back among the living, and he had ten grand in his pocket. He told his brother, "I think I got this thing beat."

DJ found himself working with mostly younger guys, recent dive school grads on their first Gulf tour. But he bonded most closely with a Tennessee diver named Bryan Ernest, who, like DJ, was in his mid-thirties. Toward the end of the year working in the Gulf, DJ made a celebratory road trip to New Orleans with some of the younger guys. The night ended with police finding DJ lying unconscious in the French Quarter, with his pocket full of heroin. He had finally succumbed to that stash in his glove compartment. He was arrested, and although he received probation rather than jail time, it cost him his dive job.

A few months later, in February 2007, when he was on his way back home to Waltham, he got a call from Bryan. His diving buddy was about to make the first saturation dive of his career. He told DJ he was so nervous that on his drive to work that morning, he had pulled his truck over and thrown up. "Do you think I'm ready?" he asked.

"Bryan, I think you're ready," DJ replied. "Just take your time."

A few days later, he got a call from his former employer. There had been an accident. Bryan had been killed.

DJ flew to Tennessee for the funeral, where he met up with some of his former supervisors. They felt awful about what had happened, and knew how close DJ had been with Bryan. They offered DJ his job back.

All his instincts told him to decline. He had never vanquished the pain from having to bring Billy and Tim out of the tunnel eight years earlier. Now, not only did he have an addiction to contend with, but he also had a fresh source of grief in having lost another good friend. Seeing no other options, though, DJ accepted the offer. Four months after his return, outside Galveston, Texas, he was arrested for possession and sentenced to several months in jail.

When he got out in February 2008, he returned to Waltham, unemployed and depressed. He reconnected with Chris Politis, who had been his closest friend in town. Chris was now an unemployed ironworker who, like DJ, had a child and, also like DJ, a raging addiction to heroin.

They took care of each other. Back when he'd still had money, DJ had shared his dope with Chris. He often let Chris crash on his mother's couch—one reason why Chris was so fond of Lorraine. Now that DJ was out of cash, Chris stepped in to fund their habit. DJ knew addicts often stole from their families to feed their addiction, but he refused to break the promise he'd made to himself never to take from his mother. For three months, he and Chris functioned in a haze of codependency. DJ knew Chris wasn't working, and eventually he learned how his friend was coming up with the money. Another friend who worked as a bank teller had once explained to Chris that employees there were trained to try to get robbers out of the bank as quickly as possible. They were told not to try to play hero by stalling or even pressing the panic button. Insurance would cover the loss of a few grand from the teller's drawer. What the bank most wanted to avoid was an escalation that could leave people dead. So tellers were drilled to hand over the bills calmly and quickly. Based on this insider knowledge, Chris had settled on bank robberies as a relatively easy way to get the cash he needed to function. He decided there was no reason to be threatening, no need to take on the added risk of using a gun, or even alluding to one in his holdup note. Across several months, he hit a string of branches, each time using nothing more than a scribbled message on the back of a deposit slip that read either "50's and 100's" or "big bills."

DJ wanted no part of this criminal behavior, and Chris never asked for his help. Still, by the spring of 2008, the banks and the cops were getting warmer on Chris's trail. Surveillance cameras had captured his image, partially concealed by a Boston Red Sox cap pulled down tight onto his head and a bulky winter coat draped over his burly body. Copies of that grainy screenshot had by now been flashed during local TV newscasts and posted in bank lobbies throughout the area. DJ feared it was only a matter of time before Chris got pinched.

His cell phone buzzed while he was lying in bed. It was early Saturday morning, May 10, and DJ's withdrawal sickness had kept him up most of the night.

"You ready?" Chris asked.

"Yeah," DJ said flatly, even if he didn't mean it. "I guess so."

"Okay, I'll call you when I get to your street."

DJ heard the hallway floor creak. His mother was approaching. Lorraine had seemed especially worried about her son lately, keeping a closer eye on him. As she popped her head into his doorway, DJ spoke casually into the phone, "Yeah, give me a call later." Then he rolled back over in bed, as though trying to get back to sleep.

Lorraine continued on into the bathroom. As soon as he heard her turn on the shower, DJ bounded out of bed and headed downstairs. Chris called again as he was pulling up in his Honda Accord in front of the duplex, and DJ hopped in.

Obligation and addiction had brought DJ to this pathetic point, poised to do something that, even in his dope-sick fog, he knew was immoral. Like any addict, he had let all the rationalizations play out in his mind. He wouldn't be armed, so no one would get hurt. He needed to assume some of the risk, since Chris had been shouldering it all for months in order to feed both their habits. DJ saw the emptiness of those justifications, but he also knew that without them, he wouldn't be able to go through with it. And then where would he be? He needed to stop being dope sick. To do that, he needed more dope.

The day before, they had stopped at a costume supply shop to snatch up a fake beard and a tube of scar makeup. Now, at 8:10 a.m., they pulled up to a Walgreens pharmacy, looking for wool caps, but they struck out. Winter hats weren't a big seller in May.

At a contractor supply shop, DJ persuaded the kid behind the counter to retrieve a box of wool hats from the storeroom. After fishing out a New England Patriots cap and another hat, he paid for them with the last few dollars he had in his pocket.

In the parking lot, Chris told DJ, "Fuck it. I'm going to do this myself."

DJ had been dreading even the thought of going through with a robbery. Still, he knew it wasn't right to continue having Chris take all the risk. "Are you sure?" he asked.

"Yeah," Chris said. "If you get caught, I'm gonna have to live with that, and your mother is gonna fuckin' disown me."

DJ felt a lot better limiting his role to getaway driver. After Chris put on the fake beard, DJ helped him apply the scar makeup to his right cheek. DJ couldn't decide if his buddy looked more like a fresh Taliban recruit or an Amish guy separated from his farm cart. "I don't know about that beard," he said, chuckling. "You look a little shady."

They started driving, still not sure where they were going to go.

Sitting behind the wheel, DJ mindlessly rubbed his hands along his thighs, as if trying to iron out the wrinkles in his jeans. He took a long drag on his Marlboro Red, then mumbled, *Jesus, what have I gotten myself into?* The Honda was now parked in a cozy business district

overlooking the Massachusetts Turnpike in Newton, not far from
the Waltham line. His cell phone rang. He looked down and saw his
mother's number. She'd called several times since he slipped out of
the house. He let this call, like the others, go straight to voice mail.

Wearing his winter hat, he sat waiting for Chris to emerge from
the Village Bank, a brick building with arched windows and a quaint
cupola. Looking to the right again, DJ spotted a Newton police cruiser
on the prowl. *Oh Jesus, this is over.*

Moments later Chris yanked open the passenger's door and hopped
in. "Let's go," he said.

DJ inched the black Honda forward.

"C'mon, dude," Chris complained. "Let's go!"

"Wait a minute," DJ shot back. "A fucking cop just went by. He
took the right. I want to let him get up the street a little. I don't want
him to see us pulling out and get a description of our car."

Carefully, DJ let the Honda creep toward the intersection. When
he saw the cop pass Dunkin' Donuts on the right, evidently on his
way to circle back around the bank, DJ banged a left. He headed along
Route 30, which trails the Charles River, to get to the Mass Pike.

Now Chris was agitated. "D, you think we're going to be all right?"

"Yeah," DJ said, peering into the rearview mirror. "I don't see anybody yet, and there's nobody behind us. Just don't turn your head
back, Chris."

Once on the Pike, heading east toward Boston, DJ looked over to
see his buddy peeling off the beard and wiping away the scar makeup.
Chris tore up the beard as best he could and stuffed it into a coat
pocket. He ripped up the deposit slip, the same one he had handed
to the teller with "big bills" written on the back. He lowered the car
window and chucked out the clump of deposit-slip bits. Instantly, the
gusts along the Pike sent the clump flying back into the Honda, landing in the rear seat.

"Chris, that shit went back inside the car," DJ said.

Chris didn't seem worried. "Yeah," he said. "We'll get it later."

DJ looked across the median strip of the divided turnpike and saw two Newton police cruisers flying along the westbound side. "Dude, we're all set," he said.

"You think so?" Chris asked.

"Yeah, they're responding to the bank, and we're already this far in the other direction. We got it. We're good." For the first time all morning, DJ felt the tension starting to lift.

Chris smiled and reached over to shake DJ's hand. As he started counting the bundles of bills to determine the size of his score, he smiled again. "Oh, Jesus, I did good on this one." When he reached $5,000, he turned to DJ and said, "Dude, there's still a couple of thousand in here. I think we might have like seven or eight grand on this one." He handed DJ a thousand-dollar bundle of fifties. "Here, D, this is for you."

"Thanks, Chris," DJ said, jamming the wad deep into the front pocket of his jeans.

"Thanks for driving."

A few moments later, Chris peeled off another $500 in twenties. "Here, D," he said, handing it over. "Take your mom out for Mother's Day."

"Really, Chris?"

"Absolutely. Your mom deserves to be taken out. We've put her through enough shit."

Chris told DJ to get off the Pike at the next exit, at the Allston-Brighton tolls, and hop onto the Storrow Drive Parkway.

DJ glanced over to see Chris pulling out his cell phone and begin dialing. "I'll be over in like ten minutes," Chris told his dealer, then hung up.

Approaching the tollbooth, DJ grabbed one of the twenty-dollar bills Chris had just handed him and paid the $1.25 toll, fishing around the ashtray for a loose quarter so he'd get an even $19.00 back in change.

At the end of the parkway, he steered the Honda onto the expressway and then into the Callahan Tunnel, a fifty-year-old tube built under Boston Harbor and best known as a route to Logan Airport. But they wouldn't be going that far. At the end of the tunnel, before the entrance to the airport, there is a right-turn exit leading to the heart of East Boston, a dense and tribal neighborhood mostly separated from the rest of the city by the harbor.

DJ looked over at Chris. "We gotta get some help after this," he said. "We can't keep doing this."

"I know, I know," Chris agreed.

Daylight met them near the end of the Callahan Tunnel, and DJ guided the Honda past a strip of asphalt, where a single police cruiser was routinely stationed. His stomach dropped when he noticed that instead of one, there were two Boston Police cruisers there, along with a Boston paddy wagon and a state police cruiser. Up ahead, toward the airport entrance, another statie was idling. There was nothing routine about this contingent.

Chris looked over at DJ, the color drained from his face. "Dude, that's not for us, is it?"

DJ was panicked, too, but somehow mustered sarcasm. "No, Chris, that's for the fucking guy who crossed the solid yellow line behind us," he barked. "Of course it's for us!"

Time now slowed. DJ swore he could hear his heart pounding, echoing throughout the car. *Ga-boom. Ga-boom*. He looked over to see a Boston cop pointing, slowly mouthing "That's them!" He saw another cop approaching with his gun drawn and, behind him, a couple of officers holding shotguns. Blue lights were flashing, sirens were blaring.

DJ looked over at Chris and thought he could tell what he was thinking: *Should we make a run for it?* But DJ could see there was no-where to go. Suddenly, he wondered if death by a barrage of bullets would be easier than the humiliation that would follow an orderly surrender.

He was scared and broken. And dumbfounded. *After all I've survived, this is where it's going to end?*

At 2:38 p.m., the bailiff's deep voice punctured the quiet in a bright courtroom paneled with warm English oak. "All rise." Everyone did as they were told, watching as the judge entered. Even in his robe, Judge Mark Wolf didn't cut an imposing figure. He was short, bald, and round-faced. But knowing how DJ's lawyer had described the judge—that he was very smart, very serious, and often very tough—Lorraine was bracing herself for the worst.

It had been a year since Lorraine had received that crippling call from DJ at the police station. Now, in June 2009, she had come to the John Joseph Moakley United States Courthouse, a handsome new building hugging the revived harbor, to hear her son's fate. She'd spoken with his court-appointed lawyer several times over the past year, while DJ had been held in a federal lockup facility an hour's ride away, in Plymouth. The lawyer, Pete Horstmann, initially thought DJ's best legal strategy would be to demonstrate that he had gotten into Chris's Honda without knowing what his buddy intended to do. But as new evidence came to light, and as DJ refused to pin everything on his friend, both DJ and Chris decided to plead guilty in the Village Bank case, and Chris also admitted to his earlier robberies.

DJ hoped his plea would result in reduced prison time, but Horstmann cautioned both him and his mother to be realistic. When it came to guilty pleas, judges usually gave the prison time that government prosecutors recommended, and on rare occasions they actually gave more. Horstmann mentioned a previous client who had pleaded guilty to armed bank robbery, only to receive an unusually stiff sentence. The judge in that case happened to be the same man who would now decide DJ's punishment.

Lorraine sat in the courtroom gallery with her sister at her side. Staring at the defendant's table in front of her, she was pleased to see her son looking more like himself. DJ had used his year in jail to get clean and return muscle to his frame. He wore an off-white, long-sleeve thermal top and green, loose-fitting pants. He had grown a goatee and wore his light brown hair high and tight, combed forward on top. There was a long scar on the left side of his forehead. Lorraine had seen it during one of her visits to Plymouth. She knew the official explanation was that he'd fallen out of his bed. She also knew enough about the Darwinian ways of prisons to realize that there was often a sizable gap between official explanations and the real story.

During the long hearing, the lawyers for DJ and Chris jousted with the assistant U.S. attorney and the probation officer over how to count the "points," under the complex federal sentencing guidelines, to determine the defendants' prison terms. Judge Wolf seemed impatient. He pounced to correct the lawyers for slight errors. Repeatedly, he puffed up his cheeks and then exhaled dramatically. Talk dragged on over how to count certain previous offenses, such as DJ's arrests in Louisiana and Texas. When all the quibbling was done, Wolf announced his reading of the range of sentence lengths called for under the guidelines. For Chris, the range was between 151 and 181 months. For DJ, it was between thirty-seven and forty-six months. Lorraine was relieved to hear the ceiling for her son was four years. Still, four years was far too long for her boy to be away.

Wolf explained to DJ and Chris that it was their right to address the court before they were sentenced, but they were not required to do so. DJ's lawyer, in fact, had strongly discouraged him from speaking. Horstmann explained to DJ that, in his experience, a defendant seldom did himself any favors by addressing the court and actually ran the risk of saying something that might unintentionally tick off the judge.

As she looked on, Lorraine had no idea what her son was going to do. The answer came soon enough. When the judge asked DJ and Chris if they were interested in speaking, both of them nodded yes. Chris went first.

"People told me I should write a letter," he said, his thick Boston accent rendering that last word as *lettah*. "But I don't forget, every waking day, the embarrassment I caused my family." As sobs came from his relatives in the gallery, Chris began to cry himself, lifting his white T-shirt up from his stomach to wipe away the tears. When he was finished, the court was completely still.

Before DJ spoke, his lawyer rose to address Wolf on his behalf. He knew the veteran judge had seen an endless parade of defendants file through his courtroom over the years, blaming all manner of bad behavior on their addiction to drugs. As Horstmann had explained it to Lorraine, however, he felt that, for perhaps the first time in his career, he had in DJ a client who wasn't simply a drug addict but someone with a legitimate, compelling reason for his addiction. He also suspected that DJ's trauma in the tunnel was so wrenching and unusual that it might appeal both to the judge's heart and to his intellectual curiosity. That DJ had never sought to use the tunnel horror to his advantage—in fact, Horstmann had found out about it only after DJ decided to plead guilty—reinforced in the lawyer's mind his client's genuineness and almost naïveté.

"This is a person with a painful addiction," Horstmann told the judge, "and an addiction that seemed to emerge after an incredibly traumatic experience." He briefly explained the tunnel accident, building on the detailed reference letter he had submitted to the judge from John Prescott, one of DJ's lawyers in the tunnel case. Horstmann told the judge how DJ had used drugs to try to run from the nightmares that had haunted him ever since the accident, and how the financial settlement he'd received had allowed him to fund a very long, drug-fueled run. Despite having hit bottom, and having made some inexcusable decisions, Horstmann said, DJ was a good man. "I like my client," he said. "He's a fighter. He was able to carry his friend's body eight miles under Boston Harbor."

Lorraine watched as Horstmann sat down and DJ stood up. Her son began by demonstrating the manners that had always served him well in life. He apologized first to his mother, "who worked so hard to raise me and my brother on her own." But he didn't stop there, apologizing to the court and to the Village Bank. He made it clear that Chris hadn't coerced him into getting involved in the robbery. "Coercion was the drug," he said.

"My drug addiction is strong, but it doesn't justify my actions," he continued, breaking down. "I'm a grown man." After steadying himself, he asked the judge to consider their state of mind. "We weren't driving to the bank to get cash so we could get high and have a good time," he said. "We were going so we could get cash and not be sick, so we could create the impression for our families that we were responsible people in control of our lives."

Lorraine's eyes welled up.

As he listened to DJ, Judge Wolf once again puffed up his cheeks and exhaled before tipping his head to the right and then to the left. After DJ sat down, the judge paused to reflect. Then, a few minutes after four o'clock, he was ready to speak. First the judge addressed Chris. Although Chris had admitted to multiple robberies, the judge said, the government would have had difficulty proving his involvement in those earlier holdups without Chris's confession. He sentenced Chris to eighty-four months in jail. Looks of relief cascaded over the faces of Chris's relatives. They were still trying to do the math in their heads, but they knew this sentence was much lighter than what they had expected.

Lorraine whispered, "How many years is eighty-four months?" and someone whispered back, "Seven."

Then Wolf turned to DJ, telling him he had clearly played only a supporting role by driving the getaway car. "Mr. Gillis, I sentence you to eighteen months in custody," the judge said, adding a recommendation that DJ spend at least three months in an inpatient drug treatment facility upon his release.

DJ's lawyer looked shocked. It was less than half the minimum sentence called for under the guidelines. Lorraine, knowing her son had already served more than a year while waiting for the wheels of justice to turn, looked overjoyed. With a few months shaved off his sentence for good behavior, she figured he might be released by the end of the summer, in time to celebrate his fortieth birthday.

The judge also ordered the pair to make restitution payments to the banks. Chris would be required to pay back $41,004, for the money he had pocketed from his six-month string of robberies. DJ would be required to pay back one dollar, for the buck he'd used at the tollbooth on their getaway ride.

Wolf acknowledged that his sentences were far lighter than the federal guidelines suggested, but he argued the variances were justified. The case, he said, was about drugs much more than it was about theft. He told DJ he thought he was a good candidate for beating drugs. "It's hard," he said, "but you've demonstrated that you can do hard things that require discipline." It all depended, Wolf told DJ, on how he responded to this break the judge was giving him. "If you blow this off, I'll see you within the year. Then I may lock you up for three years," the judge said. "But if you take advantage of the opportunity, your future will be much happier than your past."

Wolf said he had closely read Prescott's letter about DJ's role in the tunnel accident and he had been moved by it. "Your problems didn't begin when your friend died and you survived that horrible accident," the judge said. "But they got a lot worse." While stressing that the tunnel accident didn't excuse his role in the bank robbery, he said, "it helps explain it." The accident, he was suggesting, must still loom large over DJ's life. With that, the judge stood up and left.

As he was led out of the courtroom, DJ turned and winked at his mother, mouthing the words *I love you*. Underneath his off-white shirt, there was evidence to support the judge's hunch. There, on his right shoulder, was the reminder DJ saw every morning when he looked in the mirror. It was the tattoo he'd had inked in a dodgy basement shortly after the accident, when Riggs had been staying with him.

At its center, there was a large diving mask. Below it were the names Billy and Tim. Above it were the words: NOT TO BE FORGOTTEN.

EPILOGUE

I stood at the registration desk of the Washington State Convention Center in Seattle, waiting for my day pass, hoping it would lead me to him. The conference was called OCEANS '10, an international gathering of oceanic engineers and scientists. I had come in pursuit of the man I had spent so much time thinking about, and talking about, but had never met.

It was a gamble, since Harald Grob's name appeared nowhere on the conference materials. Since his return to Canada shortly after the deaths of Billy and Tim and his four-hour interview with the state police and OSHA, Harald had managed to lie low. The many investigators and lawyers who had tried to chase him down, during the years of criminal investigations and civil litigation, had come away frustrated. They warned me not to get my hopes up.

I had communicated once with Harald, however. In 2009, shortly before I wrote about the tenth anniversary of the tunnel deaths in *The Boston Globe Magazine*, I e-mailed Harald a request for an interview. He sent me this one-paragraph reply:

> *Two colleagues died on the Deer Island Outfall project and it*
> *deeply affected all of us. Over the years we have learned to live*
> *with this tragedy. For my part, I do not wish to start the healing*
> *process over again. Therefore, I respectfully decline your offer of*
> *an interview.*
>
> *Thank you*
> *H. Grob*

After I included his note in my article, some of the divers' family members told me they were offended by his comments. "It's great that he was able to finish the healing process," Billy's mother told me, "but we're still struggling to heal."

Although I already had plenty of material conveying Harald's point of view when I decided to write this book, I was determined to gather much more. Over the course of several years, I was able to amass a mountain of documents and correspondence involving Harald: transcripts of his conversations with investigators and lawyers; audio recordings of his interview with OSHA and state police; video footage of his work in and out of the tunnel; official logs and unofficial notations scratched into day planners by Kiewit managers after their phone conversations with Harald; faxed memos and e-mails between him and his equipment suppliers; and e-mails between him and Roger as they strategized their defense following the accident (including the unforgettable one describing his sightseeing trip to Borneo).

By then, I was confident that I could accurately describe Harald's perspective and the role he had played. Still, I wanted to talk to him in person. I learned he was working for a Canadian company called OceanWorks International. Based in British Columbia, OceanWorks specializes in cutting-edge diving systems, such as atmospheric diving suits that are designed to withstand underwater pressure at two thousand feet. At the OCEANS conference held in Germany in 2009, Harald had been a featured presenter, delivering a paper titled "Cabled

Observatory Technology for Ocean Renewable Energy Devices." He seemed to be getting the kind of industry recognition that investigator Joan Parker had come to believe was a far bigger motivator for him than financial gain.

Unlike the conference in Germany, this gathering in Seattle in the fall of 2010 did not list Harald as a presenter. Still, as I made my way into the exhibition hall, I figured there was a good shot he'd be in attendance. As I rounded the corner, I spotted the OceanWorks booth. There were three men standing behind a table, including a short guy who held his arms out, as though he were carrying something under each one. Aside from the gray-white hair on the side of his head, he was completely bald. He wore a blue Oxford shirt, and as I drew closer, the name tag clipped to his breast pocket came into view. *Harald.*

As I offered my hand, I watched his eyes travel to my name tag, and then saw his smile disappear. I introduced myself, although by then it hardly seemed necessary. I asked if we could speak privately. At a table in a far corner of the hall, I told him that, with this book, I hoped to understand the project from every perspective. As much documentary evidence as I had describing his role, I said, I was sure he could offer some additional insight. And I reminded him that my goal was not to look for simple black-and-whites in the story but to explore the many shades of gray. He told me he had not read my pair of magazine articles about the tunnel project. Based on his body language, I wasn't persuaded that was the case, but I took him at his word and said I would send the stories. My initial impression was that he was polite and remarkably soft-spoken—so soft-spoken, in fact, that I had to crane my head to hear him over the din of the exhibition hall.

Eventually, though, the other side of Harald—the one that the divers and investigators had described to me—came through. When I told him I would like to continue our conversation after he'd had a chance to read what I sent him, he narrowed his eyes and snapped,

"I know that's what *you* would like." The comment instantly took me back to the exchange when an OSHA investigator had pressed Harald on how using an unreliable handheld analyzer to measure the oxygen content could possibly be accurate, and Harald had huffed, "It is *certainly* accurate." I asked him to call me after he'd a chance to read the materials I sent him, and he said he would consider it.

Although I sent him the articles right away, I never heard back from him. A year after our meeting, I e-mailed Harald again, detailing for him the even bigger mountain of materials I had accumulated and explaining that I had been fortunate enough to speak with every major player on the project. I followed up my e-mail with a fax, a letter, and a voice mail message. A week later I spotted an e-mail reply from Harald in my inbox. As soon as I read the first sentence, I knew I didn't need to read further. He had simply cut-and-pasted his reply to me from two years earlier, without adding a single word.

For years, a debate has persisted among the people who examined the diver deaths about who had been most responsible. Was it Harald, for the decisions he made in conceiving and implementing the breathing system, and for his insistence on pushing ahead despite all the warning signs? Or were the various bigger players more to blame, since they had let the project get to such a crisis point that they needed to call in Harald for the rescue mission—and then had signed off on his plan without giving it the scrutiny it desperately needed?

Plaintiffs' attorneys John Prescott, Nina Pelletier, and Bob Norton, perhaps not surprisingly, argue that more responsibility rested with the big players, since without their failings, there would have been no Harald. On the flip side, the AG's Joan Parker and the union's Dan Kuhs maintain that Harald was far more to blame, because his actions were so willful. Clearly, he had not intended for Tim and Billy to die. But, they point out, he had controlled the complex breathing system that was the divers' means of staying alive in the tunnel. He had chosen to keep critical information from other players and had disregarded the divers' concerns.

And how about Roger Rouleau, Norwesco's owner and Harald's boss? Parker had kept him in her sights because he was the only one from Norwesco who had stood to gain financially from the job in any meaningful way. Ultimately, though, she concluded that Harald was far more responsible because of the way he had hoarded both information and operational control.

Three years after the diver deaths, Roger closed down Norwesco and, with a colleague, formed a new business that same year, doing the same kind of diving work but under a new name, Associated Underwater Services. In the summer of 2007, at a BP oil refinery in Washington State, a piling detached from a vibrating hammer and killed one of Roger's employees, a Massachusetts native by the name of Christopher Primeau. OSHA levied a "serious" safety-violation fine against Associated Underwater and several more against the general contractor. Roger told me the incident dredged up bad memories from Deer Island and contributed to his decision to leave the dive business. Primeau's family members remain furious with him and his business partner for safety lapses they contend were inexcusable. In 2009, around the tenth anniversary of the deaths of Billy and Tim, Roger sold his share of the business to his partner.

In retrospect, Roger told me, he had failed to supervise Harald adequately on the Deer Island project, and failed to heed the concerns that Hoss and others had raised with him. "There were a lot of things going on that I didn't understand fully, but I accepted the fact that Harald had his bases covered," he said. Only after the accident did Roger come to believe something important about Harald: "He didn't know what he was doing." Yet Roger argues that the bulk of blame ultimately should rest with the players who were more powerful than either him or Harald. The designer, contractor, construction manager, and tunnel owner all knew about the plug problem for years but failed to deal with it, instead sloughing it off on a small diving subcontractor. (Of course, Norwesco had chosen to take on the job even though two other established dive companies had declined to bid because of safety concerns.) OSHA officials had raised no objections when Roger presented Harald's plan to them, even though it blatantly violated the agency's requirement for mechanical ventilation in the tunnel. Looking at it now, Roger said, one thing is clear. "We should have never been in there."

After spending years marinating in this project and its aftermath, I now find Mary McCauley's argument for shared responsibility to be the most persuasive. The state police detective maintains that Harald must shoulder the heaviest blame because of his reckless actions. However, she stresses that significant responsibility also rests with the project's memo-warring big players, with Roger, and even with OSHA officials, who all became disturbingly hands-off when it was time to review the plan and supervise Harald.

Whenever a worker dies, there is a natural inclination to hunt for a huge, single failure that can be blamed. In reality, a worker's death is usually caused by a series of small, bad decisions made by many individuals, none of which, on its own, would have been enough to produce a fatality. Disaster strikes only when all the holes in the Swiss cheese line up.

But the fact that all those holes did line up in the tunnel case forces a larger, lingering question. How could this idea of sending divers to a place as remote as the moon, asking them to entrust their lives to an improvised, untested breathing system, have ever made sense to sensible people? The answer lies in the dangerous cocktail of time, money, stubbornness, and frustration near the end of the over-budget, long-delayed job. The major players desperately needed the project to surmount its last enormous hurdle. It's almost as if, amid all the fatigue, expense, and mutual distrust that had built up like plaque in the arteries, these players looked at Harald's dazzling plan, then closed their eyes and hoped that it made sense. If they had kept them open, they would have had to confront the many ways in which it didn't. They also might have hatched a better plan, like the "steel straw" vent solution that was ultimately used to get the plugs out so efficiently.

Even Dave Corkum, Kaiser's memo-writing manager, while not walking away from his insistence that Harald's plan seemed reasonable, acknowledged to me that as distrustful as he and everyone else had become by the end of the seemingly interminable tunnel project, he probably wanted to believe the plan was sounder than it actually was. "Was this like our fairy godmother coming in? Like manna from heaven?" he asked. "Maybe."

The deaths of Billy and Tim had been more than just preventable. They had been more than just predicted. Given all the bad decisions—by all the players—the deaths had effectively been preordained.

Because of its untried nature, the divers' plug-pulling mission appeared to be in a category of its own. However, when I consulted several specialists in workplace safety and organizational behavior, they found in Billy and Tim's deaths important and widely applicable lessons for avoiding all sorts of bad outcomes on the job—and in life in general.

Injuries and deaths tend to happen late in projects, when confidence runs high and tolerance for delay dips especially low. This springs from a phenomenon known as normalization, which, in this case, allowed people to accept looser standards in the name of greater speed. The more people do something without suffering a bad outcome, the harder it becomes for them to remain aware of the risks associated with that behavior. This lesson has implications for any task involving some level of hazard, from cleaning gutters to chopping vegetables. If it was worth climbing down the ladder to move it a few feet at the beginning of the gutter job, it should be worth taking that same sensible precaution near the end.

Another important lesson: a poorly thought-out safety measure is often worse than no safety measure at all. This explains how child-safety products that careful parents once viewed as necessities—from baby walkers to bath seats—can end up leading to infant injuries and deaths. It also explains how early requirements for ironworkers to be tied off with a rope when working at high elevations unwittingly introduced serious choking hazards, a problem that continued until the requirement was upgraded to include a harness. Remember that when it came to the tunnel, the fifty-five plugs were a *secondary* safety measure. By failing to spell out how those plugs could be safely removed after the tunnel was finished, the tunnel designers and managers baked into the project an enormous and unnecessary risk. Like Prohibition spawning organized crime, it was a cruel case of unintended consequences. Along the same lines, all the redundancies that Harald had touted in his plan not only failed to safeguard the lives of the divers but made things worse by creating a false sense of security.

Finally, there is a bucket of lessons that all deal in some way with group dynamics. The fact that emotional intelligence is often a better predictor of a person's effectiveness than IQ is something most of us confront early in our working lives. As brilliant as Harald may have been in many respects, he appeared to have suffered from a staggering deficit of so-called EQ—the ability to read, process, and manage the emotions of the people around you (as well as your own). Harald's tendency to bristle when the divers raised concerns about his plan, and to stress his own rank in the org chart, made matters much worse.

Then there's the seductive appeal of the outsider—or more broadly, the consultant—which helps explain why the contractor, construction manager, designer, and tunnel owner were all eager to turn to

Harald and Norwesco to solve their problem rather than relying on their own collective brainpower. These weary and wary players put a higher value on Harald's "fresh eyes" and can-do attitude than on their own common sense and intimate knowledge of the project. Turning to an outsider also afforded them more distance from the true risks associated with the mission. This helps explain why corporate executives looking to downsize their workforce will often call in high-priced management consultants, seeking a report that comes to the same conclusion. It also explains how sports apparel companies, by outsourcing the assembly of their sneakers to third-world subcontractors, can insulate themselves from any mistreatment of those workers. The farther managers are from the actual decision making, the greater their tolerance for risk.

But distance can be overrated. In the end, the brilliant "steel straw" plan was hatched and refined by people who'd been toiling on the tunnel job for years, the same guys who had apparently exhausted all their original ideas prior to Harald's arrival. As it turned out, the fresh eyes they needed were staring right at them. Agency chief Doug MacDonald used his status as a nonengineer to ask a basic question, which got the engineers to look at the problem from a new angle. They were then able to devise a solution themselves.

Why did it take the tragedy of Billy and Tim's deaths for them to find their creative spark? Part of the answer was that the battling parties had become so fixed in their positions that they could no longer trust the intentions of the other side. That's how the designer and construction manager could have refused to give serious consideration to Kiewit's promising proposal to yank out the plugs just before removing the bag line, install a temporary dam, and then institute a policy of evacuating all workers from the tunnel whenever a major storm was forecast. Once the main players realized that they had become incapable of settling on a solution themselves and began canvassing for outside help, they fell prey to something called the availability bias.

They based their decisions on what was most available to them—in this case, the dazzling plan that Harald presented—which was itself an extension of the only other dive plan that Kiewit had considered, the one Tap had submitted and that Kiewit had asked Harald to review. The "steel straw" solution wasn't considered because it wasn't yet available. It didn't become an option until, in desperation after the deaths, everyone stepped back, took financial concerns off the table, and explored the problem with genuinely open minds.

In the face of all the mounting warning signs—the blazing red flags that the divers had waved in front of Harald and others about how the breathing system was failing on that Monday and Tuesday—why in the world had the divers been sent back into the tunnel on Wednesday and made to rely on that same failing system? The significance of those red flags appears to have been masked by more powerful crosscurrents: The financial pressures that the contractor felt, fueled by enormous cost overruns. The schedule pressures that the owner felt, fueled by the ticking clock of a court-ordered deadline. The client-relation pressures that Norwesco and Black Dog felt, in trying to please Kiewit, their most important, rainmaking customer. The reputational pressure that Harald presumably felt, to live up to his own inflated billing and preserve his standing as the ultimate fixer. All these forces combined to create an overwhelming pressure to carry on.

Like stepping on the gas when running late for an appointment, only to create a bigger delay by getting pulled over, the act of giving in to all those pressures had backfired in every conceivable way. While the original Norwesco bid that Harald submitted for the job came in at just under $600,000, the cost of making things right after everything had gone so wrong turned out to be fifty to sixty times that sum.

Financially, the toll for just the "steel straw" solution and the civil settlement with the divers and their families approached $25 million.

Add to that another $5 million to $10 million in estimated costs associated with legal fees, insurance and administrative matters, and lost time. The rush to meet the court-ordered deadlines had ended up adding a full year to the tunnel schedule. The relationships that Norwesco and Black Dog and Harald personally had with Kiewit were shattered, as all three companies became embroiled in litigation over who should shoulder the blame for the diver deaths. The Deer Island project even held lessons for the general public. After all, the tunnel job showed how measures that are often popular with taxpayers when it comes to major public works projects—things like low-bid contracts, staggered payments, and late-fee penalties for missed deadlines—can work against their interests by incentivizing risky behavior that can ultimately cost those same taxpayers dearly.

The tunnel ended up altering the way many of the players behind it approached future projects. The most lasting lesson that both Kiewit boss Ken Stinson and Doug MacDonald took away was that when problems arise on a megaproject, as they inevitably do, the priority has to be on solving them. Figuring out who's to blame and who's going to pay can wait. Otherwise the project suffers, and the problem gets a lot harder and more expensive to fix. "Once relationships get poisoned," Stinson told me, "it's very hard to unpoison them."

In recent years, the trend in the tunneling industry has been toward using a new contract structure that aims to reduce the kind of toxic antagonism that stymied the Deer Island job. Under this "design-build" approach, the project designer functions as a subcontractor to the main contractor. Although the approach can introduce its own complications, and the enormous unknowns of doing work underground always inject great risk into a project, design-build can significantly reduce the amount of finger-pointing while streamlining areas of responsibility. That, in turn, encourages the contractor and owner to focus on keeping their relationship strong. As Stinson put it to me, "There are no good stories that come out of jobs that

have bad relationships with the owner." The shame of the Deer Island project, he says, was that it was "a terrific story of engineering
success." But because of the tragic deaths, "no one came out of this
feeling like a winner."

That includes the surviving divers. "We were wronged, but I don't
feel like the world owes me anything," Dave Riggs told me. "I feel
like the world owes itself the truth. I feel like the powers that be that
brought this whole situation into play should be held responsible. And
I have no feeling whatsoever that that's happening or going to happen. I'm not saying that Roger and Harald are the only ones. But I
certainly feel that Roger and Harald have a lot more to answer for
than they have." Riggs, whose relationship with his wife became increasingly strained, said he's careful not to pin all his problems on
Harald and Roger and the big project players. "It's not their fault that
my marriage is a wreck. I don't blame them," he said. "I blame them
for killing Billy and Tim—and damn near killing me."

It's a gorgeous, cloudless day in 2013 as I type this, much like that
radiant Wednesday in 1999 when five divers ventured into the tunnel and only three returned. I'm sitting on a wrought-iron bench
on Deer Island, looking out into Massachusetts Bay. In the distance,
past the white sailboats bobbing on the gentle blue waves, I can just
barely make out a spire of drab granite. Although it appears to be
standing sentinel at the far edge of the horizon, the lighthouse is actually only about five miles from me. For more than a century, Graves
Light served as a signpost for seafarers sailing from the bay to enter
the shipping channel into Boston Harbor. But I've come to see it as a
different kind of signpost, since the Deer Island tunnel runs directly
below it. Even though I have to squint to bring that lighthouse into
view, I know that the spot near the end of the tunnel where Billy and

Tim took their final breaths is almost twice as far away, not to mention hundreds of feet below those blue waves.

I'm sitting on Tim's bench, with the plaque that reads: IN MEMORY OF TIM NORDEEN 1960–1999. Twenty feet to my right, there's an identical black bench featuring a plaque with Billy's years: 1965–1999. A few hundred feet behind me on the island stands another memorial. On a replica of one of those fifty-five Apollo-module-shaped diffuser heads that sit on the ocean floor, the MWRA has affixed a plaque, preserving the same words that the *Boston Globe* letter-writer had chosen to extol the unknown divers: ORDINARY HEROES.

The parade of people strolling, running, and biking past me along the popular public walkway that rings Deer Island is just one testament to the rebirth of Boston Harbor and Massachusetts Bay. Another is the steady stream of fishermen setting up their poles near me on the walkway or floating by on boats, getting their daily catch of striped bass and bluefish, which no longer come blanketed in cancerous tumors. It's an even more remarkable sight when you consider that they are fishing the same waters into which sewage from the old Deer Island plant once expectorated without embarrassment. Other native fish have returned in significant numbers, along with seals and porpoises. Fears that the tunnel would harm the whales and other marine mammals of the bay have so far proved unfounded.

The other side of Deer Island looks out onto Boston Harbor, where the daily catch is flounder, and the view across the blue water is the glistening skyline of downtown Boston. The bustle along the 2.6-mile walkway ringing Deer Island is repeated all along the new 39-mile HarborWalk, which makes the bulk of the harbor perimeter accessible to the public. That walkway now runs from the northern edge of the city, where the USS *Constitution* sits on the Charlestown coast, to the Seaport District, a former wasteland of muddy parking lots that now buzzes with top restaurants and hotels, a sprawling convention

center, and the gleaming new Institute of Contemporary Art building. And it keeps going south, past the city beaches of South Boston
and Dorchester, which attract throngs of families and preening teens
now that they no longer face routine closures because of elevated bacteria levels. The renaissance of the harbor that the federal Environmental Protection Agency calls "a great American jewel" has been so
profound that Boston's center of gravity is moving inexorably from
the Back Bay to the waterfront. The mammoth cleanup of Boston
Harbor is a reminder that, even during an era when government has
been diminished and derided, it remains capable of doing great things
to improve the public good. Still, this postcard-perfect image of re-
born Boston also reminds me of what Billy's mom said during an earlier visit I'd made with her to the benches: "People here should never
forget the lives that were sacrificed to give them a clean harbor."

Around the same time as that conversation with Olga on Deer Island,
I had a similar one with a different diver's mother while sitting on a
different bench. After the bailiffs led DJ out of the federal courtroom
following his sentencing, I walked out of the gallery with Lorraine
and her sister. The corridor outside the courtroom offers a sweeping
view of Boston Harbor through a massive, curved glass wall, which
stands seven stories high and stretches for nearly four hundred feet.
The $228 million Moakley Courthouse is one of Boston's newest
showpieces, designed by the same firm that handled the glass pyra-
mid addition to the Louvre in Paris. As we rode the elevator down to
the lobby, Lorraine told me she could feel the worry inside her simply
wash away. She knew her son still had a long road ahead of him if he
was going to put drugs fully in his past. But she was hopeful, because
he'd been clean for a year now. "I need a cigarette," she said. I followed her as she walked out the front door, turned left, and found

a curved granite bench in the grassy park that snakes between the courthouse and the harbor.

Turning away from the harbor to look in the other direction, she spied the top of the fifty-two-story Prudential Tower rising above the Back Bay. Her carpenter father had helped build that landmark while protected by little more than the Virgin Mary medal he kept in his pocket, the same medal that DJ had tied to his hat on the day Billy and Tim were killed and he barely survived.

As she quickly smoked her cigarette down to the filter, Lorraine said, "DJ told me once, 'Ma, you have to be on these jobs to see the things that we do. You can't imagine them.'" She paused to blow a plume of smoke out of the side of her mouth. "Well, take a look around us," she told me, gesturing to all the high-rises soaring above downtown and rimming the harbor. "Who put all these things together but people like him? When you think about it, it's not doctors and lawyers. It's these guys that make things happen. The buildings are built, the trains run, the planes fly, because of people like my son who get out there and can put things together."

Lorraine paused to light up another Benson & Hedges, and when she looked up, she saw the taillights of an armored van heading out of the court parking lot, toward the highway. "I think DJ was in there," her sister Ann said, waving hopefully. Lorraine nodded, knowing it would be delivering her son back to his cell.

DJ was released from prison a few months after that conversation, and about one month after the tenth anniversary of Billy and Tim's deaths. He told me the magazine articles I had written about the tunnel incident helped ease his transition out of jail. While he had feared old friends and acquaintances around Waltham might cross the street to avoid him, on account of the bank robbery and the drugs, he was relieved to be met with the opposite reaction. People sought him out, telling him that reading about his ordeal in the tunnel helped them finally understand his descent during the years that followed the accident. I was grateful that they saw in DJ what I had seen, a guy who had gone through hell and made a host of horrible decisions but remained a good person to his core.

In the months and years after his release from prison and an inpatient drug treatment program, I was able to spend countless hours with DJ. He was back living with his mother in their duplex apartment, and we often talked there. Other times we got in the car and visited the haunts of his past, including Deer Island. Along the way, I got to know his family and friends. Eventually, though, I told him there was someone else I needed to talk to, and I asked for his help in connecting us. He agreed, reluctantly.

On a cold winter afternoon at the start of 2011, I followed DJ into a swank hair salon in the shadow of Fenway Park. A blonde stood at the reception desk, wearing her coat and holding her purse, giving an employee some last-minute instructions before heading out the door. As she looked up and her eyes met DJ's, she unleashed a big smile. "You're not seriously showing up here when I have to leave, are you?" she shouted, then charged over to him to deliver an exuberant hug and a kiss.

"How are you, Lisa?" DJ said, returning her welcome with a broad smile and warm hug of his own. The nervousness he'd had about how he would be greeted, showing up here after so many years, instantly evaporated.

Lisa looked to the back of the salon, where a black-haired stylist stood cutting a client's hair, and shouted, "Dana! Look who's here!"

Dana walked toward DJ slowly and also leaned in for a hug, though hers was more restrained than her sister's.

Dana and Lisa were still working together at their uncle's salon, and Dana was still single, though she now had a young daughter. Lisa had to leave for an appointment, but she and Dana both agreed to meet with me later, and they both exchanged numbers with DJ.

That night Dana called him, and she called him every night after that for several weeks. She told him that seeing him walk into the salon reminded her how much she missed and loved him. After the accident, his intensifying addiction had scared her. So she had left. But, she said, she never stopped caring about him. Soon she came over to his mother's place to visit, bringing along the engraved Jesus medal she'd bought him for his thirtieth birthday. Now she handed it back to him. Within a few weeks, DJ and Dana were once again dating.

Other reunions didn't go as well. During DJ's wild period, his son's mother had understandably shut off DJ's access to Cody. She continued to refuse him visitation now, although she did allow Lorraine to have contact with her grandson. Since Cody was now a teenager, DJ worried that the window on rebuilding their relationship was closing fast. He sometimes found himself dejected at essentially having to start from scratch around the age of forty, with no job, no car, no house of his own, and a past that always demanded an explanation. At one point, he let self-pity get the better of him, leading to a failed drug test that landed him in a halfway house for a spell. But he forged ahead, returning to Local 56 as a union diver. Most days he's able to remind himself of his good fortune. Yes, he was forced to start his life over. But there were two men he thought about every day who would never have that chance.

APPENDIX

UPDATES

DJ Gillis continues to work as a diver, pile driver, and welder. His renewed relationship with Dana lasted nearly two years, before she once again broke it off.

Dave Riggs remains a certified welding inspector in California. He and his wife, Karen, divorced in 2011.

Donald "Hoss" Hosford travels the globe working as a saturation diver and superintendent. He and Karolina married in 2010, but later divorced. In between jobs, he returns home to Coeur d'Alene, Idaho.

Judy Milner continues her child psychiatry practice in Washington. For several years, she traveled extensively to countries ravaged by war and natural disasters, volunteering her services. More recently she has focused on working with veterans who have PTSD. She has dated, but not seriously, realizing how hard it would be to find what she had with Tim. When his parents, **Bob** and **Shirley Nordeen,** visited her in 2011, they all found a beautiful spot in front of a lighthouse near Puget Sound. There they finally scattered Tim's ashes, along with those of his dog Shackles. The closet where Judy still keeps some of Tim's clothes remains their cat Matta's favorite spot.

Michelle Rodrigue earned her college degree and now works as an X-ray technician at a hospital in Wolfeboro, New Hampshire. Like Judy, she has dated, but not seriously. She treasures the bond she had with Billy, and works with **Deb** and **Ken Jones** to keep his spirit alive through the nonprofit William "Billy" Juse Memorial Fund, which supports community projects and awards scholarships to local students. (For more information, write to the fund at P.O. Box 1482, Wolfeboro, NH 03894.)

Tap Taylor, who is also involved with the memorial fund, still runs Black Dog Divers. But he says Billy's death forced him to reevaluate his priorities. He drastically cut back his hours to spend more time with his kids.

Olga Juse and **Jolene Juse-Paige** visit New Hampshire every July 21, attending a memorial mass for Billy, leaving flowers at his gravesite, and sharing a meal with Michelle. During their visit to Billy's grave in 2011, Jolene called her father, **Bill Juse,** with good news. At age forty-three, she was pregnant. In February 2012, Jolene delivered her first child, a six-pound, ten-ounce boy. His middle name is Billy.

Doug MacDonald left his MWRA post after nine years to become secretary of transportation for the state of Washington. The highlight of his tenure there was the construction of the new Tacoma Narrows Bridge, the longest suspension bridge built in the United States in more than forty years. The project, structured as a design-build contract, came in on schedule, a hair under its $850 million budget, and without a single worker fatality. The contractor was a Kiewit joint venture. At the ribbon cutting, he posed for photos with his old sparring partner, Kiewit chairman **Ken Stinson**. Both MacDonald and Stinson have since retired.

Dave Corkum became a partner with the law firm that represented his former employer on the Deer Island project. The former engineer was part of a team that received the 2001 NOVA Award, a prestigious honor presented by the Construction Innovation Forum, for the design and execution of the tunnel's "steel straw" vent solution. In 2014, he returned to engineering for a new tunnel project.

Mary McCauley was transferred in 2007 to a special state police detective unit attached to the attorney general's office. Promoted from detective to sergeant the following year, she now works for the narcotics and organized crime unit.

Joan Parker, following her resignation as the attorney general's director of safety, worked for a while as a consultant. In 2007 she returned to her horticulture roots, opening her own landscaping business, Parkwood Garden Restoration.

John Prescott, Nina Pelletier, and **Bob Norton** continue their law work, though Prescott is now general counsel for a private company.

Dan Kuhs, who became a regional manager for the union, vigorously lobbied Massachusetts lawmakers to follow the lead of twenty-seven other states and add a criminal negligence statute to the books. Had that lesser charge been available at the time of the tunnel investigation, he argues, there would likely have been a successful prosecution. In July 2004, on the fifth anniversary of Billy and Tim's deaths, Kuhs testified before the legislature's joint committee on criminal justice. The bill died. "This kind of accident could happen again tomorrow," he says. "No laws have been changed."

ACKNOWLEDGMENTS

I never met Billy Juse or Tim Nordeen. But in the last five years, I feel as though I've spent untold hours in their presence. I'm grateful that they made such good companions. These were not just brave and honorable men, but bright and funny guys, too.

I'm fortunate that their family and friends turned out to be such warm individuals whose company I enjoyed greatly, and whose trust I was honored to earn. I'm indebted to Tim's wife, Judy, his parents, Bob and Shirley, and his brother, Todd, as well as Billy's partner, Michelle, his parents, Olga and Bill, his sister, Jolene, and his close friends Deb and Ken Jones. Wonderful people, all of them.

This book would not have been possible without the cooperation and candor of DJ Gillis, Donald "Hoss" Hosford, and Dave Riggs. These three remarkable men welcomed me into their lives and, across hundreds of hours of interviews, allowed me into the recesses of their minds. I know some of the wrenching memories described in the preceding pages are ones they would have preferred to have left alone. But I'm enormously grateful that they trusted me to tell this story, and that they intuitively understood the importance of telling it completely and honestly. I learned so much from each of them.

Their families also played critical roles in my understanding of this project and the personal toll it exacted. DJ's mother, Lorraine Jones, is one of those special people who, within a few minutes of meeting her, feels as familiar as a childhood friend. I appreciate all the stories she told me—and all the tea she served me. Thanks also to DJ's brother, David; his aunt Ann Maillet, and the pair of sisters who played important roles in his life, Dana and Lisa. I also benefited from the hospitality and insights of Riggs's former wife, Karen, and of Hoss's former wives, Heather and Karolina.

Tap Taylor, whose life had for so long been intertwined with Billy's, was giving of both his time and valuable perspective. It's nice to see the bonds remain strong between him and Billy's New Hampshire family, which also includes Thomas Wachsmuth. Deb, who met Billy shortly after losing her brother, once told me: "I always felt like God took my brother and gave me Billy. And then when Billy died, I always felt like God took Billy and gave me Tap."

Roger Rouleau and his wife, Tawnie, welcomed me into their home, even though he made clear his initial reluctance. It's worth noting how remarkably close Roger has become with Tim's parents in the years since the accident. Roger began by dutifully calling them once a week and sending them cards for Mother's Day and Father's Day. Bob and Shirley told me they could sense in Roger a need to atone for whatever responsibility he had in Tim's death. Bob described Roger's thinking this way: *I took your son, maybe I can somehow replace him.* Even if Roger's attentiveness began out of obligation and remorse, a genuine friendship developed. Bob and Shirley told me they now see Roger as a member of their extended family.

Many people played crucial roles in helping this book become a reality. At Crown, I've been fortunate to work with a wonderful team. I simply can't heap enough praise on my world-class editor, Amanda

Cook. She possesses a rare ability to sink deeply into a manuscript and emerge knowing precisely what needs to be done to improve it. Her intelligence, vision, conviction, and caring made this book profoundly better in so many ways. The lineup that the brilliant Maya Mavjee has assembled at Crown brims with talent. Special thanks to David Drake, Molly Stern, Jill Flaxman, Penny Simon, the indispensable Emma Berry, Min Lee, Domenica Alioto, Eric White, Annsley Rosner, Elina Nudelman, Jay Sones, Linnea Knollmueller, Mark McCauslin, Janet Biehl, Julie Cepler, Sarah Pekdemir, and Matthew Martin.

My agent, Sarah Chalfant, is an extraordinary advocate and by now a trusted friend. I leaned on her extensively for her wise counsel, impeccable judgment, and uncanny ability to see around corners. Thanks also to Andrew Wylie and everyone at the Wylie Agency who pitched in, notably Alba Ziegler-Bailey and Rebecca Nagel.

I owe a huge debt to the incomparable Stephanie Vallejo, who served as my fact-checker, transcriber, research assistant, and effectively my left brain on this project. Her commitment was astounding, and her contributions invaluable.

This book has its roots in the pages of *The Boston Globe Magazine*, where I've had the good fortune to write for more than a decade. The gifted editor of the magazine, Susanne Althoff, sharpened both my writing and thinking in the pair of articles I wrote about the tunnel accident. Doug Most, her predecessor who now oversees the paper's special sections, provided important encouragement all along.

I'm particularly grateful for the crucial support I received from Marty Baron, who led the *Globe* to remarkable heights as top editor for nearly a dozen years, and to current editor Brian McGrory, who is taking it to new ones. Thanks to them and to publisher John Henry and CEO Mike Sheehan for their inspired leadership.

I benefited greatly from the talents of several colleagues and friends with whom I worked on those original stories, notably Suzanne Kreiter, Scott LaPierre, Josue Evilla, Barbara Pattison, John Burgess, Brendan Stephens, Ann Silvio, Greg Klee, Mary Zanor, Christine Makris, research guru and private eye nonpareil Lisa Tuite, and graphics star Javier Zarracina, who also created the illustrations for this book. Others provided key assists as I transitioned to book mode, including Ellen Clegg, Paula Nelson, Christine Chinlund, Francis Storrs, Veronica Chao, Janice Page, Wanda Joseph-Rollins, Stan Grossfeld, and Helen Morissette. I learned from the *Globe*'s earlier reporting on the harbor cleanup and tunnel project, especially from Scott Allen and Beth Daley. Also helpful was coverage from the industry bible *Engineering News-Record* as well as from the *Boston Herald* and the Quincy *Patriot-Ledger*. And thanks to my friend Tony Marcus, for all his jokes and encouragement from behind the counter of the *Globe* cafeteria.

My first draft of this manuscript was half again as long as the final product. Three accomplished friends from the creative world— Micheal Flaherty, Matt Bai, and BJ Roche—plowed through those bloated pages and offered shrewd advice for where I should aim the knife. Thanks for the inspiration from writer friends Mitch Zuckoff, Josh Wolk, Pete Chianca, and especially Michael Holley, who served as a sort of marathon training partner when we were both immersed in our own book projects. Michael kept checking in on me long after he had crossed the finish line.

When it came to understanding the complicated investigation into the diver deaths, I benefited enormously from several people who knew this case cold and were passionate and indefatigable in their pursuit of answers. Mary McCauley, Joan Parker, and Dan Kuhs were generous with their time and deep knowledge, and if they became annoyed by the volume of my follow-up questions, they never let on. I also appreciated the patient tutoring by Dr. Marie Knafelc, who schooled me in matters of diving, oxygen deprivation, and physiology. Thanks also for the work of others who helped sort out what caused the accident, notably John O'Leary, Pat Haggan, Kirk Van-Zandt, Jake Jacob, Maria-Lisa Abundo, and Elena Finizio.

Many divers and pile drivers shared their expertise, beginning with Ron Kozlowski, who periodically checked in from ports all over the world. Dave Woodman was Dan Kuhs's full partner in Local 56's push for accountability. Also helpful were Jerry Frongillo, Patrick Seiden, Tracy Markham, Jim Clark, Jay Brinton, and John C. Roat.

For the diffuser riser installation project, Dave Beck, with his encyclopedic knowledge of the job, was a font of information. I'm grateful for all the time and insights that Ricky Spears's widow, Cathy, gave me, as well as the help of his brother Joe Maloney and former colleague Greg Kolacz.

And now I come in praise of lawyers. The profession is often derided, but the attorneys I encountered on this project were impressive in their commitment to their clients—and to the truth. I wouldn't have known about this stunning story were it not for my friend Michael Riseberg, an accomplished defense attorney who was involved in the civil lawsuit that grew out of the tunnel accident. Michael has a writer's eye for a great story, and I'm so appreciative to him for directing my eyes to this one.

Plaintiffs' attorneys John Prescott, Nina Pelletier, and Bob Norton turned out to be indispensable guides for me to the contours of the case. I was impressed with their investigative prowess in unearthing key documents. Even more impressive was how deeply they cared about the divers and their families. These are advocates in the truest sense of the word. As DJ's court-appointed lawyer in his criminal case, Pete Horstmann showed not just legal acumen but great concern for his client. He also filed the motion that paved the way for my first interview with DJ, while he was still in jail. U.S. District Court Judge Mark Wolf, who approved that motion, struck me as someone who brought remarkable care to his deliberations.

It was critical to me that I get the full complexity of the case and not see it simply through one side. So I'm grateful for the defense attorneys who spoke candidly to me, particularly Pat Riley, John Lawler, and Fran Meaney.

Then there were the lawyer friends I leaned on for essential guidance as I worked my way through this book. David McCraw, assistant general counsel for the New York Times Company, possesses one of the nation's sharpest minds in media law, but he also has an outstanding feel for narrative writing. Susan Amster is a gifted publishing lawyer and a terrific person.

I'm grateful for the cooperation—and exhaustive record keeping—of the staff of the Massachusetts Water Resources Authority. Ria Convery, who is unfailingly honest and helpful, should be leading workshops nationally on the right way to be an agency communications director. Doug MacDonald and his wife, Lynda Mapes, were generous with their time, and I benefited from Doug's intelligence, deep caring, and love of history. Several former MWRA officials played critical roles in my understanding of the project, even though their names do not appear in the preceding pages. Paul Levy had the vision to put the massive cleanup on the right path, under trying circumstances. Walter Armstrong, Ralph Wallace, and Charlie Button played key roles in carrying out that vision and in helping me grasp it in all its complexity. I'm grateful for the assistance of several others at the agency, notably Fred Laskey, John Vetere, Steven Remsberg, Tom Lindberg, Barbara Allen, Marianne Connolly, Joe Duplin, Rebecca Kenney, and former spokesperson Tom Lee.

The cooperation of all the major contractors and consultants to the project made a huge difference in my understanding. From Kiewit, big thanks to Ken Stinson, Kirk Samuelson, Bill Currier, and Bob Kula. Thanks also to Dave Corkum, who served as Kaiser's tunnel point man and who indulged an endless stream of follow-up questions from me. The command he continues to have of the details surrounding this project is remarkable. At PB, thanks to the very knowledgeable

Eldon Abbott, Brian Van Weele, and Judy Cooper. During my visit to PBI Dansensor's headquarters in Denmark, Steen Hansen schooled me on the inner workings of the MAP Mix 9000. Bobby Malkasian and Mike Kanash were entertaining guides into the fascinating and foreign world of the sandhogs. And Dick Robbins gave me an essential tutorial on the art and science of tunnel-boring machines. I gained important historical insights about Boston Harbor from the late Charles Haar's book *Mastering the Harbor* and Eric Jay Dolin's *Political Waters,* as well as from Bill Golden, whose filthy running shoes helped get the harbor clean.

Others providing assists were the Boston Society of Civil Engineers Section, paramedic Keith Wilson, Julie Primeau, Ann Scales, Lisa Fliegel, mine rescue specialists Rob McGee and Curt Ahonen, the late Ed Willwerth, John Chavez of OSHA, Tom Botts of Shell, Sue Dong of the Bureau of Labor Statistics, Christina Spring of the CDC-NIOSH, Don Gibson of the Fairfield University School of Business (and Katy Mathias for connecting us), Larry Palinkas of the University of Southern California, David Procopio of the Massachusetts State Police, Don Abbott of the Texas Tower Association, and two of the brightest minds in the country on PTSD: Terry Keane of the VA, and Charles Figley of Tulane University. I learned lots about how NASA prepares for high-risk missions from retired astronaut Shannon Lucid, NASA psychologist Al Holland, and psychiatrist Chris Flynn. For insight into the Apollo 13 mission, Jim Lovell's book *Lost Moon* (cowritten by Jeffrey Kluger) and a 2005 report from *IEEE Spectrum* were essential reading. For insight into dive culture, Robert Kurson's book *Shadow Divers* was indispensable.

Help with archives, photo and otherwise, came from Katrina Scott, Kevin Kirwin, Dale Freeman and Joanne Riley of the University of Massachusetts Boston, Scott Lewis of *Engineering News-Record,* Bill Kole of the Associated Press, TJ Walsh of NECN, Michael Spear, and the staff of the Franklin Trask Library. Thanks also to Sister Peggy Rooney, Chris and Kevin Balfe, Caroline Bruce Macaulay, Nils Bruzelius, Heather Barry, Dan McGinn, and the Martha's Vinyard Writers Residency. Special thanks to the members of the Alray Scholars family, who are my partners in the nonprofit scholarship and mentoring organization that grew out of my first book. (Check it out at alray.org.)

Finally, my biggest debt of personal gratitude goes to my family. My amazing parents, Sam and Mary Swidey, have been a source of bottomless encouragement—and top flight copyediting. As a reporter, I learned from them how to make genuine connections with people from all walks of life. As a writer, I was blessed to learn the craft from my dad, a master teacher and the finest man I've ever known. A small consolation of his passing in 2014 was knowing he had been around to see the publication of this book, which he had done so much to help shape. Most of the important life lessons that didn't come from my parents were ones I learned from my sensational siblings—Eric Swidey, Maryann Jackson, Bob Swidey, and Judy D'Angelo—who continue to serve as my support system, right down to Eric (our family's MacGyver) teaching himself HTML so he could build a Web site for me. That support system has only been strengthened by the addition of their wonderful spouses and kids: Kristin, John, Lynn, Michael, Kelsea, Brett, Eddy, James, Zachary, Mia, Samantha, and Ava.

My outstanding in-laws, Herb and Sara Drower, have added so much to my life with their restless minds and warm support. Many thanks for the encouragement from my entire extended family, with special nods to Mary Hayes and Jayne Paris for being such faithful fans.

This book owes much of its existence to my awe-inspiring wife, Denise Drower Swidey, and the other three beautiful women in my home, our daughters Sophia, Nora, and Susanna. (Ollie, the other male in our house, deserves recognition for his loyalty and low-key acceptance of everything besides passing trucks and bigger dogs.) In many ways, Denise's fingerprints are on every page. As always, she was my sounding board, first reader, and first editor, diplomatically but firmly challenging me to make the prose on every page clearer. More than that, she gave of herself so I could give this book the time it needed. And somehow she resisted the understandable urge to complain about the state of my home office, with its walls covered with charts and timelines and its floor impassable with boxes of documents and depositions. For too long, it looked like the lair of an obsessed detective or deranged villain from one of those TV crime procedural shows. As for our girls, I can't thank them enough for their love, patience, and daily inspiration, and especially for all the laughs they gave me to lighten the dreariest parts of the writing process. All for a book that they have no desire to read! They often ask, "When are you going to write something *interesting?*" Give me time, girls.

As for me, I found the world of commercial diving and megaproject infrastructure endlessly fascinating. I emerge with immense respect for the work done by Billy, Tim, DJ, Riggs, Hoss, and others like them. The high-risk assignments that so many anonymous workers take on every day are essential to the progress the rest of us enjoy.

NOTES

I conducted hundreds of interviews in the course of researching this book and reviewed thousands of pages of documents as well as extensive video footage, audio files, and photographs. But in the interest of space, I have generally not included citations for my interviews in this notes section, except in cases where clarity demands it. In most of the citations that follow, I source particular documents or materials, or interviews conducted by someone other than me. All quoted material and dialogue comes from transcripts, written documents, or audio recordings, or from the specific recollections of at least one—usually two—of the people being quoted. In cases where people's memories on a certain point did not agree, I made my best judgment on whose was likely the most accurate. I placed the greatest emphasis on the memories that were supported by contemporaneous notes, or were corroborated by others, or came from someone who in my judgment had had a superior vantage point on a particular event. In the notes that follow, I have, where appropriate, pointed out instances where people's memories do not align.

PROLOGUE

5   **Each egg was fourteen stories high:** Although different sources report varying heights for these tanks, the Massachusetts Water Resources Authority (MWRA) says 140 feet (or roughly the equivalent of fourteen stories) is the accurate measurement.

6   **billions of gallons of treated sewer water:** The technical term for this is *effluent.* The tunnel was designed to be able to discharge up to 1.3 billion gallons of effluent a day, though the average daily flow was expected to be about one-third that.

8   **in the days of the Hoover Dam's:** The U.S. Bureau of Reclamation reports the official number of industrial deaths during construction of the dam at ninety-six, but it acknowledges that the true number was likely much higher. For instance, some workers who were injured on the job but who did not die on site were likely not counted in the official number. For more, see www.usbr.gov/lc/hooverdam/History/essays/fatal.html.

8   **In 2010 alone, eleven oil rig workers lost their lives:** David Barstow, David Rohde, and Stephanie Saul, "Deepwater Horizon's Final Hours," *New York Times,* December 25, 2010.

8   **the *Las Vegas Sun* won a Pulitzer Prize:** The Pulitzer board gave the newspaper the award for public service but singled out the courageous reporting by Alexandra Berzon. "The 2009 Pulitzer Prize Winners: Public Service," www.pulitzer.org/citation/2009-Public-Service.

9   **The report from the Centers for Disease Control:** Division of Safety Research, National Institute for Occupational Safety and Health, CDC, "Deaths Associated with Occupational Diving—Alaska 1990–1997," *Morbidity and Mortality Weekly Report (MMWR) Weekly,* June 12, 1998 / 47(22): 452–55, www.cdc.gov/mmwr/preview/mmwrhtml/1153331.htm.

11   **Not long after Riggs:** Details are drawn from contemporaneous accident notes written by Donald "Hoss" Hosford and Dave Riggs as well as from diver interviews by the author.

CHAPTER 1: DJ

20   **This job in the fall of 1993:** DJ Gillis's employment history from Local 56 union records.

23   **Native Americans used to gather:** Marilyn Ledoux, "Bellows Falls Attraction: The Fish Ladder," *Eagle Times,* July 18, 2008.

24   **riverside justice:** While Ron Kozlowski recalls that the dispute focused primarily on the wet gear, DJ, who witnessed the confrontation, remembers it concerning the young diver taking credit for work completed by Kozlowski.

28   **During the storm:** Dave Beck (Kaiser manager for the diffuser riser installation project), interview by author, April 2011.

CHAPTER 2: ISLAND SECRETS, 1675–1995

35   **"Harbor of Shame,"** *Boston Herald,* April 28, 1987.

36   **angry protesters camped:** Paul Levy (former MWRA executive director), interview by author, November 2010. The protesters were opposing the controversial decision to site a sewage landfill in Walpole, Massachusetts.

36   **A century earlier:** "The Boston Sewer System and Main Drainage Works," *Scientific American,* December 3, 1887, as well as *Scientific American* supplement 524. Also, Levy interview.

36   **About 140,000 pounds:** Walter G. Armstrong and Ralph Wallace, "A Case Study of Construction Management on the Boston Harbor Project: Reflections at Project Completion," CM eJournal, January 2001.

36   **Even worse, as the population grew:** Norman Boucher, "The Dirtiest Job: Not Everyone Would Take on the Task of Cleaning Up the Nation's Most Polluted Harbor and Wrestling with the Political Fallout. Enter Paul Levy," *Boston Globe Sunday Magazine,* May 8, 1988.

37   **On Christmas Day:** Charles M. Haar, *Mastering Boston Harbor: Courts, Dolphins, and Imperiled Waters* (Cambridge, Mass.: Harvard University Press, 2005), 147–48.

37   **To the Smithsonian's astonishment:** Norman Lockman, "Dukakis Proposes Sewer Authority as New MDC Unit," *The Boston Globe,* April 20, 1984, referenced in Eric Jay Dolin, *Political Waters: The Long, Dirty, Contentious, Incredibly Expensive but Eventually Triumphant History of Boston Harbor—A Unique Environmental Success Story* (Amherst: University of Massachusetts Press, 2004), 115.

37   **In the harbor's Dorchester Bay:** Haar, *Mastering Boston Harbor,* 71–74, 326.

37   **Studies found that the organic material:** Ibid., 74.

37   **They rebranded the tampon applicators:** Scott Allen, "Closing In on a Healthy Harbor," *Boston Globe,* March 15, 2000.

38   **By then, the cleanup was already under way:** "During the 1970s, the federal government provided up to 75 percent of the funds necessary to construct wastewater treatment facilities [to comply with the Clean Water Act]. Throughout the 1980s, however, the total amount of federal grants declined steadily. By the end of the decade, federal grants for constructing sewage treatment facilities ceased." Armstrong and Wallace, "Case Study of Construction Management."

39   **more than three thousand construction workers:** Shani Wallis, "Boston Activity: Boston Harbor, Massachusetts," *World Tunnelling and Subsurface Excavation,* February 1, 1994; Walter Armstrong and Ralph Wallace, interviews by author, 2010 and 2013.

39   **"the most prolific issuer of municipal revenue bonds":** John J. Doran, "Massachusetts Is About Ready to Start Cleaning Up Its Dirty Harbor," *Bond Buyer,* June 5, 1989.

40   **some of the biggest megaprojects:** "Pay Controversy Fogs Denver Airport Project," *Engineering News-Record,* December 9, 1991; "Big Bay Expansion Ahead," *Engineering News-Record,* December 23, 1991; "Water Supply

Project Now into Second Stage," *Engineering News-Record,* August 5, 1991; other publications.

40 **360 different companies:** Robert A. Grace, *Marine Outfall Construction: Background, Techniques, and Case Studies* (Reston, Va.: American Society of Civil Engineers, 2009), 294.

40 **serving as a manager for the first time in his life:** MacDonald benefited greatly from the expertise of his deputies Dick Fox, Walter Armstrong, and Ralph Wallace, as well as their deputy for construction, Charlie Button. They handled much of the day-to-day management of the massive project. In addition, MacDonald's predecessor as executive director, Paul Levy, was a planning whiz who had tackled many of the knottiest planning issues around the project and put it on solid footing.

40 **"Demand a recount":** David L. Chandler, "Frugal MWRA Head May Be What Ratepayers Seek," *Boston Globe,* February 2, 1992.

41 **a Puritan minister named John Eliot:** Jill Lepore, *The Name of War: King Philip's War and the Origins of American Identity* (New York: Random House, 1999), 138.

41 **about half of the five hundred:** Lepore, 139, 141.

41 **city officials sent many of them:** "Island Facts: Deer Island," National Park Service, accessed 2010, www.nps.gov/boha/historyculture/facts-deer. htm.

43 **The $15 million TBM was carving:** Dick Robbins (former CEO of the Robbins Company, which made the TBM), interview by author, September 2010.

43 **Although there were three names on the letterhead:** For the sake of clarity, I refer to the Kiewit-Atkinson-Kenny joint venture as Kiewit, the name of the lead partner, in subsequent references. Many of the project players did the same.

44 **pay a penalty of $30,000:** MWRA contract with Kiewit. The liquidated damages provision appears in Section 5.2.3 (500).

45 **concrete segments cracked and chipped:** Rodney Garrett, "Boston Harbor Outfall; Construction of Wastewater Treatment Plant System in Boston Harbor, Massachusetts," *World Tunnelling and Subsurface Excavation,* March 1, 1993.

45 **The tool was called a jackleg:** For a video clip of a sandhog using a jackleg, see www.youtube.com/watch?v=1b8LdQ9SzN8.

46 **better known by his nickname:** Bobby Malkasian, interviews by author, 2010; Mike Kanash, interview by author, July 2010.

46 **The local's recording secretary:** Daniel Golden, "Under the Waterfront: From the Harbor Cleanup to the Airport Tunnel, Boston's Building Boom Has Moved Underground," *Boston Globe,* July 25, 1993.

47 **Before government safety standards:** Malkasian and Kanash interviews.

49 **Gravity, rather than engines or pumps:** Project documents; Charlie Button and various project officials, interviews by author, 2010–11.

49 **Three days after Christmas:** Bill Currier, interview by author, April 2011;

Malkasian interview and photos. Also, Paul Zick, "One Step Closer: A Major Milestone," Deer Island Job News from December 28, 1995, Hard Hat News (project newsletter), 1996.

53 **Around lunchtime:** OSHA report on death of Michael Lee; "Michael R. Lee, 40, Harbor Tunnel Worker," *Patriot Ledger* (Quincy, Mass.), July 2, 1992.

53 **the most dangerous day:** Doug MacDonald, interview by author, April 2011.

54 **"It's a boy":** Cathy Spears, interview by author, April 2011; Greg Kolacz, interview by author, May 2011; Dave Beck, interviews by author; OSHA report on death of Ricky Spears.

54 **Although he had lived to see his baby:** Cathy Spears, interview by author, 2011. Cathy explained that even though Ricky had picked up her and their baby son, Cody, at the hospital and driven them home, Cody was sleeping a lot, and Ricky didn't want to wake the baby. Ricky, a decorated Vietnam vet, left for work offshore that same afternoon. The next time Cathy saw her husband, he was on life support at Boston City Hospital.

54 **well-liked character:** OSHA report on death of Dick White; Dave Corkum, interview by author, November 2011.

CHAPTER 3: MEMO WARS

56 **Shoppers settled on Christmas gifts:** James Varney and Petula Dvorak, "Rammed: Runaway Freighter Smashes Riverwalk; Scores Injured, Several Believed Missing," *Times-Picayune* (New Orleans), December 15, 1996.

56 **A red and black beast of a freighter:** National Transportation Safety Board, *Marine Accident Report: Allision of the Liberian Freighter* Bright Field *with the Poydras Street Wharf, Riverwalk Marketplace, and New Orleans Hilton Hotel in New Orleans, Louisiana, December 14, 1996* (Washington, D.C.: U.S. Government Printing Office, 1998), www.ntsb.gov/investigations/summary/MAR9801.htm.

57 **"It felt just like an earthquake":** Chris Gray, "Workers, Shoppers Run for Safety in Chaos," *Times-Picayune* (New Orleans), December 15, 1996. Also, Rick Bragg, "A Nightmare Along the Mississippi," *New York Times,* December 16, 1996.

58 **Hydraulic studies predicted:** Eldon Abbott, memo to Paul Zick and Dave Corkum, "Outfall Tunnel—Tunnel Flooding Due to Loss of Diffuser Head," December 15, 1997; Dave Sinsheimer, memo to Ken Stinson and Ron Minarcini, "Deer Island Hydraulic Study," December 19, 1997.

58 **It would resemble a flash flood:** One hydraulic study estimated that the wave would race through the entire 9.5 miles of the tunnel and make it to the base of the shaft in about three and a half hours. It would likely take workers a lot longer than that to get out of the tunnel since, during a flood, the lokey would be of no use. Within another eight hours, the height of the wave would reach the top of the tunnel.

59 **2.4 million tons of rock:** Eric Niiler, "MWRA Finishes Drilling Sewage Outfall Tunnel," *Patriot Ledger* (Quincy, Mass.), September 20, 1996.

59 **Kiewit chose a three-pronged approach:** Tom Corry (Kiewit executive and project "sponsor") deposition in civil lawsuits filed by divers and families, June 2000, 72–79.

60 **Those elbows would create:** These side tunnels, which the sandhogs dug mostly with their jacklegs, ranged in length from nine feet to thirty-five feet, depending on how far that particular riser was located from the main tunnel. After digging their way out to each vertical riser, the sandhogs would attach the prefabricated, concrete-encased elbow piece to the bottom of that pipe. Once the elbow and the horizontal connector pipe were in place inside each of these side tunnels, crews would pump in concrete to backfill around the connector and riser pipes. With that, the width of each of the fifty-five openings, from the main tunnel all the way up to the ocean floor, was no wider than thirty inches.

61 **Metcalf & Eddy produced:** Metcalf & Eddy, "Conceptual Design: Design Package 6, Effluent Outfall Tunnel and Diffusers," final report, May 31, 1989.

62 **In 1990, after winning the bid:** "Boston Harbor Project—Effluent Outfall Tunnel, CP-282," MWRA Contract No. 5637, official notice to proceed given by MWRA on August 9, 1990. The specifications concerning "scuba gear" appear in Part 3, Section 3.01, "Equipment."

63 **To the extent that Kiewit executives:** Ken Stinson and Kirk Samuelson, interview by author, September 2011.

63 **Kiewit was already in the red:** "Boston Harbor Project, CP-282 Effluent Outfall Tunnel, Status Report for the Office of the State Auditor," May 25, 2000; also, project players representing the contractor, designer, owner, and construction manager, interviews by author

64 **He lived and breathed:** Stinson and Samuelson interview; Dick Robbins interview.

66 **Kiewit fired the first shot:** Robert Regazzini (Kiewit construction manager), memo to Dave Corkum (ICF Kaiser Engineers area manager), "Change/Clarification Request Form No. 96," March 10, 1997.

66 **MacDonald and his deputies relied:** Dave Corkum was directly employed by Stone & Webster, a subcontractor for Kaiser. In most tunnel correspondence, however, parties involved identify him as a Kaiser employee, and he essentially functioned as one.

66 **He firmly rejected:** Dave Corkum, memo to contractor Kiewit, "Subject: CCR 282-96—Reuse of Tunnel Offtake Plugs," April 4, 1997.

67 **asking a worker to crawl:** Robert Regazzini, memo to Dave Corkum, "Subject: FTM 282-2164, FTM 282-2165: Reuse of Tunnel Offtake Plugs," April 8, 1997.

67 **"only the vaguest idea":** Dave Corkum, memo to contractor Kiewit, "Subject: Bulkhead at Venturi," September 3, 1997.

68 **"time being of the essence":** Robert Regazzini, memo to Dave Corkum,

"Subject: Removal of Secondary Riser Pipe Safety Plugs," October 16, 1997.

69   **In his reply, Corkum said:** Dave Corkum, memo to contractor Kiewit-Atkinson-Kenny, "Subject: Removal of Diffuser Safety Plugs," November 22, 1997.

69   **Corkum invoked the freighter crash:** Corkum noted, both at the time and in subsequent interviews with the author, the likelihood that many captains sailing into Boston Harbor might not understand English. But in his letter to Kiewit invoking the New Orleans incident, he did not specifically mention the role that language may have played.

70   **"the risk of catastrophe":** Robert Regazzini, memo to Dave Corkum, "Subject: Removal of Secondary Riser Pipe Safety Plugs," October 16, 1997.

70   **"enormous in terms of both money":** Fred Anderson, occupational safety consultant hired by Kiewit, memo to Robert Regazzini, November 28, 1997.

70   **an "emergency" in the contract:** Robert Regazzini, memo to Dave Corkum, "Subject: Removal of Diffuser Safety Plugs, FTM No. 282-2304, Notice of Defective Specification and Emergency Situation Under Article 6.21 of the General Conditions," December 16, 1997.

71   **"to not remove or otherwise compromise":** Dave Corkum, memo to contractor Kiewit, "Subject: Response to CTM 282-1136," December 19, 1997.

71   **A high-ranking Kiewit executive:** Kiewit-Atkinson-Kenny, AJV, "Minutes of Hemispherical Plug Removal Meeting—MWRA Headquarters," January 19, 1998. The Kiewit executive was Jerry Toll. The MWRA's Charlie Button ran the meeting.

71   **Researchers in organizational behavior:** I'm particularly grateful to Don Gibson, dean of the Fairfield University School of Business, for the insight he provided during a January 2011 interview.

73   **one Kiewit executive argued:** Kiewit, "Minutes of Hemispherical Plug Removal Meeting," January 19, 1998. Kirk Samuelson was the Kiewit executive who made this point.

73   **an entirely different legal front:** Fran Meaney, interview by author, May 2010.

74   **"both difficult and expensive":** Robert Regazzini, memo to Dave Corkum, "Subject: Diffuser Plug Removal, FTM No. 282-2335," February 11, 1998.

74   **"Mr. Corkum responded":** Ibid.

74   **he received a startling call:** Meaney interview; Corry deposition; Stinson interview, 2011.

CHAPTER 4: ARRANGED MARRIAGE

79   **because she preferred it that way:** Olga doesn't recall being the one who came up with this alternate pronunciation of the surname, but Bill is sure

she was. He said he pronounced it to sound like *juice,* the same way his
father had.

80   **He told Tap:** Tap Taylor, interview by author, February 2011.

81   **On the last Thursday in May:** Tap Taylor interviews by author; Kiewit
executives, interviews by author at the company's Omaha headquarters;
September 2011; Tom Corry, deposition, June 27, 2000.

81   **winning the contract would generate:** Tap Taylor interviews.

82   **They were both representatives:** Descriptions in this section are from
Tap Taylor, interviews by author, June 2009 and February 2011; Roger
Rouleau, interviews by author, June 2009 and September 2010; and
Rouleau deposition, OSHA case, November, 283.

83   **Success soon attracted:** Roger Rouleau, interview by author, 2010.

83   **Can-Dive sought legal protection:** Leslie Ellis, "Entrepreneurs Revisited:
Up from the Depths; March 1991; Can-Dive Services, Inc., and International
Hard Suits, Inc.," *Profit,* Fall 1992.

85   **Norwesco divers helped install:** Paul Rosta, "Outfall Goes Deep in San
Diego," *Engineering News-Record* 242, no. 4 (1999): 22; Jack Burke, "South
Bay Outfall; South Bay Ocean Outfall Project," *World Tunnelling and
Subsurface Excavation* 10, no. 2 (1997): 47(5); Kim Crompton, "Diving for
Dollars in the Deep," *Spokane Journal of Business,* July 15, 1999.

86   **Creating the water tap required:** Kiewit's Bob Kula to author, September
20, 2011.

86   **"We do not possess the engineering prowess":** Specialty Diving Services
of North Kingstown, R.I., to KAK's law firm, declining to bid, May 18,
1998. A similar letter from the Massachusetts firm CLE Engineering,
dated March 16, 1998, had read, "Our decision not to bid is based on our
site visit, safety and operating constraints, and thorough review of the
proposed project documents."

87   **he included this warning:** Harald Grob to Robert Fitzgerald, "MWRA
Contract No. 5637, Re: Effluent Outfall Tunnel, Diffuser Bulkhead
Removal—Revision I," May 21, 1998.

88   **Tap's plan involved a diesel flatbed truck:** Tap Taylor to Robert Fitzgerald,
"MWRA Contract No. 5637, Effluent Outfall Tunnel Proposal," March 11,
1998.

88   **"overly complex":** Harald Grob to Tom Corry, March 22, 1998.

89   **Harald had determined:** Harald Grob to Robert Fitzgerald, May 21, 1998.

89   **He'd get the same amount of breathing gas:** Calculations have been
checked against breathing equipment suppliers and confirmed by Dan
Kuhs of Local 56 and regional manager, Specialty Trades, New England
Regional Council of Carpenters.

90   **Waiting in the car for him was Tom Corry:** Roger Rouleau, interview by
author, 2010; Rouleau and Corry depositions.

90   **He knew there were plans:** Rouleau interviews; Norwesco subcontract
with KAK.

91   **The Norwesco plan had already won plaudits:** Dave Corkum (Kaiser area

manager), memo to contractor Kiewit, "Subject: Diffuser Plug Pulling Operation—Submittal 282-01650," July 21, 1998.

91 **a series of safety violations:** The fines included $250,000 that OSHA levied in 1992 for ten workplace safety violations that the agency said had contributed to the death of worker Michael Lee. See Steve Adams, "Two Workers Die in MWRA Tunnel Under Bay," *Patriot Ledger* (Quincy, Mass.), July 22, 1999; OSHA report.

93 **Harald had taken the lead:** Footage of the fact-finding mission, Mazzone Collection, Joseph P. Healey Library, University of Massachusetts Boston; Hoss, interview by author; Tap Taylor, interview by author; diver Patrick Seiden, interview by author, February 2011.

94 **oxygen would be displaced:** U.S. Department of Labor, "OSHA Fact Sheet: Hydrogen Sulfide," October 2005, www.osha.gov/OshDoc/data_Hurricane_Facts/hydrogen_sulfide_fact.pdf; U.S. Department of Labor, "OSHA Fact Sheet: Carbon Monoxide Poisoning," 2002, www.osha.gov/OshDoc/data_General_Facts/carbonmonoxide-factsheet.pdf; and U.S. Department of Labor, Occupational Health and Safety Administration, "Confined Spaces," www.osha.gov/SLTC/confinedspaces/index.html.

94 **wielding a radio antenna:** In addition to conducting interviews with multiple participants, I viewed video footage of this fact-finding operation.

CHAPTER 5: MINE RESCUE

95 **a NASA manager pulled Jim Lovell aside:** Jim Lovell and Jeffrey Kluger, *Lost Moon: The Perilous Voyage of Apollo 13* (New York: Houghton Mifflin, 1994), 56.

95 **a treacherous Appalachia-like range:** Ibid., 76.

96 **NASA had always taken great care:** Al Holland (NASA psychologist), interview by author, July 2012; Shannon Lucid (retired astronaut), interview by author, June 2012; Chris Flynn (psychiatrist and former NASA flight surgeon), interview by author, June 2012.

96 **Lovell could detect every nuance:** Lovell and Kluger, *Lost Moon*, 83.

96 **Mattingly's voice crackled:** Ibid., 78–80.

97 **a different, nonsimulated crisis:** Ibid., 80–81.

97 **Forty-eight hours before launch:** Ibid., 83.

97 **Fifty-five hours, fifty-four minutes, and fifty-three seconds:** Ibid., 126.

97 **an unholy number of setbacks:** Stephen Cass, "Apollo 13, We Have a Solution: Rather than Hurried Improvisation, Saving the Crew of Apollo 13 Took Years of Preparation," *IEEE Spectrum,* April 2005.

98 **"This crew is coming home":** Ibid.

98 **a pile driver on a Big Dig–related project:** DJ's union employment records.

99 **Harald phoned Kiewit executive Tom Corry:** Tom Corry, notations in his day planner, May 13, 1999, produced as part of discovery in the subsequent court proceedings.

100 **Harald's primary equipment supplier:** Jerry Anderskow (of A-L Compressed Gases) to Customer Service, MVE Cryogenics, May 17, 1999.

101   **just under $5,000 per setup:** Jerry Anderskow to Harald Grob, May 26,
      1999.

104   **After DJ awoke in the intensive care unit:** " 'I Thought I Was Going
      to Die': How DWH Staff Saved a 28-Year-Old Man's Life," Deaconess
      Waltham Hospital *Today*, 1998.

107   **"Don't try to bluff your way through":** U.S. Department of Labor, Mine
      Safety and Health Administration (MSHA), "Mine Rescue Training
      Module," West Virginia University Mine Extension Service, 18. I
      obtained from DJ the actual copy that he used during the training in New
      Hampshire.

107   **A mine rescue association:** Rob McGee, "Miners Killed During Recovery
      Operations," U.S. Mine Rescue Association, Uniontown, Pennsylvania,
      www.usmra.com/rescuer_deaths.htm.

108   **graphic descriptions in the training manual:** MSHA, "Mine Rescue
      Training Module," 15.

108   **It does nothing to support human breathing:** Ibid., 28.

108   **consequences of falling oxygen concentration:** Ibid., 27.

109   **The two biggest concerns were:** Dr. Marie E. Knafelc, interviews by
      author; Christopher Swan, *The History of Oilfield Diving: An Industrial
      Adventure* (Santa Barbara, Calif.: Oceanaut Press, 2007); NAUI Worldwide,
      "Dive Tables," www.scubadiverinfo.com/images/Dive_tables_NAUI
      .jpg; and Divers Emergency Service UK, "Types of Decompression
      Illness," www.londonhyperbaric.com/decompression-illness/types-of-
      decompression-illness.

111   **a minimum day rate of $742:** Black Dog Divers subcontract to Norwesco,
      June 23, 1999, Attachment A.

CHAPTER 6: MOBILIZATION

117   **His frustration increased:** Although Riggs said he generally doesn't recall
      having been offended by profanity other than *goddamn* and the like,
      DJ and Hoss had vivid recollections of Riggs asking them to curb their
      cussing.

121   **the one time he had complained to a Norwesco supervisor:** Hoss,
      interviews by author, 2009 and 2010.

122   **On a salvage dive years earlier:** Judy Milner, interview by author,
      September 2010; Bob and Shirley Nordeen, interview by author, 2010.

125   **the MAP Mix 9000 was coming from Europe:** DJ, Hoss, and Riggs,
      interviews by author; Drybanksi deposition; Corry day planner.

125   **Harald had searched the Internet:** John Prescott, memo to file "Gillis/
      Juse—Conversation with Harald Grob," September 13, 2000, following his
      questioning of Harald.

127   **the combined East-West crew in Tap's yard:** Trooper Mary T. McCauley,
      "Interview of Harald Grob," Massachusetts State Police Suffolk County
      Detective Unit, January 7, 2000, 4; DJ, interviews by author.

127 **Tap went so far as to report DJ's absence:** Tap Taylor, deposition; Taylor, interview by author.
129 **"The system's been approved":** Taylor deposition, 140.
129 **project sponsor Tom Corry had flown in:** Ibid; Corry day planner.
129 **"Is everything tested?":** Rouleau deposition, 115.

CHAPTER 7: THE CAVALRY
132 **"in small submarines":** Dave Corkum, deposition, 184.
132 **he didn't push it:** Ibid., 159.
132 **"As far as I knew":** Ibid., 187
133 **About as far as Corkum had been willing to probe:** Ibid., 212–13.
133 **"How are you defendants doing":** Tom Long, "A. D. Mazzone, Judge Who Led Harbor Effort, Dies at Age 76," *Boston Globe,* October 27, 2004.
133 **After the divers parked the vehicles:** This scene was captured on raw video footage I viewed from the mission. Also, Dave Corkum, interviews by author, 2009, 2011.
135 **Harald kept the radar monitor:** Raw video footage from the mission.
135 **When the doors to both trailers were open:** Tap Taylor, interview by author.
135 **damaged in a roadway accident:** Tom Corry, day planner, July 12, 1999.
138 **to get to the island either by bus or barge:** Paul Levy, Walter Armstrong, and Ralph Wallace (former MWRA executives), interviews by author, 2010.
139 **they had to call a welding company:** Harald Grob, topside log, July 13, 1999.
139 **"Lost one day":** Corry day planner.
141 **The divers spent the day:** Trooper Mary T. McCauley, "Interview of Harald Grob," Massachusetts State Police Suffolk County Detective Unit, January 7, 2000, 3.
141 **the engines would likely stall:** Curt Ahonen (mine safety trainer), interview by author, 2012; Georgia Institute of Technology, Health and Safety Office, "Fire Safety," www.ehs.gatech.edu/fire/.
142 **roughly cut, unpainted plywood:** I observed video and photographic footage of the setup, taken inside the tunnel during the plug-pulling operation.
143 **now those tires were flat:** Corry day planner; Grob topside log, July 14, 1999.
144 **DJ had to back the gray Humvee:** I observed this scene from a video recording.
146 **it was now after eleven:** Grob topside log, July 14, 1999.
147 **the electronics of the vehicle's oxygen analyzer:** Ibid., July 15, 1999.
148 **By ring 7500:** Kelly Mills (Kiewit/KAK manager), day planner, July 15, 1999.
149 **a consciousness-robbing 9 percent:** Grob topside log, July 15, 1999.
149 **to be filled with a special foam:** Corry day planner.

149   **The elusive MAP Mix 9000:** Tony Drybanski, deposition, 131.
150   **eight inches tall, nine inches wide, and sixteen inches deep:** "MAP Mix
       9000 User's Manual," PBI Dansensor A/S, Ringsted, Denmark, 1996. I've
       converted the measurements from the metric.
151   **Just three hoses:** Ibid.
152   **Norwesco had paid more than $10,000:** A-L Compressed Gases of
       Spokane, invoice, June 30, 1999. Each mixer cost $5,062.50.
152   **Most of its customers:** Steen Hansen, interview by author, Ringsted,
       Denmark, June 2011.
152   **During the long delay, Harald had warned:** Trooper Mary McCauley,
       "Interview of PBI Dansensor Company (TOPAC GAS MIXER) in
       Reference to Tunnel Death Investigations at Deer Island," Massachusetts
       State Police Suffolk County Detective Unit, October 9, 1999, 3.
153   **"Just tie it back up":** Dave Riggs, deposition.

CHAPTER 8: MONDAY
154   **Billy sat by the lake in New Hampshire:** Michelle Rodrigue, Ken Jones,
       and Deb Jones, interviews by author in August 2010 and after.
157   **At seven-thirty a.m. on Monday:** Harald Grob, topside log, July 19, 1999.
158   **At nine-thirty, with all the preliminaries done:** Kaiser, shift report, July
       19, 1999.
159   **after that air had exited the MAP Mix 9000:** Hoss, DJ, and Riggs,
       interviews by author, 2009–12; Hoss's accident notes from July 23, 1999:
       "On Monday, I asked Harold if the gas mixer was meant for breathing and
       why it didn't have an analyzer built into it. He said, 'I guess this model
       doesn't have one'" (p. 11).
160   **Inside the 3 million-gallon egg-shaped digester tanks:** MWRA, *The State
       of Boston Harbor: The New Treatment Plant Makes Its Mark,* 1995,
       www.mwra.state.ma.us/harbor/enquad/pdf/1996-06.pdf.
160   **electricity to help power the plant:** MWRA, "The Deer Island Sewage
       Treatment Plant," September 2, 2009, www.mwra.state.ma.us/03sewer/
       html/sewditp.htm.
160   **divers arrived at riser number 12:** Grob topside log, July 19, 1999.
162   **shrinking from seven feet to six feet to five:** Tom Corry, deposition, 75.
162   **The same dynamics are at play:** Dr. Marie E. Knafelc, interview by author,
       May 2012.
162   **the first sign is usually frustration:** G. S. Tune, "Psychological Effects of
       Hypoxia," *Perceptual and Motor Skills* 19 (1964): 551–62.
162   **"one of the most intense and happiest moments":** Ernst A. Rodin, "The
       Reality of Near-Death Experiences: A Personal Perspective," *Journal of
       Nervous and Mental Disease* 168, no. 5 (1980): 259–63.
163   **under water to the point of near-drowning:** Calvert Roszell, *The Near-
       Death Experience: In the Light of Scientific Research and the Spiritual Science of
       Rudolf Steiner* (Hudson, N.Y.: Anthroposophic Press, 1992), 24–26.
163   **switch to their backup supply of HP air:** Trooper Mary McCauley,

"Sudden Death/Deer Island, Winthrop/07/21/99," Massachusetts State
Police Suffolk County Detective Unit, January 14, 2000, 6–7.

163   **By 6:25 p.m., Tim informed Mars:** Grob topside log, July 19, 1999.

164   **"No plugs pulled":** Kaiser shift report, July 19, 1999.

165   **In a feature story about Norwesco:** Kim Crompton, "Diving for Dollars in
the Deep," *Spokane Journal of Business,* July 15, 1999.

CHAPTER 9: TUESDAY

166   **"Every time Harald called":** Tony Drybanski, deposition, 134.

167   **about 100 pounds per square inch (PSI):** Ibid., 137; Trooper Mary T.
McCauley, "Interview of Harald Grob," Massachusetts State Police Suffolk
County Detective Unit, January 7, 2000, 7.

167   **gas going out into the divers' umbilicals:** Gunnar Bak Rasmussen to Tony
Drybanski, June 29, 1999.

167   **The test showed:** Harald Grob, topside log, July 20, 1999.

167   **It was just after nine-thirty a.m.:** Humvee log, July 20, 1999.

168   **There was no gauge or monitor on the mixer:** McCauley, "Interview with
Grob," 5.

169   **The three divers on the excursion crew:** Ibid., 8.

169   **Back during setup in New Hampshire:** Tap Taylor, deposition, 247.

171   **Opening the valve to that port:** McCauley, "Interview with Grob," 6.

173   **Just before midnight, a navy:** Kurt Andersen, "Night of Flaming Terror,"
Time, June 8, 1981; Associated Press coverage and video footage of the
crash at www.youtube.com/watch?v=IEnqR86TJ1k.

175   **But those bolts:** Dave Riggs to author, June 2009.

176   **first safety plug successfully removed:** Humvee log, July 20, 1999,
4:05 p.m.

176   **His mask started free-flowing:** Humvee log, July 20, 1999, 5:40 p.m.

CHAPTER 10: WEDNESDAY

187   **Riggs walked over to the lunch truck:** I'm grateful to Riggs and Hoss for
committing detailed notes to paper immediately after the accident, while
the events were still fresh in their minds, and for sharing these notes with
me. Dave Riggs, accident notes, 1.

187   **He borrowed a pair of scissors:** Ibid., 2.

188   **This relatively inexpensive gauge:** Dr. Marie E. Knafelc, "Investigation of
Deaths Within the Boston Harbor Outflow Tunnel," Navy Experimental
Diving Unit, September 3, 1999, 3.

188   **"Hey, Harald," Riggs said:** Riggs accident notes, 2.

189   **flying through the premission checklists:** Ibid., 3.

189   **They could use the Humvee's other handheld:** Trooper Mary T.
McCauley, "Interview of Harald Grob," Massachusetts State Police Suffolk
County Detective Unit, January 7, 2000, 8. Of course, the use of this
additional handheld addressed none of the fundamental problems with
the monitoring system.

189 **He told him his back was hurting:** Tap Taylor, interviews by author.

190 **The small team of remaining sandhogs:** Dave Riggs to author, June 2009.

191 **cruising into the tunnel at 8:20 a.m.:** Harald Grob, topside log, July 21, 1999. Riggs accident notes, 4.

191 **Then he squatted:** Riggs accident notes, 5.

192 **Tim turned to him:** Ibid., 4.

192 **Hoss, Riggs, and DJ began trudging east:** Humvee log, July 21, 1999, 11:14 a.m.

193 **"The pressure's down,":** Hoss accident notes, 2.

194 **He was still smarting:** Tap Taylor, deposition, 224.

194 **Their sole communication with topside:** Ibid., 230–31.

195 **off the coast of Martha's Vineyard:** On the contemporaneous media accounts of the JFK Jr. recovery, see "The Life of JFK Jr.: Special Report: Coverage of the JFK Jr. Tragedy," *Washington Post*, 1999, www.washingtonpost.com/wp-srv/national/longterm/jfkjr/stories.htm.

196 **Riggs began his crawl in:** Humvee log, July 21, 1999, 12:18 p.m.

196 **Once he removed the bolts:** Riggs accident notes, 7.

196 **The whole process:** Humvee log, July 21, 1999, 12:39 p.m.

196 **So he took a breather:** Riggs accident notes, 7.

197 **"before we pull number 4":** Hoss accident notes, 2; Riggs accident notes, 7.

198 **DJ just stopped walking:** Riggs accident notes, 8; Hoss accident notes, 3.

199 **Hoss reached Tim:** Hoss accident notes, 3. In his accident notes, Hoss writes that Tim said, "Shit, it's 8.9%." Although he acknowledges in his notes that Tim may have reported 9.8%, Hoss writes, "I think 8.9%."

200 **the oxygen demand spikes:** "Stress—The Body's Response," University of Maryland Medical Center, February 13, 2009, www.umm.edu/patiented/articles/what_biological_effects_of_acute_stress_000031_2.htm#ixzz2N7Ep1WvW.

200 **Looking at the packed vessel now:** Hoss accident notes, 4.

201 **Riggs tried adjusting his regulator:** Riggs accident notes, 9.

203 **mouth-to-mouth by remote control:** Trooper John O'Leary, "Interview of: Donald J. Gillis," Massachusetts State Police Suffolk County Detectives, July 28, 1999, 7; DJ Gillis, interview by author.

204 **DJ picked up the receiver:** Harald Grob, topside log, July 21, 1999, 1:45 p.m.; also DJ Gillis, interviews with author.

204 **"Don't plug into that!":** Hoss accident notes, 6; O'Leary, "Interview of Gillis," 7.

CHAPTER 11: TRAPPED IN BLACK

207 **He was startled to find the cylinder of liquid oxygen:** Trooper Mary McCauley, "Sudden Death/Deer Island, Winthrop/07/21/99," Massachusetts State Police Suffolk County Detective Unit, January 14, 2000, 6–7.

209 **Government incident reports:** U.S. Mine Rescue Association, "Miners Killed During Recovery Operations," Uniontown, Pennsylvania.

209   operating three "scoop" vehicles: U.S. Mine Rescue Association, "Frank
       Crawford, Jr., Coal Company No. 1 Mine Explosives Accident,"
       www.usmra.com/saxsewell/frank_crawford.htm.

209   Two of the miners: "Around the Nation: 3 Killed and One Hurt in
       Kentucky Coal Mine," United Press International, April 16, 1985.

209   An investigation determined: Ibid.

209   "It's immoral": Curt Ahonen, interview by author, July 2012.

210   Hoss turned to Riggs: Riggs accident notes, 11.

212   Riggs had managed with one hand: Ibid. .

212   asked him to do the cutting: Hoss accident notes, 7.

213   Hoss did what he sensed DJ could not: Ibid., 8; DJ Gillis, interview by
       author, June 2009.

214   He knew there was a standby team: Trooper Mary T. McCauley,
       "Interview of Harald Grob," Massachusetts State Police Suffolk County
       Detective Unit, January 7, 2000, 9.

216   alert a Kiewit supervisor about the potential problem: Ibid.

216   two men were down: Harald Grob, topside log, July 21, 1999, 1:45 p.m.;
       Riggs accident notes, 11.

216   "still unconscious": Grob topside log, July 21, 1999, 1:55 p.m.

220   As she made her way down the slopes: "Skier Revived from Clinical
       Death," BBC News, January 28, 2000, news.bbc.co.uk/2/hi/
       health/620609.stm.

220   stopped for nearly three hours: John Naish, "Can Humans Hibernate?"
       Daily Mail (London), February 21, 2012.

220   The mammalian diving reflex protected her: John Rennie, "How the
       Dive Reflex Extends Breath Holding," *Scientific American* (Web exclusive),
       March 22, 2012, www.scientificamerican.com/article.cfm?id=breath-
       holding-dive-reflex-extends.

220   a month breathing off a ventilator: "Skier Revived from Clinical Death,"
       BBC News.

221   Hoss got out and picked up the receiver: Grob topside log, July 21, 1999,
       2:40 p.m.

227   At 3:36 p.m., Riggs navigated the Humvee: Ibid., 3:36 p.m.

228   The stopwatch now read: Hoss accident notes, 10.

CHAPTER 12: "WHY, GOD?"

231   Mary McCauley hung up: Trooper Mary McCauley, "Sudden Death/Deer
       Island, Winthrop/07/21/99," Massachusetts State Police Suffolk County
       Detective Unit, January 14, 2000, 4; McCauley, interviews with author.

232   they found lots of commotion: Ibid.

234   Word came in from both Mass General and Boston Medical: Ibid., 5.

242   the Associated Press had moved a story: "Two Workers Contracted by
       MWRA Die in Sewage Tunnel," Associated Press, July 21, 1999, 6:16
       p.m. At my request, Boston Bureau chief Bill Kole searched AP archives
       of alerts and determined that this was the first story about the accident

that the wire service transmitted. Even in the next day's papers, coverage
of the accident was extremely limited. I believe there are two reasons for
that. First, the media's focus remained squarely on JFK Jr., a story that was
drawing international attention. The next day's *Boston Globe* featured an
enormous photograph, spread across all six columns of the front page, that
captured family patriarch Ted Kennedy grimly looking on as several of his
kin carried the casket containing the remains of his nephew. The article
about the deaths of the divers in the tunnel, meanwhile, was relegated
to the paper's section for local news. Second, the then-spokesman for the
MWRA told me that at that point, even the people at the agency weren't
sure exactly what had happened. Those who knew the most, the three
surviving divers, did not speak publicly about the accident until ten years
later, when they shared their stories with me for the *Globe Magazine*. And
even they had been kept in the dark about many of the events leading up
to the disaster. Despite those limitations, two *Globe* reporters were able to
produce a pair of informative stories in the days following the accident:
Jordana Hart, "Questions in Tunnel Deaths," *Boston Globe*, July 23, 1999;
and Joanna Weiss, "Two Divers Who Died in Deer Island Prepared
Extensively for Perilous Job," *Boston Globe*, July 24, 1999.

243 **They started with DJ:** Trooper John O'Leary, "Interview with Donald J.
Gillis," Massachusetts State Police, Suffolk County Detectives, July 28,
1999; DJ and Riggs, interviews by author.

243 **the state police agreed to interview them together:** McCauley, "Sudden
Death/Deer Island," 5–7.

244 **frozen as an ice sculpture:** Ibid., 6; Riggs, interviews by author.

247 **So he was crushed:** Bill and Olga said they learned about Billy's sale of his
partnership sometime after the funeral, when they began going over his
financial papers.

250 **Roger would later say:** "I have no recollection of something like that,"
Roger told me. "Good God. I mean I paid Harald all the way through the
whole thing. Maybe he heard something out of context." Hoss, however,
told me his recollection on this point is crystal clear, and Riggs, with
whom he discussed it at the time, confirms Hoss's account.

250 **The reflection had actually been published:** Parker T. Pettus, "A Salute to
the Ordinary Heroes," Letter to the Editor, *Boston Globe*, July 23, 1999.

251 **two lawyers who had flown in from Boston:** Bob Norton and John
Prescott, interviews by author, 2009–12.

253 **There were also everyday artifacts:** Trooper John O'Leary, "Inventory
of Evidence from Outfall Tunnel," Massachusetts State Police, Suffolk
County Detectives, September 20, 1999; also, video footage.

CHAPTER 13: THIS CRAZY IDEA
255 **essentially odorless pump-house building:** Ray McEachern, "Houghs
Neck Advocate, Project Praised; Nut Island Officially Joins National
Park," *Patriot Ledger* (Quincy, Mass.), September 11, 1999.

255   **running for more than a year:** MWRA, "A History of the Sewer System: Massachusetts Water Resource Authority," www.mwra.state.ma .us/03sewer/html/sewhist.htm.

256   **"This place is like an instant hospital":** McEachern, "Houghs Neck Advocate, Project Praised."

256   **on track to cost $100 million more:** Kiewit executives, interviews by author, 2011; Ria Convery (MWRA communications director), interview by author, 2010.

256   **Shellfish beds:** Beth Daley, "For Harbor Cleanup, a Light at the End of the Tunnel; Sewage Outfall Pipe Prepared for Use," *Boston Globe,* July 18, 2000.

256   **leaving it cloudy and producing algae blooms:** David Holmstrom, "Pulling the Plug," *Christian Science Monitor* (Boston), September 9, 1999.

257   **the light rain that had been falling turned into a downpour:** McEachern, "Houghs Neck Advocate, Project Praised."

257   **"We may be many weeks away":** Quoted in Holmstrom, "Pulling the Plug."

257   **What's more, the MWRA:** Peter J. Howe, "Permit Clears the Way for Mass. Bay Sewage Outfall Pipe," *Boston Globe,* May 21, 1999. Also David Wedge, "EPA Removes Final Obstacle to MWRA Sewage-Flow Tunnel," *Boston Herald,* July 13, 2000; Associated Press, "Nearly 10-mile-long, $390-million Tunnel Now Carries Boston's Waste Out to Sea," *Providence Journal-Bulletin,* September 6, 2000; and Holmstrom, "Pulling the Plug." To oversee this monitoring program, the federal EPA and the Massachusetts Department of Environmental Protection created the independent Outfall Monitoring Science Advisory Panel, made up of scientists and engineers with expertise in the field. See the panel's Web site at www.epa.gov/region1/omsap/index.html. Also see MWRA, "Boston Harbor and Massachusetts Bay: MWRA Environmental Quality Department," www.mwra.state.ma.us/harbor/html/outfall_update.htm.

259   **one of his top engineers said the idea:** MacDonald says Ken Chin, senior construction manager for the MWRA, played a crucial role in taking his offhand idea to the next level.

260   **one of Kiewit's junior partners:** "Bankruptcy Claims on Guy F Atkinson," *Tunnels and Tunnelling International,* December 1997.

260   **the contractor was perhaps $80 million in the hole:** MWRA, "Boston Harbor Project CP-282 Effluent Outfall Tunnel Status Report for the Office of the State Auditor," 4–5.

262   **Abbott told him there were three ideas:** Tom Walsh, "No Quick Fix for Troubled Tunnel; Completing Sewage Outfall Pipe Where 2 Workers Died Will Add Millions to Project's Cost," *Patriot Ledger* (Quincy, Mass.), February 11, 2000; Beck and Abbott, interviews by author.

262   **"I need a plumber,":** *United States vs. Metropolitan District Commission et al.,* Civil Action No. 85-489-ADM, U.S. District Court, State of Massachusetts, hearing, March 30, 2000.

263 **the Sludge Judge:** Tom Long, "A. D. Mazzone, Judge Who Led Harbor Effort, Dies at Age 76," *Boston Globe,* October 27, 2004.

263 **The hearing began with a briefing:** *United States vs. Metropolitan District Commission et al.,* Civil Action No. 85-489-ADM, hearing, March 30, 2000.

263 **An engineer for the MWRA:** Charlie Button, who had left his post as the MWRA's construction point man to work for a private company, had by now returned as a consultant to help the agency find a way out of its plug jam. As he took the stand, the unassuming Button was nervous. Judge Mazzone startled him with a question, "I don't have to put you under oath, do I, Mr. Button?" and then followed that up with a comment, "You went to Everett High School." Button's nerves tightened until he saw the judge smile. Then he recalled that the judge had graduated from the same school. Meeting the judge's grin with one of his own, Button said, "Yes. We learned to tell the truth."

263 **Installing a new bag line:** *United States vs. Metropolitan District Commission et al.,* Civil Action No. 85-489-ADM, March 30, 2000, transcript, 17–18.

264 **"How do you get air out nine miles?":** Ibid., 29.

264 **three-quarters of a million dollars to reserve the IB-909:** Ibid., 36.

266 **Popeo cited a host of unknowns:** Ibid., 42–44.

267 **he retrieved his video camera:** Videotape recording by Ron Kozlowski.

270 **A team of sandhogs:** "Removing Safety Caps—Inside the Tunnel and Under the Sea," *Under Construction: Building the Boston Harbor Project 1992–2001* (Boston: MWRA, 2001), 166.

270 **The jet-engine fan on the barge:** Dave Corkum, interview by author, March 2013.

270 **Workers used treated sewer water:** Ibid.

271 **took only about three days:** "Removing Safety Caps," *Under Construction.*

271 **clocking in at $15 million:** Walter Armstrong and Ralph Wallace, interviews by author, 2010. See also the July 21, 2000, issue of *Engineering News-Record.*

CHAPTER 14: JUSTICE

274 **OSHA's Finizio asked him to elaborate:** Trooper Mary T. McCauley, "Interview of Harald Grob," Massachusetts State Police Suffolk County Detective Unit, January 7, 2000. In addition to Finizio, industrial hygienist Maria-Lisa Abundo was also heavily involved in the investigation for OSHA.

275 **"It's *certainly* accurate":** Ibid., 5.

276 **The autopsies found gas bubbles:** Commonwealth of Massachusetts, Office of the Chief Medical Examiner, Timothy Nordeen Autopsy Report, July 22, 1999.

276 **Reading rattled off a host of examples of its poor design:** Timothy Reading (Airgas Northeast safety and compliance manager) to Mary McCauley, July 30, 1999.

278   **Kiewit and Norwesco had met with several OSHA officials:** Brenda
       Gordon (OSHA area director) to Mary McCauley, September 15, 1999.
279   **McCauley fired off an angry letter:** Mary McCauley to Brenda Gordon,
       August 12, 1999.
279   **Gordon's defense:** Brenda Gordon to Mary McCauley, September 15, 1999.
280   **Billy and Tim had died as a result of asphyxia:** Massachusetts Chief
       Medical Examiner, Nordeen Autopsy Report.
281   **with the help of her state police colleagues:** McCauley stressed the key
       roles played by her partner, Trooper John O'Leary; their supervisor, Sgt.
       Randy Cipoletta; and Trooper Wes Wanagal of the State Police Crime
       Scene Services Section.
281   **"The design of the breathing apparatus":** Dr. Marie E. Knafelc,
       "Investigation of Deaths Within the Boston Harbor Outflow Tunnel,"
       Navy Experimental Diving Unit, September 3, 1999, 6.
281   **For short money:** Ibid., 5.
282   **Knafelc noted Harald's failure:** Ibid., 3–4. The twelve-hundred-foot
       umbilical connecting the manifold in the Humvee to the manifold in the
       boat consisted of four three-hundred-foot hoses strung together. From
       there, each diver on the excursion team had his own three-hundred-foot
       umbilical between him and the manifold in the boat.
282   **But in the process of being vaporized:** Dr. Marie E. Knafelc, interviews
       by author, May 2012; Canadian Centre for Occupational Health and
       Safety, "Cryogenic Liquids—Hazards," www.ccohs.ca/oshanswers/
       chemicals/cryogenic/cryogen1.html; and Air Products, "Cryogenic Liquid
       Containers," Safetygram-27, www.airproducts.com/~/media/Files/PDF/
       company/safetygram-27.pdf.
283   **"Using liquid gases requires":** Knafelc, "Investigation of Deaths," 5.
283   **Breathing directly off a mixer:** Ibid., 4–6.
283   **The second missing device:** Ibid., 6.
285   **impaired coordination, perception and judgment:** Ibid., 2.
286   **He insisted to McCauley:** Timothy Reading to Mary McCauley, July 30,
       1999.
287   **"authorised personnel *only*":** "MAP Mix 9000 User's Manual," 2.
287   **That led Hansen to deliver a bombshell:** Trooper Mary McCauley,
       "Interview of PBI-Dansensor Representatives in Reference to Tunnel
       Death Investigation at Deer Island," Massachusetts State Police Suffolk
       County Detective Unit, October 9, 1999, 10.
287   **Hansen said he and his colleagues:** Steen Hansen, interview by author,
       June 2011. Hansen confirmed what he had told Mary McCauley: that it
       was inconceivable to them that anyone would have ever used the device
       for human breathing air. During a tour of the PBI Dansensor plant in
       Ringsted, Denmark, he showed me the warning labels that were added to
       the devices as a result of the tunnel deaths.
288   **In her view, he had:** Contributing to McCauley's thinking was a letter
       from Airgas Northeast safety manager Tim Reading, who provided a

record of Harald's contact with his company and Jerry Anderskow of supplier A-L Compressed Gases. Reading stressed that Airgas was never informed that its liquid gases would be used to provide breathable air. In her report McCauley wrote that Reading "indicated that the gas supply industry in the United States does not sell liquid gas for the producing [of] breathing air in the field. This is why the mixer that was purchased by Norwesco was not sold in the United States." McCauley, "Sudden Death/Deer Island," 14. Also, Anderskow, in his May 16, 1999, memo to Harald, wrote that after consulting with the nation's largest manufacturer of cryogenic storage systems, "We cannot find a breathing air approved mixer/blender." Harald later told plaintiffs' lawyers that's when he took to the Internet and found the overseas device.

288   **Norwesco received the biggest ticket:** OSHA Boston Area Office South to Norwesco Marine, Inc., Citation and Notification of Penalty, Inspection site: CP-282 Deer Island, issuance date January 14, 2000.

288   **Kiewit and Kaiser each received:** OSHA Boston Area Office South to Kiewit-Atkinson-Kenny, Joint Venture, Citation and Notification of Penalty, Inspection site: CP-282 Deer Island, issuance date January 18, 2000; OSHA Boston Area Office South to Black Dog Divers, Inc., Citation and Notification of Penalty, Inspection site: CP-282 Deer Island, issuance date January 14, 2000.

289   **all four of them quickly signaled publicly:** Tom Walsh, "4 Companies Appeal Worker Deaths Fines," *Patriot Ledger* (Quincy, Mass.), February 11, 2000.

295   **I don't know, but the excursion crew did:** McCauley, "Interview of Harald Grob," 14.

295   **notably in "small submarines":** Dave Corkum, deposition, 184.

296   **"in general caused the failure":** Kirk W. VanZandt, "An Evaluation of the Breathing System Used on the Boston Harbor Outflow Tunnel Project," Code A52 Coastal Systems Station, Department of the Navy, March 16, 2000, 6.

296   **with surprising frequency:** Ibid., 2.

297   **"misinformed or inexperienced":** Ibid., 5.

297   **"Skiing has been good?":** Roger Rouleau to Harald Grob, February 7, 2000.

298   **"I'm trying to get an idea":** Roger Rouleau to Harald Grob, February 16, 2000.

298   **"we should be able to get a good resale price":** Harald Grob to Roger Rouleau, February 21, 2000.

298   **"Just finished a trek in the jungles":** Harald Grob to Roger Rouleau, March 31, 2000.

298   **OSHA ultimately reduced the fines:** U.S. Occupational Safety and Health Review Commission, Settlement and Agreement Between Secretary of Labor and Norwesco Marine, Inc.

299   **"Because I don't know how to do that stuff":** Roger Rouleau, deposition, 2000.

300 **during the Great Plague of 1665:** Daniel Defoe, *A Journal of the Plague Year* (New York: Modern Library, 2001), 42. Defoe's work, first published in 1722, is credited with popularizing the connection between the six-foot grave edict and the Great Plague. However, it is an historical novel. The *Encyclopedia of the Black Death* notes that some towns in Italy and other parts of Europe tried to enforce a six-foot grave rule as far back as the fourteenth century, to stop the stink. Joseph P. Byrne, *Encyclopedia of the Black Death* (Santa Barbara, Calif.; ABC-CLIO, 2012), 301.

300 **"there was no way out at all":** Joan Parker, speech at event organized by the Massachusetts Coalition for Occupational Safety and Health in belated recognition of Workers' Memorial Day, delivered at the University of Massachusetts Lowell, May 24, 2000. Text and notes from Parker. Note that she used a depth of 400 feet for the shaft. For consistency with the rest of this book, I substituted the actual depth of 420 feet and corresponding number of six-foot graves.

301 **Other lawyers were representing Tim's widow:** Bob Norton represented Judy Milner; Warren Fitzgerald represented Hoss and Riggs.

301 **Accordingly, they had sued:** *Milner, Admx, et al. v. MWRA et al.,* Suffolk Superior Court Case Summary, Civil Docket, Massachusetts Administrative Office of the Trial Court.

303 **Because tattoo parlors:** A Massachusetts Superior Court judge had actually lifted the state's long-standing ban on tattoo parlors just one week before the mediation session, though DJ appears to have gotten inked before the ruling. As it turned out, the ban was reinstated for several months to allow time for the drafting of regulations. "Massachusetts Judge Temporarily Revives Tattoo Ban," Associated Press, November 21, 2000.

304 **A memo from the tunnel designer:** Eldon Abbot (Parsons Brinckerhoff project manager) to Dave Corkum, February 17, 1998.

306 **the value of the wrongful death claims for Billy and Tim:** John Prescott, memo to John Fitzgerald, October 26, 2000.

308 **No one was entitled:** Olga told me that even if the law didn't recognize Michelle as Billy's life partner, she did. That's why, she said, she took the $75,000 payout she received from Billy's insurance policy and turned it over to Michelle.

310 **the tax-free settlements were not huge:** Joint Petition for Approval of Third-Party Settlement and Dismissal of Plaintiffs' Claims, Suffolk, Superior Court, Commonwealth of Massachusetts. The order for these records to remain impounded expired in January 2011. Figures come from these court papers. The parties did not disclose settlement amounts.

311 **Then those insurers went to court:** *Milner, Admx, et al. v. MWRA et al.,* Suffolk Superior Court Case Summary, Civil Docket, Massachusetts Administrative Office of the Trial Court.

312 **"was fortunately a relief":** *United States v. Metropolitan District Commission*

et al., Civil Action No. 85-489-ADM, U.S. District Court, State of Massachusetts, hearing, March 30, 2000.

312 **In that case:** The case involved Tewksbury Industries, Inc., and its CEO, Thomas E. Bowley. See Henry Goldman, "Death in the Workplace: It's Becoming a Crime," *Philadelphia Inquirer,* May 18, 1997; and Armando Roggio, "Former Scrap Exec Enters Plea Accord," *American Metal Market,* October 1, 1998.

315 **This delayed the final decision:** Commonwealth of Massachusetts, Office of the Attorney General, Harald Grob & Norwesco Marine, Case status: Closed on September 17, 2001.

CHAPTER 15: THE LONG TAIL OF TRAUMA

318 **vets felt helpless to stop reliving a traumatic event:** Charles Figley, *Stress Disorders Among Vietnam Veterans: Theory, Research and Treatment* (New York: Brunner-Routledge, 1978).

318 **Figley opened all his files:** Charles Figley and Terence Keane both confirmed these details in separate interviews with me in 2012.

320 **is now believed to affect 24 million Americans:** Keane, interview by author; U.S. Department of Labor, "Frequently Asked Questions About Post-Traumatic Stress Disorder (PTSD) & Employment," America's Heroes at Work, www.americasheroesatwork.gov/forEmployers/factsheets/FAQPTSD/.

320 **That psychiatrist:** Dr. Jonathan Shay connected PTSD to Achilles and Odysseus. See Anna Badkhen, "Psychiatrist Treated Veterans Using Homer Work Made Him MacArthur Fellow," *Boston Globe,* September 25, 2007.

325 **Riggs thought to himself:** Dave Riggs, interview by author, July 2010. I drew the details about this memorial service from interviews with numerous attendees as well as photographs, the ceremony program, and other firsthand documents. An additional account was Adrianne Appel, "Ceremony Honors 2 Divers Who Died in Outfall Pipe," *Boston Globe,* September 15, 2002.

326 **Their worst fight had taken place:** Dave Riggs and Karen Riggs, (separate) interviews by author, July 2010.

327 **"Take some ibuprofen and drink a couple of beers":** Dave Riggs, deposition, 101.

332 **Klonopin and other antianxiety medication:** *United States v. Donald J. Gillis,* U.S. District Court, District of Massachusetts, presentence report, May 27, 2009, 12.

334 **He rounded up a stack:** DJ's Army Ranger application.

334 **OxyContin had arrived on the scene only in 1996:** Donna Gold, "A Prescription for Crime: Abuse of 2 Painkillers Blamed for Rise in Violence in Maine's Poorest County," *Boston Globe,* May 21, 2000. For more on opiates, see www.nlm.nih.gov/medlineplus/ency/article/000949.htm.

338   **The night ended with police finding DJ:** *United States v. Donald J. Gillis,* Presentence Report, 3.

338   **Bryan had been killed:** "Deaths: Selmer—Bryan Lee Ernest," *Commercial Appeal* (Memphis, Tenn.), February 9, 2007.

339   **Chris had settled on bank robberies:** Geoffrey Kelly (FBI special agent), affidavit, June 4, 2008. Based on a tip, authorities began watching Chris, attaching a GPS device to his car.

341   **they pulled up to a Walgreens:** Ibid; also, DJ, interviews by author.

346   **During the long hearing:** I attended this sentencing hearing, so these observations were gathered firsthand. I drew some details about Chris Politis's role from his statements in court as well as from those made by others about him; also from court documents and extensive interviews with DJ. As of this writing, Chris remains incarcerated.

346   **sentence lengths called for under the guidelines:** *United States v. Christopher A. Politis and Donald J. Gillis,* U.S. District Court, District of Massachusetts, June 3, 2009, clerk's notes.

349   **an inpatient drug treatment facility upon his release:** *United States v. Donald J. Gillis,* U.S. District Court, District of Massachusetts, judgment in a criminal case, June 6, 2009. In addition to the prison time, the judge ordered DJ and Chris to each serve three years of supervised release, which is something like probation but served after a prison sentence rather than in lieu of it.

349   **DJ would be required to pay back one dollar:** Ibid.

EPILOGUE

352   **Two colleagues died:** Harald Grob to author, June 23, 2009.

354   **He had simply cut-and-pasted his reply:** Harald Grob to author, November 7, 2011.

355   **OSHA levied a "serious":** In December 2009 an administrative law judge vacated the serious violation and $2,500 fine against Associated Underwater Services (AUS), the successor company to Norwesco. But in February 2012 that decision was overturned and the citation reinstated. AUS paid the fine the following month, according to Patricia Drummond of the U.S. Department of Labor. Primeau's sister Julie argues that the kind of company recklessness seen in the tunnel deaths was repeated in her brother's death.

355   **In 2009, around the tenth anniversary:** After Roger Rouleau sold his interest in the company, his former business partner Kerry Donohue and Donohue's wife took control. Rouleau, interviews by author, 2009–10; licensing records.

358   **the ability to read, process, and manage:** For more on EQ, see Daniel Goleman, *Emotional Intelligence: Why It Can Matter More Than IQ,* 10th anniversary edition (New York: Bantam Books, 2005).

358   **the seductive appeal of the outsider:** On this point, I benefited from the insights of Don Gibson, Larry Palinkas, and Tom Botts, among others.

359   **insulate themselves from any mistreatment of those workers:** Benjamin
      Powell and David Skarbeck, "Sweatshops and Third World Living
      Standards: Are the Jobs Worth the Sweat?" Independent Institute
      Working Paper 53, Department of Economics, San Jose State University,
      2004; Michael Clancy, "Sweating the Swoosh: Nike, the Globalization of
      Sneakers, and the Question of Sweatshop Labor," Institute for the Study
      of Diplomacy, School of Foreign Service, Georgetown University, 2000, 5.

360   **Financially, the toll:** The $30 to $35 million estimate takes documented
      dollar figures for such items as the design and construction costs for the
      vent solution and the legal settlements with the divers and their families,
      and combines them with estimates from multiple project players for
      such line items as legal fees, fines, insurance and administrative costs,
      and lost time. This newspaper account from prior to the vent solution
      decision in 2000 estimated that the additional costs to complete the
      tunnel following the accident could be as high as $30 million. Tom Walsh,
      "No Quick Fix for Troubled Tunnel," *Patriot Ledger* (Quincy, Mass.),
      February 11, 2000. Several years before his death in 2013, Ed Willwerth,
      a noted industrial hygienist and marine chemist, told me he had closely
      examined the tunnel case in preparation for presentations he delivered
      at industry conferences, including discussions with one of the key OSHA
      representatives in the case. In his talks, Willwerth said if that $30 million
      published estimate was correct, it would make the diver deaths "the most
      expensive confined space accident not involving a fire or explosion in
      history."

361   **Although the approach:** Dave Corkum went on to coauthor a book
      for specialist readers on the trend toward design-build. This approach
      has by now become mainstream in tunneling and other underground
      projects. He says it can speed up the process by which a project goes from
      concept to reality, and the project owners tend to like it because it offers
      streamlined responsibility. If there is a problem, the owner doesn't have to
      figure out whether it's related to design or to construction. Says Corkum:
      "He can point to his design-builder and say, 'It's all yours.'" But Corkum
      cautions that the approach can be misused by owners to dump all the risk
      onto the design-builder. While it can reduce the likelihood of the kind of
      dysfunction seen on the Deer Island job, Corkum says he doesn't know if
      a design-build approach would have eliminated the fundamental standoff
      over the timing of the plug removal. For more technical perspective on
      this trend, see Gary S. Brierley, David H. Corkum, and David H. Hatem,
      Design-Build Subsurface Projects (Littleton, Colo.: Society for Mining,
      Metallurgy and Exploration, 2010).

APPENDIX

371   **for the union:** Kuhs is the regional manager for the specialty trades of the
      New England Regional Council of Carpenters.

INDEX

high price of, 152

lack of gauge/monitor, 159, 168–69, 188–90, 276–77

manual "authorised personnel use" warning, 166–67, 287

McCauley's investigation of, 254, 286–87

oxygen content measurement protocol, 187–88

placement in Humvee of, 151, 152, 161, 168

pressure gauge adjustment, 166–67, 287

pressure reading, 193, 284

risks of, 282–85

as unsuitable for human use, 286–88

marine life, 37, 39, 48, 256, 257, 363

Markham, Tracy, 147, 161

Mars, Mike, 147, 150, 152, 161, 163, 164, 167, 174, 176–77, 189, 197

Martini's Law, 109

Massachusetts Bay, 28, 39, 48, 50, 54, 269, 362

rebirth of, 363

shipping lane, 57–58, 69, 113, 134

treated wastewater release into, 257

Massachusetts General Hospital, 233, 234, 235, 236, 238, 240, 242, 307

Massachusetts Water Resources Authority. See MWRA

Mattingly, Ken, 96, 97, 98

Mazzone, Armando David, 133, 134, 225, 262–63, 265–66, 270, 312

McCauley, Mary, 231–35, 273–81, 292–93, 294, 298

background/appearance/personality of, 231, 232, 292

criminal charges and, 280, 288, 289, 290–91, 311

diver death investigation by, 231, 232, 243–44, 252–54, 272–81

Harald interview by, 273–76

update on, 371

Meaney, Fran, 73, 74–75

medical examiner's office, 276, 279–80

Metcalf & Eddy, 61, 62

methane, 94, 132, 158, 160

Milner, Judy (Tim's wife), 122, 123, 124, 170

fears about Tim's safety of, 158–59, 180

Tim's death and, 238–40, 249, 250, 322–23

Tim's gun collection and, 119, 252

update on, 369–70

wrongful death suit and settlement, 251, 301–4, 307–8, 310

mine rescue training, 105–9, 112–13, 115, 127, 210, 214

fallen comrade protocol, 101, 107, 206, 208–9

Mintz Levin (law firm), 73, 265

mixed-gas breathing system, 9, 62–63, 118, 139, 161, 170–71, 210–11, 261, 275–79, 286, 295, 301

approval by major parties of, 86–92, 101, 126–27, 131–34, 278–79, 280, 357

backup systems for (see HP)

components of (see liquid oxygen/liquid nitrogen)

concerns about, 113–14, 116, 121, 124–29, 135, 146, 162–65, 167, 179

cost of, 89, 90, 360

deaths of Billy and Tim and, 204, 207–8, 221, 225, 244, 249, 253, 278, 281–82, 294, 354

diagram of, 140

divers' anger about, 179, 180, 221

effect of greater demand on, 282, 296–97, 328–29

fundamental flaws in, 142, 179, 191–92, 281–86, 296–97, 327

Harald's secrecy/reassurances about, 129, 132, 151, 221, 225, 274–75, 294–97, 313

as improvised and untested, 7, 126, 142, 254, 276–78, 288, 356, 357

inadequacy of, 100–101, 125, 162, 166–67, 176, 179, 193, 199–200, 207, 244, 281–85, 288–89, 294–97, 305, 322, 356, 358, 360

McCauley's assessment of, 254

mixer/blender as centerpiece of (see MAP Mix 9000)

monitoring of, 161, 168–69, 190, 276–77, 281–82

Riggs's in-line analyzer suggestion for, 187–88, 210, 244, 281

Tap's plan vs., 88–89, 146

umbilical lengths and, 142, 161–62

Murphy's Law, 113

MVE (cryogenics supplier), 100–101

MWRA (Massachusetts Water Resources Authority), 38–39, 62–71, 73, 134, 139, 257–66, 304

ABOUT THE AUTHOR

NEIL SWIDEY is author of *The Assist*, a *Boston Globe* bestseller that was named one of the best books of the year by *The Washington Post,* and a coauthor of the *New York Times* bestselling *Last Lion: The Fall and Rise of Ted Kennedy*. A staff writer for *The Boston Globe Magazine*, Swidey has been a finalist for the National Magazine Award and has won the Sigma Delta Chi Award from the Society of Professional Journalists three times. His work has been featured in *The Best American Science Writing*, *The Best American Crime Writing*, and *The Best American Political Writing*. As an outgrowth of his first book, he founded the nonprofit Alray Scholars Program, which helps Boston students return to college. He lives outside Boston with his wife and three daughters. For more on Neil Swidey, and to see photos and illustrations of the events described in this book, please visit www.neilswidey.com.